T0136997

Alternating Direction Method of Multipliers
for Machine Learning

Zhouchen Lin • Huan Li • Cong Fang

# Alternating Direction Method of Multipliers for Machine Learning

Zhouchen Lin
Key Lab. of Machine Perception (MoE)
School of Artificial Intelligence
Peking University
Beijing, China

Huan Li
Institute of Robotics and Automatic
Information Systems
College of Artificial Intelligence
Nankai University
Tianjin, China

Cong Fang
Key Lab. of Machine Perception (MoE)
School of Artificial Intelligence
Peking University
Beijing, China

ISBN 978-981-16-9842-2      ISBN 978-981-16-9840-8    (eBook)
https://doi.org/10.1007/978-981-16-9840-8

This Springer imprint is published by the registered company Springer Nature Singapore Pte Ltd.
The registered company address is: 152 Beach Road, #21-01/04 Gateway East, Singapore 189721, Singapore

*To our families. Without your great support, this book would not exist and even our careers would be meaningless.*

# Foreword

Alternating direction method of multipliers (ADMM) is an important algorithm for solving constrained optimization problems. It particularly fits well for the machine learning community because the latter basically favors algorithms with low per-iteration cost and does not need high numerical precision. Due to its versatility and high usability, I would not hesitate to make it one of the top recommendations if one wants to develop a general-purpose optimization library or AI chip. So there has been renewed interest on ADMM since its successful application in solving low-rank models around 2010. Since then, ADMM has been extended significantly, going far beyond the traditional setting: deterministic, convex, two-blocks of variables, and centralized. Unfortunately, the vast literature on ADMM is scattered across various sources, making it difficult for non-experts to track the advances in this important optimization technique. This book resolves this issue in a timely manner. Its materials are quite comprehensive, covering ADMM for various situations: convex (and deterministic), nonconvex, stochastic, and distributed. It is self-contained, with detailed proofs, so that even a beginner can grasp the state-of-the-art quickly, not just the pseudo-codes but also the proof techniques. More importantly, this book has not simply compiled various papers together. It has actually completely rewritten the materials so that the notations are consistent and the deductions are smooth, removing the major obstacle of reading existing literatures. In addition, the book puts more emphasis on convergence *rates* rather than only convergence. This makes the theoretical analysis extremely informative to practitioners. I would say that this book is definitely a valuable reference for researchers and practitioners from multiple areas, including optimization, signal processing, and machine learning.

The authors, Zhouchen Lin, Huan Li, and Cong Fang, are experts in the intersection of optimization and machine learning. Besides contributing greatly to this field with technical papers, they have also endeavored a lot in sorting out valuable algorithms that are fit for engineers, making another kind of good contribution to the community. After their previous book, *Accelerated Optimization for Machine Learning: First-Order Algorithms*, which I like very much, I am happy to advocate their book once more.

Xi'an Jiaotong University, Xi'an, China                                    Zongben Xu
October 2021

# Foreword

With the advance of sensor, communication, and storage technologies, data acquisition has become more ubiquitous than any time in the past. This has enabled us to learn a considerable amount of valuable information from big diverse data sets for effective inference, estimation, tracking, and decision-making. Learning from data requires the proper modelling and analysis of big data sets, which are usually formulated as optimization problems. Consequently, large-scale optimization involving big data has attracted significant attention from various areas, including signal processing, machine learning, and operations research.

To minimize a cost function involving a large number of variables, the most popular approach is block coordinate descent/minimization (sometimes also called alternating optimization). However, if variables are coupled linearly, the block coordinate descent/minimization method no longer works. The alternating direction method of multipliers (ADMM) can be considered an extension of block coordinate descent/minimization method for linearly constrained optimization problems. Given the abundance of application problems that can be cast in the form of linearly constrained optimization problems (convex/nonconvex, smooth/nonsmooth), ADMM has been the method of choice for machine learning and signal processing problems involving big data. It is widely (sometimes wildly) applied in many different contexts, often times without sufficient theoretical underpinning on its convergence.

This book provides an excellent summary of the state of the art for the theoretical research on ADMM. It introduces the basic mathematical form of ADMM as well as its variations. The core material is on the convergence analysis of ADMM for different classes of linearly constrained optimization problems, including convex, nonconvex, deterministic, stochastic, and centralized/distributed. The mathematical treatment is concise, up to date, and rigorous. A nice bonus is the last chapter where the practical aspects of ADMM are discussed, which should be very valuable for practitioners or first-time users of ADMM.

The first author is a well-known researcher in optimization, particularly on optimization methods for machine learning applications. The text is written in a reader friendly manner, complete with appendices that introduce the mathematical tools and background for the convergence analysis covered in the book. It should be a valuable reference for both researchers and users on ADMM and will be a great read for graduate students in optimization, statistics, machine learning, and signal processing.

The Chinese University of Hong Kong, Shenzhen, China          Zhi-Quan Luo
November 2021

# Preface

Alternating direction method of multipliers (ADMM) is a magic algorithm to me. In my biased opinion, it is more or less a universal method for solving a wide range of constrained problems that ordinary practitioners in machine learning may encounter. Unlike gradient descent, which is roughly a universal method for unconstrained problems, ADMM appears to be more elegant yet less transparent. The secret lies in the Lagrange multiplier, which temporarily makes the constrained problem unconstrained, not only removing the difficulty in handling the constraints but also overcoming some inherent defects of the penalty method and the projected gradient descent, while non-experts are much easier to think of the latter two methods. The Lagrange multiplier also plays a central role in the proofs of convergence and convergence rate of ADMM. With possible exaggeration, I would say that one who cannot appreciate the beauty of ADMM cannot be a good researcher in optimization.

Since my first encounter with ADMM around 2009, I have seen that more and more machine learning people are using ADMM and extending its scope of applications. I also benefited a lot from and contributed a bit to the new developments. Yet, I also found that many engineers are not using ADMM correctly (the most notable example is to naively apply the ADMM for two blocks of variables to problems with more than two blocks). There has been an excellent tutorial on ADMM, *Distributed Optimization and Statistical Learning via the Alternating Direction Method of Multipliers*, written by Boyd et al. in 2011. Nonetheless, it is now 10 years old and does not cover the new developments, which I actually think are more important than the traditional ones for the wider applications of ADMM, because the new developments were done out of demands from real applications in signal processing and machine learning. So, I think that writing a new book on ADMM will be very useful for many people, including myself when teaching and advising students. My goal is to incorporate the most important aspects of the new developments in ADMM, rather than being confined to the traditional materials, which are typically for convex and two-block cases. Clearly, I am unable to review

Prof. Bingsheng He

all papers on ADMM. So, the strategy is to choose representative algorithms by their types (e.g., convex, nonconvex, stochastic, and distributed) instead. As a result, one should not be surprised that some variants of ADMM are not included (because they are not the most representative ones of their types but just discuss in more depth, or are too complex to use or analyze) while some variants of ADMM appear to be rather crude but they are still included (because that type of ADMM is less explored). Of course, personal flavor and limited knowledge also matter greatly. Finally, being self-contained is also very important. So, I also want to present proofs in detail.

I truly feel lucky as my former PhD students, Huan Li and Cong Fang, agreed to join this task even though they have graduated, and I have tortured them in the previous book, *Accelerated Optimization for Machine Learning: First-Order Algorithms*. I am also very lucky that more students contributed to the proofreading, including checking the proofs thoroughly (most of the proofs have undergone rewriting, rather than being directly copied from corresponding papers), which made the work less daunting. Nonetheless, the book is still far from being perfect. One of the major regrets is that using adaptive penalty is critical for speeding up convergence (see Sect. 7.1.2), but all the algorithms introduced in this book use a fixed penalty. Actually, most of the literatures do not consider adaptive penalty. Although it is quite possible that some of the algorithms introduced in this book may also work with adaptive penalties, we are unable to test which adaptive penalty strategy to use and then rewrite the proofs for adaptive penalties (actually drastic changes in the proofs may be necessary). The other regret is that we have to leave out learning-based ADMM, an emerging yet immature type of ADMM, as the theoretical guarantees are weak.

I expect that there will still be errors in the book despite the great efforts from my students and myself. So, if the readers detect any problem, please feel free to write an email to zclin2000@hotmail.com.

Finally, I would like to pay tribute to Prof. Bingsheng He. He has devoted most of his life to ADMM and contributed significantly to the research and education of ADMM. This book also introduces many of his works. I am glad to see that he has been well recognized in China, manifested by winning the "Operations Research" Research Award of the Operations Research Society of China in 2014. However, he is much less recognized internationally. I hope that my advocacy here could add to his credit.

Peking University, Beijing, China Zhouchen Lin
October 2021

# Acknowledgments

We would like to thank all our collaborators and friends, especially Bingsheng He, Licheng Jiao, Junchi Li, Qing Ling, Guangcan Liu, Risheng Liu, Yuanyuan Liu, Canyi Lu, Zhi-Quan Luo, Yi Ma, Fanhua Shang, Zaiwen Wen, Xingyu Xie, Zongben Xu, Shuicheng Yan, Wotao Yin, Xiaoming Yuan, Xiaotong Yuan, Yaxiang Yuan, Tong Zhang, and Pan Zhou. We also thank Yiming Dong, Dr. Yuqing Hou, and Prof. Hongyang Zhang for proofreading, Yikang Li, Qiuhao Wang, Zhoutong Wu, Tong Yang, and Pengyun Yue for carefully checking all the proofs (in particular, Qiuhao Wang gave a much simpler proof for Lemma 3.16) and Celine Chang from Springer for helping the publication of the book. The authors are truly honored to have forewords written by Prof. Zongben Xu and Prof. Zhi-Quan Luo. This book is supported by National Natural Science Foundation of China under Grant Nos. 61625301 and 61731018.

# About the Book

This book introduces the basic concepts of ADMM and its latest progress. Specifically, it introduces various ADMMs under different scenarios: convex and deterministic ADMM (Chap. 3), nonconvex and deterministic ADMM (Chap. 4), stochastic ADMM (Chap. 5), and distributed ADMM (Chap. 6). To make the book self-contained, it gives the detailed proofs of the convergence rates for most of the introduced algorithms.

This book serves as a useful reference to the recent advances in ADMM. It is appropriate for graduate students and researchers who are interested in optimization or practitioners who seek a powerful tool for optimization.

# Contents

# Acronyms

| | |
|---|---|
| AAAI | Association for the Advancement of Artificial Intelligence |
| Acc-SADMM | Accelerated Stochastic Alternating Direction Method of Multipliers |
| ADMM | Alternating Direction Method of Multipliers |
| ALM | Augmented Lagrangian Method |
| DRS | Douglas-Rachford Splitting |
| KKT | Karush-Kuhn-Tucker |
| LADMM | Linearized Alternating Direction Method of Multipliers |
| MISO | Minimization by Incremental Surrogate Optimization |
| RPCA | Robust Principal Component Analysis |
| SADMM | Stochastic Alternating Direction Method of Multipliers |
| SAG | Stochastic Average Gradient |
| SDCA | Stochastic Dual Coordinate Ascent |
| SGD | Stochastic Gradient Descent |
| SPIDER | Stochastic Path-Integrated Differential EstimatoR |
| SVD | Singular Value Decomposition |
| SVRG | Stochastic Variance Reduced Gradient |
| VR | Variance Reduction |

# Chapter 1
# Introduction

Optimization plays a central role in fields that use mathematical models, such as signal processing and machine learning [14]. The celebrated formula [14]:

$$\text{machine learning} = \text{representation} + \text{optimization} + \text{evaluation},$$

proposed by P. Domingos, an AAAI Fellow and a Professor at University of Washington, dictates the importance of optimization in machine learning.

Besides the wide applications of unconstrained optimization (see [42] for example), a lot of mathematical models can also be formulated or reformulated as constrained optimization. The constraints either add more prior information to the mathematical models (e.g., nonnegativity and boundedness) or are introduced when reformulating the original optimization problem (e.g., using auxiliary variables) so that the reformulated problem can be solved more easily. The alternating direction method of multipliers (ADMM) is a powerful tool for solving constrained problems, ranging from the classic linearly constrained problems with separable objectives to general problems with nonlinear and inequality constraints. Due to our limited knowledge, we focus on ADMMs studied in the machine learning community.

## 1.1 Examples of Constrained Optimization Problems in Machine Learning

We provide three representative examples of constrained optimization in machine learning, all of which can be solved by ADMM. The first one is the famous robust principal component analysis (RPCA) model [8], the second one is the widely used consensus problem in distributed optimization, and the third one is the non-negative matrix completion [63].

The RPCA model [8] decomposes an observation matrix into a low-rank matrix and a sparse matrix. It is cast as the following linearly constrained convex problem with separable objectives:

$$\min_{\mathbf{X},\mathbf{Z}} \left( \|\mathbf{X}\|_* + \tau \|\mathbf{Z}\|_1 \right), \quad s.t. \quad \mathbf{X} + \mathbf{Z} = \mathbf{M}, \tag{1.1}$$

where $\| \cdot \|_*$ is the nuclear norm, defined as the sum of singular values, and $\| \cdot \|_1$ is the $\ell_1$ norm,[1] defined as the sum of absolute values of the entries, which are the convex relaxations of the rank function and the $\ell_0$ (pseudo-)norm, respectively. In practice, the observation matrix $\mathbf{M}$ may be corrupted by noise. Accordingly, people often solve the following model instead:

$$\min_{\mathbf{X},\mathbf{Z},\mathbf{Y}} \left( \|\mathbf{X}\|_* + \tau \|\mathbf{Z}\|_1 + \frac{\mu}{2} \|\mathbf{Y}\|^2 \right),$$
$$s.t. \quad \mathbf{X} + \mathbf{Z} + \mathbf{Y} = \mathbf{M}, \tag{1.2}$$

where we use $\| \cdot \|$ to denote the Frobenius norm of a matrix and the Euclidean norm of a vector in a unified way. In real applications, the observation matrix $\mathbf{M}$ may be huge and it is unrealistic to compute the singular value decomposition (SVD, Definition A.1) of a huge matrix. To reduce the computational cost, people often factorize the low-rank matrix $\mathbf{X}$ as a product of two much smaller matrices and solve the following model instead:

$$\min_{\mathbf{U},\mathbf{V},\mathbf{Z},\mathbf{Y}} \left[ \frac{1}{2} \left( \|\mathbf{U}\|^2 + \|\mathbf{V}\|^2 \right) + \tau \|\mathbf{Z}\|_1 + \frac{\mu}{2} \|\mathbf{Y}\|^2 \right],$$
$$s.t. \quad \mathbf{U}\mathbf{V}^T + \mathbf{Z} + \mathbf{Y} = \mathbf{M}, \tag{1.3}$$

due to $\|\mathbf{X}\|_* = \min_{\mathbf{U}\mathbf{V}^T=\mathbf{X}} \frac{1}{2} \left( \|\mathbf{U}\|^2 + \|\mathbf{V}\|^2 \right)$. On the other hand, sometimes we want to use the rank function and the $\ell_0$ norm directly, rather than their convex relaxations, then we have the following nonconvex model:

$$\min_{\mathbf{X},\mathbf{Z},\mathbf{Y}} \left( \text{rank}(\mathbf{X}) + \tau \|\mathbf{Z}\|_0 + \frac{\mu}{2} \|\mathbf{Y}\|^2 \right),$$
$$s.t. \quad \mathbf{X} + \mathbf{Z} + \mathbf{Y} = \mathbf{M}. \tag{1.4}$$

Models (1.1) and (1.2) are convex problems with two blocks and three blocks of variables, respectively, while (1.3) and (1.4) are nonconvex ones. All of them can be solved by particular forms of ADMM with convergence guarantee.

---

[1] Please see the table in Sect. A.1 for commonly used notations in this book.

The second example is the consensus problem [7], where we want to solve the following general finite-sum problem in a distributed environment:

$$\min_{\mathbf{x}} f(\mathbf{x}) \equiv \sum_{i=1}^{m} f_i(\mathbf{x}), \qquad (1.5)$$

in which $m$ agents form a connected and undirected network and the local function $f_i$ is only accessible by agent $i$ due to storage or privacy reasons. When the network has a central (or called master) agent, we can reformulate the above problem as the following linearly constrained one:

$$\min_{\{\mathbf{x}_i\}, \mathbf{z}} \sum_{i=1}^{m} f_i(\mathbf{x}_i), \quad s.t. \quad \mathbf{x}_i = \mathbf{z}, \ i \in [m],$$

where the central agent is responsible for updating $\mathbf{z}$ while each worker agent is responsible for updating $\mathbf{x}_i$. When the network is decentralized, we cannot use the constraints $\mathbf{x}_i = \mathbf{z}$ since there is no central node to compute $\mathbf{z}$. Instead, we associate each edge in the network with a variable $\mathbf{z}_{ij}$ if it connects agents $i$ and $j$, and reformulate Problem (1.5) as follows:

$$\min_{\{\mathbf{x}_i\}, \{\mathbf{z}_{ij}\}} \sum_{i=1}^{m} f_i(\mathbf{x}_i),$$

$$s.t. \quad \mathbf{x}_i = \mathbf{z}_{ij}, \ \mathbf{x}_j = \mathbf{z}_{ij}, \ \forall (i, j) \in \mathcal{E} \text{ with } i < j,$$

where $\mathcal{E}$ is the set of edges. Both the above two reformulations can be solved by ADMM efficiently (see Sects. 6.1 and 6.2).

The third example is the non-negative low-rank matrix completion problem [63], which is a useful model for dimensionality reduction, text mining, collaborative filtering, and clustering. It is originally formulated as:

$$\min_{\mathbf{X}} \left( \|\mathbf{X}\|_* + \frac{1}{2\mu} \|\mathbf{b} - \mathcal{P}_{\Omega}(\mathbf{X})\|^2 \right), \quad s.t. \quad \mathbf{X} \geq \mathbf{0}, \qquad (1.6)$$

where $\mathbf{b}$ is the observed data in the matrix $\mathbf{X}$ contaminated by noise and $\mathcal{P}_{\Omega}$ is a linear mapping that selects the entries whose indices are in $\Omega$. The above problem is not easy to solve directly, so we reformulate it as the following one:

$$\min_{\mathbf{X}, \mathbf{Y}, \mathbf{e}} \left( \|\mathbf{X}\|_* + \frac{1}{2\mu} \|\mathbf{e}\|^2 + \chi_{\mathbf{Y} \geq 0}(\mathbf{Y}) \right),$$

$$s.t. \quad \mathbf{b} = \mathcal{P}_{\Omega}(\mathbf{Y}) + \mathbf{e}, \ \mathbf{X} = \mathbf{Y}, \qquad (1.7)$$

by introducing auxiliary variables $\mathbf{Y}$ and $\mathbf{e}$, where $\chi$ is the indicator function:

$$\chi_{\mathbf{Y} \geq 0}(\mathbf{Y}) = 0 \text{ if } \mathbf{Y} \geq 0, \text{ and } \infty \text{ otherwise.}$$

After reformulation, (1.7) can be easily solved by multi-block ADMM.

## 1.2   Sketch of Representative Works on ADMM

ADMM was originally proposed by Glowinski and Marrocco [22] and Gabay and Mercier [19] in the mid-1970s. ADMM did not catch much attention in the machine learning community until people started using it to solve low-rank problems around 2010, such as RPCA [8] and low-rank representation [43], when sparse and low-rank learning was the hottest research topic of machine learning at that time. Early practice includes [8, 39, 40, 43, 56]. The tutorial-like booklet by Boyd et al. [7] also greatly contributed to the popularity of ADMM in the machine learning community.

The convergence of ADMM for convex problems was proved by many researchers, including Gabay [18] and Eckstein and Bertsekas [15]. However, the convergence rate remains an open problem until He and Yuan [26] proved the $O(1/k)$ rate in the ergodic sense (i.e., considering an average of the past iterates) in 2012 via variational inequality, where $k$ is the number of iterations. To make the subproblems easily computable, many authors extended ADMM to its linearized variants by linearizing the augmented term, as well as the smooth objectives. See [24, 40, 52, 58] for example. Some researchers (see [48] and [37] for example) combined linearized ADMM with Nesterov's acceleration. However, the convergence rate is not improved for generally convex problems. It is still $O(1/k)$. When smoothness and strong convexity is assumed, faster convergence rates can be proved. Compared to the ergodic convergence rate, we may be more interested in the non-ergodic (i.e., considering the latest iterate) convergence rate. It was first studied by He and Yuan [29] in 2015, and then extended by Davis and Yin [12] in 2016. They proved the $O(1/\sqrt{k})$ convergence rate for generally convex problems, and the latter further proved that this rate is tight. ADMM was originally designed to solve two-block problems, i.e., the variables are grouped into two blocks and variables in the same block are updated simultaneously, but many problems in machine learning are modeled with multi-blocks. However, the convergence of the direct extension of two-block ADMM for multi-block convex programming is unknown, although it appeared to be effective on some problems, e.g., [56]. Later, Chen et al. [11] gave a counter-example to show that such a direct extension of two-block ADMM is not necessarily convergent for multi-block convex programming. Thus, to guarantee the convergence, some modifications have to be made, such as adding the Gaussian back substitution [25] or the contractive step [27] or using the parallel splitting instead,[2] e.g., [28, 41, 44]. ADMM was also originally designed to solve problems with linear equality constraints. For broader applications, some researchers have extended it to general constraints, including linear equality constraint, linear inequality constraint, as well as nonlinear constraint. See [21] for example. Learning-based ADMM is a recent interesting topic by treating the iterative algorithm as a structured deep neural network, and relaxing the parameters of ADMM to be learnable such that they are optimal to specific data or problems. See [62, 65] for example, but learning-based ADMM is rather immature and only [62] has some theoretical guarantees.

Nonconvex ADMM, as a new topic of interest, has been studied by a lot of researchers (for example, see [5, 30, 33, 36, 60]), especially the proximal ADMM with Bregman distance [59]. Zhang and Luo [67] proposed another proximal ADMM, which uses the Bregman distance from an exponential averaging of all history iterates, rather than just the previous one. The methods in the above references are designed to solve problems with nonconvex objectives and linear constraint. On the other hand, Gao et al. [20] proposed a nonconvex ADMM to solve problems with multilinear constraints, which has broad applications in machine learning, such as non-negative matrix factorization [6, 23], RPCA [8], and the training of neural networks [38, 57]. The result on problems with general nonlinear and nonconvex constraint is much scarcer, and the previous studies mainly focus on the augmented Lagrangian method. See [50] and the references therein.

In machine learning and statistical models, the learning objectives are often in the form of an expectation over a large number of individual functions, each of which is associated with a datum. Under this regime, modern algorithms can be designed in a stochastic fashion, which per-iteratively only randomly access one (or several) individual function(s) serving as an estimator of the full counterpart, thereby reducing the overall computation costs. The first works for stochastic ADMM may come from [47, 55], which achieved an $O(1/\sqrt{k})$ rate in the convex setting. Azadi et al. [2] further combined it with Nesterov's acceleration and obtained a similar convergence rate. On a different research line, for unconstrained problems where the objectives are of an expectation over a finite number of individual functions, it was known that a useful technique termed Variance Reduction (VR) can control the variance of the stochastic gradient and provably accelerate the convergence rate when the number of individual functions is not too large (for example, see [13, 34, 51]). This idea has been successfully introduced into stochastic ADMM in [66, 68], and the resultant algorithms achieved a rate of $O(1/k)$ in the ergodic sense. Later, Fang et al. [16] and Liu et al. [45] fused this technique with Nesterov's acceleration and achieved faster rates by order under suitable conditions. In the nonconvex setting, stochastic ADMM was studied in [31]. Recently, Huang et al. [32] and Bian et al. [4] considered acceleration by applying a new powerful VR technique called Stochastic Path-Integrated Differential Estimator (SPIDER) [17]

---

[2] In this book, we adopt a more general sense of ADMM. Namely, the blocks of variables need not be updated sequentially.

in this setting. Note here, faster rates were achievable no matter the number of individual functions.

As for distributed optimization, it has a long history to use ADMM to solve centralized consensus problems, which may date back to 1980s [3]. A detailed introduction can be found in the popular review [7]. For the decentralized consensus problems, Shi et al. [53] proved that ADMM is equivalent to the linearized augmented Lagrangian method. Generally speaking, ADMM is not the first choice for decentralized optimization and people often prefer the primal-dual method [35], the linearized augmented Lagrangian method [1, 54], and some other methods like gradient tracking [46, 49, 64]. Asynchrony is an interesting topic in practical distributed optimization. Wei and Ozdaglar [61] modeled asynchrony by randomized ADMM, which activates only a subset of agents at each iteration. Chang et al. [9, 10] gave the convergence and convergence rate analysis on the fully asynchronous ADMM for centralized consensus optimization. Both [61] and [9, 10] focused on centralized optimization, and the literature on the asynchronous decentralized optimization is scarcer.

# References

1. S.A. Alghunaim, E.K. Ryu, K. Yuan, A.H. Sayed, Decentralized proximal gradient algorithms with linear convergence rates (2020). ArXiv:1909.06479
2. S. Azadi, S. Sra, Towards an optimal stochastic alternating direction method of multipliers, in *International Conference on Machine Learning* (2014), pp. 620–628
3. D.P. Bertsekas, J.N. Tsitsiklis, *Parallel and Distributed Computation: Numerical Methods* (Prentice Hall, Hoboken, 1989)
4. F. Bian, J. Liang, X. Zhang, A stochastic alternating direction method of multipliers for non-smooth and non-convex optimization. Inverse Prob. **37**(7), (2021)
5. R.I. Bot, D.-K. Nguyen, The proximal alternating direction method of multipliers in the nonconvex setting: convergence analysis and rates. Math. Oper. Res. **45**(2), 682–712 (2020)
6. S. Boyd, L. Vandenberghe, *Convex Optimization* (Cambridge University Press, Cambridge, 2004)
7. S. Boyd, N. Parikh, E. Chu, B. Peleato, J. Eckstein, Distributed optimization and statistical learning via the alternating direction method of multipliers. Found. Trends Mach. Learn. **3**(1), 1–122 (2011)
8. E.J. Candes, X. Li, Y. Ma, J. Wright, Robust principal component analysis? J. ACM **58**(3), 1–37 (2011)
9. T.-H. Chang, M. Hong, X. Wang, Asynchronous distributed ADMM for large-scale optimization – part I: algorithm and convergence analysis. IEEE Trans. Signal Process. **64**(12), 3118–3130 (2016)
10. T.-H. Chang, W.-C. Liao, M. Hong, X. Wang, Asynchronous distributed ADMM for large-scale optimization - part II: linear convergence analysis and numerical performance. IEEE Trans. Signal Process. **64**(12), 3131–3144 (2016)
11. C. Chen, B. He, Y. Ye, X. Yuan, The direct extension of ADMM for multi-block convex minimization problems is not necessarily convergent. Math. Program. **155**(1–2), 57–79 (2016)

12. D. Davis, W. Yin, Convergence rate analysis of several splitting schemes, in *Splitting Methods in Communication, Imaging, Science, and Engineering* (Springer, Berlin, 2016), pp. 115–163

13. A. Defazio, F. Bach, S. Lacoste-Julien, SAGA: a fast incremental gradient method with support for non-strongly convex composite objectives, in *Advances in Neural Information Processing Systems* (2014), pp. 1646–1654

14. P.M. Domingos, A few useful things to know about machine learning. Commun. ACM **55**(10), 78–87 (2012)

15. J. Eckstein, D.P. Bertsekas, On the Douglas-Rachford splitting method and the proximal point algorithm for maximal monotone operators. Math. Program. **55**(1), 293–318 (1992)

16. C. Fang, F. Cheng, Z. Lin, Faster and non-ergodic $O(1/k)$ stochastic alternating direction method of multipliers, in *Advances in Neural Information Processing Systems* (2017), pp. 4476–4485

17. C. Fang, C.J. Li, Z. Lin, T. Zhang, SPIDER: near-optimal non-convex optimization via stochastic path-integrated differential estimator, in *Advances in Neural Information Processing Systems* (2018), pp. 689–699

18. D. Gabay, Applications of the method of multipliers to variational inequalities, in *Augmented Lagrangian Methods: Applications to the Solution of Boundary-Value Problems* (1983)

19. D. Gabay, B. Mercier, A dual algorithm for the solution of nonlinear variational problems via finite element approximations. Comput. Math. **2**(1), 17–40 (1976)

20. W. Gao, D. Goldfarb, F.E. Curtis, ADMM for multiaffine constrained optimization. Optim. Methods Softw. **35**(2), 257–303 (2020)

21. J. Giesen, S. Laue, Distributed convex optimization with many convex constraints (2018) ArXiv:1610.02967

22. R. Glowinski, A. Marrocco, Sur l'approximation par éléments finis d'ordre un, et la résolution, par pénalisation-dualité d'une classe de problèmes de Dirichlet non linéaires. Rev. fr. autom. inform. rech. opér., Anal. numér. **9**(R2), 41–76 (1975)

23. D. Hajinezhad, T.-H. Chang, X. Wang, Q. Shi, M. Hong, Nonnegative matrix factorization using ADMM: algorithm and convergence analysis, in *IEEE International Conference on Acoustics, Speech, and Signal Processing* (2016), pp. 4742–4746

24. B. He, L.-Z. Liao, D. Han, H. Yang, A new inexact alternating directions method for monotone variational inequalities. Math. Program. **92**(1), 103–118 (2002)

25. B. He, M. Tao, X. Yuan, Alternating direction method with Gaussian back substitution for separable convex programming. SIAM J. Optim. **22**(2), 313–340 (2012)

26. B. He, X. Yuan, On the $O(1/t)$ convergence rate of the Douglas-Rachford alternating direction method. SIAM J. Numer. Anal. **50**(2), 700–709 (2012)

27. B. He, M. Tao, M.-H. Xu, X.-M. Yuan, Alternating directions based contraction method for generally separable linearly constrained convex programming problems. Optimization **62**, 573–596 (2013)

28. B. He, M. Tao, X. Yuan, A splitting method for separable convex programming. IMA J. Numer. Anal. **35**(1), 394–426 (2015)

29. B. He, X. Yuan, On non-ergodic convergence rate of Douglas-Rachford alternating directions method of multipliers. Numer. Math. **130**(3), 567–577 (2015)

30. M. Hong, Z.-Q. Luo, M. Razaviyayn, Convergence analysis of alternating direction method of multipliers for a family of nonconvex problems. SIAM J. Optim. **26**(1), 337–364 (2016)

31. F. Huang, S. Chen, Mini-batch stochastic ADMMs for nonconvex nonsmooth optimization (2018). ArXiv:1802.03284

32. F. Huang, S. Chen, H. Huang, Faster stochastic alternating direction method of multipliers for nonconvex optimization, in *International Conference on Machine Learning* (2019), pp. 2839–2848

33. B. Jiang, T. Lin, S. Ma, S. Zhang, Structured nonconvex and nonsmooth optimization: algorithms and iteration complexity analysis. Comput. Optim. Appl. **72**(1), 115–157 (2019)

34. R. Johnson, T. Zhang, Accelerating stochastic gradient descent using predictive variance reduction, in *Advances in Neural Information Processing Systems* (2013), pp. 315–323

35. G. Lan, S. Lee, Y. Zhou, Communication-efficient algorithms for decentralized and stochastic optimization. Math. Program. **180**(1), 237–284 (2020)
36. G. Li, T.K. Pong, Global convergence of splitting methods for nonconvex composite optimization. SIAM J. Optim. **25**(4), 2434–2460 (2015)
37. H. Li, Z. Lin, Accelerated alternating direction method of multipliers: an optimal $O(1/K)$ nonergodic analysis. J. Sci. Comput. **79**(2), 671–699 (2019)
38. J. Li, M. Xiao, C. Fang, Y. Dai, C. Xu, Z. Lin, Training deep neural networks by lifted proximal operator machines. IEEE Trans. Pattern Anal. Mach. Intell. **44**(6), 3334–3348 (2022)
39. Z. Lin, M. Chen, Y. Ma, The augmented Lagrange multiplier method for exact recovery of corrupted low-rank matrices (2010). ArXiv:1009.5055
40. Z. Lin, R. Liu, Z. Su, Linearized alternating direction method with adaptive penalty for low-rank representation, in *Advances in Neural Information Processing Systems* (2011), pp. 612–620
41. Z. Lin, R. Liu, H. Li, Linearized alternating direction method with parallel splitting and adaptive penalty for separable convex programs in machine learning. Mach. Learn. **99**(2), 287–325 (2015)
42. Z. Lin, H. Li, C. Fang, *Accelerated Optimization in Machine Learning: First-Order Algorithms* (Springer, Berlin, 2020)
43. G. Liu, Z. Lin, Y. Yu, Robust subspace segmentation by low-rank representation, in *International Conference on Machine Learning* (2010), pp. 663–670
44. R. Liu, Z. Lin, Z. Su, Linearized alternating direction method with parallel splitting and adaptive penalty for separable convex programs in machine learning, in *Asian Conference on Machine Learning* (2013), pp. 116–132
45. Y. Liu, F. Shang, H. Liu, L. Kong, L. Jiao, Z. Lin, Accelerated variance reduction stochastic ADMM for large-scale machine learning. IEEE Trans. Pattern Anal. Mach. Intell. **43**(12), 4242–4255 (2021)
46. A. Nedić, A. Olshevsky, W. Shi, Achieving geometric convergence for distributed optimization over time-varying graphs. SIAM J. Optim. **27**(4), 2597–2633 (2017)
47. H. Ouyang, N. He, L. Tran, A. Gray, Stochastic alternating direction method of multipliers, in *International Conference on Machine Learning*, pp. 80–88 (2013)
48. Y. Ouyang, Y. Chen, G. Lan, E. Pasiliao Jr., An accelerated linearized alternating direction method of multipliers. SIAM J. Imaging Sci. **8**(1), 644–681 (2015)
49. G. Qu, N. Li, Harnessing smoothness to accelerate distributed optimization. IEEE Trans. Control Netw. **5**(3), 1245–1260 (2018)
50. M.F. Sahin, A. Eftekhari, A. Alacaoglu, F.L. Gómez, V. Cevher, An inexact augmented Lagrangian framework for nonconvex optimization with nonlinear constraints, in *Advances in Neural Information Processing Systems* (2019), pp. 13943–13955
51. M. Schmidt, N. Le Roux, F. Bach, Minimizing finite sums with the stochastic average gradient. Math. Program. **162**(1–2), 83–112 (2017)
52. R. Shefi, M. Teboulle, Rate of convergence analysis of decomposition methods based on the proximal method of multipliers for convex minimization. SIAM J. Optim. **24**(1), 269–297 (2014)
53. W. Shi, Q. Ling, K. Yuan, G. Wu, W. Yin, On the linear convergence of the ADMM in decentralized consensus optimization. IEEE Trans. Signal Process. **62**(7), 1750–1761 (2014)
54. W. Shi, Q. Ling, G. Wu, W. Yin, EXTRA: an exact first-order algorithm for decentralized consensus optimization. SIAM J. Optim. **25**(2), 944–966 (2015)
55. T. Suzuki, Stochastic dual coordinate ascent with alternating direction method of multipliers, in *International Conference on Machine Learning* (2014), pp. 736–744
56. M. Tao, X. Yuan, Recovering low-rank and sparse components of matrices from incomplete and noisy observations. SIAM J. Optim. **21**(5), 57–81 (2011)
57. G. Taylor, R. Burmeister, Z. Xu, B. Singh, A. Patel, T. Goldstein, Training neural networks without gradients: a scalable ADMM approach, in *International Conference on Machine Learning* (2016), pp. 2722–2731

58. X. Wang, X. Yuan, The linearized alternating direction method for Dantzig selector. SIAM J. Sci. Comput. **34**(5), 2792–2811 (2012)
59. F. Wang, W. Cao, Z. Xu, Convergence of multi-block Bregman ADMM for nonconvex composite problems. Sci. China Inf. Sci. **61**(12), 1–12 (2018)
60. Y. Wang, W. Yin, J. Zeng, Global convergence of ADMM in nonconvex nonsmooth optimization. J. Sci. Comput. **78**(1), 29–63 (2020)
61. E. Wei, A. Ozdaglar, On the $O(1/k)$ convergence of asynchronous distributed alternating direction method of multipliers, in *IEEE Global Conference on Signal and Information Processing* (2013), pp. 551–554
62. X. Xie, J. Wu, G. Liu, Z. Zhong, Z. Lin, Differentiable linearized ADMM, in *International Conference on Machine Learning* (2019), pp. 6902–6911
63. Y. Xu, W. Yin, Z. Wen, Y. Zhang, An alternating direction algorithm for matrix completion with nonnegative factors. Front. Math. China **7**(2), 365–384 (2012)
64. J. Xu, S. Zhu, Y. C. Soh, L. Xie, Augmented distributed gradient methods for multi-agent optimization under uncoordinated constant stepsizes, in *IEEE Conference on Decision and Control (CDC)* (2015), pp. 2055–2060
65. Y. Yang, J. Sun, H. Li, Z. Xu, Deep ADMM-Net for compressive sensing MRI, in *Advances in Neural Information Processing Systems* (2016), pp. 10–18
66. S. Zheng, J. Kwok, Fast-and-light stochastic ADMM, in *International Joint Conference on Artificial Intelligence* (2016), pp. 2407–2613
67. J. Zhang, Z.-Q. Luo, A proximal alternating direction method of multiplier for linearly constrained nonconvex minimization. SIAM J. Optim. **30**(3), 2272–2302 (2020)
68. W. Zhong, J. Kwok, Fast stochastic alternating direction method of multipliers, in *International Conference on Machine Learning* (2014), pp. 46–54

# Chapter 2
# Derivations of ADMM

In this chapter, we introduce how to derive ADMM from the Lagrangian viewpoint and the operator splitting viewpoint, respectively. Especially, the former one provides some useful background and motivation.

## 2.1  Lagrangian Viewpoint of ADMM

We first briefly introduce two methods based on the Lagrangian function and then give the ADMM.

### 2.1.1  Dual Ascent

Consider the following linearly constrained convex problem:

$$\min_{\mathbf{x}} f(\mathbf{x}), \quad s.t. \quad \mathbf{Ax} = \mathbf{b}, \tag{2.1}$$

where $f(\mathbf{x})$ is proper (Definition A.25), closed (Definition A.12) and convex (Definition A.4).[1] We can solve it by dual ascent. Introduce the Lagrangian function (Definition A.17)

$$L(\mathbf{x}, \boldsymbol{\lambda}) = f(\mathbf{x}) + \langle \boldsymbol{\lambda}, \mathbf{Ax} - \mathbf{b} \rangle,$$

---

[1] As we are not interested in pathological functions, for brevity in this book we do not emphasize properness and closedness when a function is convex, especially the former.

© The Author(s), under exclusive license to Springer Nature Singapore Pte Ltd. 2022
Z. Lin et al., *Alternating Direction Method of Multipliers for Machine Learning*,
https://doi.org/10.1007/978-981-16-9840-8_2

where $\boldsymbol{\lambda}$ is the Lagrange multiplier. The dual function (Definition A.18) is

$$
\begin{aligned}
d(\boldsymbol{\lambda}) &= \min_{\mathbf{x}} L(\mathbf{x}, \boldsymbol{\lambda}) \\
&= \min_{\mathbf{x}} \left( f(\mathbf{x}) + \langle \boldsymbol{\lambda}, \mathbf{A}\mathbf{x} - \mathbf{b} \rangle \right) \\
&= -\max_{\mathbf{x}} \left( -f(\mathbf{x}) - \left\langle \mathbf{A}^T \boldsymbol{\lambda}, \mathbf{x} \right\rangle \right) - \langle \boldsymbol{\lambda}, \mathbf{b} \rangle \\
&= -f^*(-\mathbf{A}^T \boldsymbol{\lambda}) - \langle \boldsymbol{\lambda}, \mathbf{b} \rangle ,
\end{aligned}
\tag{2.2}
$$

where $f^*$ is the conjugate function (Definition A.16) of $f$. $d(\boldsymbol{\lambda})$ is concave (Definition A.5) and the domain of $d(\boldsymbol{\lambda})$ is $\mathcal{D} = \{\boldsymbol{\lambda} | d(\boldsymbol{\lambda}) > -\infty\}$. The dual problem (Definition A.19) is

$$
\max_{\boldsymbol{\lambda} \in \mathcal{D}} d(\boldsymbol{\lambda}). \tag{2.3}
$$

Denote $\boldsymbol{\lambda}^*$ as the optimal solution of the dual problem. We can recover the optimal solution of the primal problem (2.1) as

$$
\mathbf{x}^* = \operatorname*{argmin}_{\mathbf{x}} L(\mathbf{x}, \boldsymbol{\lambda}^*),
$$

as the strong duality (Proposition A.13) holds for Problem (2.1). When $f$ is strictly convex (Definition A.6), due to Danskin's theorem (Theorem A.1) and Proposition A.7, we know that $d(\boldsymbol{\lambda})$ is differentiable and $\nabla d(\boldsymbol{\lambda}^k) = \mathbf{A}\mathbf{x}^{k+1} - \mathbf{b}$, where $\mathbf{x}^{k+1}$ is the minimizer of $L(\mathbf{x}, \boldsymbol{\lambda}^k)$. So we can use the gradient ascent method to solve the dual problem (2.3), which consists of the following iterations:

$$
\mathbf{x}^{k+1} = \operatorname*{argmin}_{\mathbf{x}} L(\mathbf{x}, \boldsymbol{\lambda}^k), \tag{2.4}
$$

$$
\boldsymbol{\lambda}^{k+1} = \boldsymbol{\lambda}^k + \alpha_k (\mathbf{A}\mathbf{x}^{k+1} - \mathbf{b}), \tag{2.5}
$$

where $\alpha_k$ is the step size which needs to be chosen appropriately. The first step is a minimization step in the primal space, while the second step is the update in the dual space.

### 2.1.2   Augmented Lagrangian Method

The disadvantage of the dual ascent method is that to make the dual function differentiable, we require $f$ to be strictly convex. Otherwise, (2.5) is a subgradient ascent of the dual function, and the resulted dual subgradient ascent converges much slower. Even worse, the subproblem (2.4) may not have a solution, e.g., when $f$ is an affine function of $\mathbf{x}$. To address these issues, we can use the augmented Lagrangian method [6]. Introduce the augmented Lagrangian function

$$L_\beta(\mathbf{x}, \boldsymbol{\lambda}) = f(\mathbf{x}) + \langle \boldsymbol{\lambda}, \mathbf{A}\mathbf{x} - \mathbf{b} \rangle + \frac{\beta}{2} \|\mathbf{A}\mathbf{x} - \mathbf{b}\|^2,$$

where $\beta > 0$ is called the penalty parameter. The associated dual function is

$$d_\beta(\boldsymbol{\lambda}) = \min_{\mathbf{x}} L_\beta(\mathbf{x}, \boldsymbol{\lambda}).$$

For any $\boldsymbol{\lambda}$, we have $d(\boldsymbol{\lambda}) \leq d_\beta(\boldsymbol{\lambda})$. Moreover, for any $\boldsymbol{\lambda}$, we have $d_\beta(\boldsymbol{\lambda}) \leq f(\mathbf{x}^*)$. Since $d(\boldsymbol{\lambda}^*) = f(\mathbf{x}^*)$, we know $d(\boldsymbol{\lambda}^*) = d_\beta(\boldsymbol{\lambda}^*) = f(\mathbf{x}^*)$. Thus introducing the augmented term $\frac{\beta}{2}\|\mathbf{A}\mathbf{x}-\mathbf{b}\|^2$ does not change the solution. Another way to draw this conclusion immediately is to regard $L_\beta(\mathbf{x}, \boldsymbol{\lambda})$ as the Lagrangian function associated with the following problem:

$$\min_{\mathbf{x}} \left( f(\mathbf{x}) + \frac{\beta}{2} \|\mathbf{A}\mathbf{x} - \mathbf{b}\|^2 \right), \quad s.t. \quad \mathbf{A}\mathbf{x} = \mathbf{b}.$$

However, using the augmented Lagrangian function brings great benefits: for $d_\beta(\boldsymbol{\lambda})$ to be differentiable we only require $f$ to be convex, rather than being strictly convex, as shown by the following lemma.

**Lemma 2.1** *Let $\mathcal{D}(\boldsymbol{\lambda})$ denote the optimal solution set of $\min_{\mathbf{x}} L_\beta(\mathbf{x}, \boldsymbol{\lambda})$. Then $\mathbf{A}\mathbf{x}$ is invariant over $\mathcal{D}(\boldsymbol{\lambda})$. Moreover, $d_\beta(\boldsymbol{\lambda})$ is differentiable and*

$$\nabla d_\beta(\boldsymbol{\lambda}) = \mathbf{A}\mathbf{x}(\boldsymbol{\lambda}) - \mathbf{b},$$

*where $\mathbf{x}(\boldsymbol{\lambda}) \in \mathcal{D}(\boldsymbol{\lambda})$ is any minimizer of $L_\beta(\mathbf{x}, \boldsymbol{\lambda})$. We also have that $d_\beta(\boldsymbol{\lambda})$ is $\frac{1}{\beta}$-smooth (Definition A.9), i.e.,*

$$\|\nabla d_\beta(\boldsymbol{\lambda}) - \nabla d_\beta(\boldsymbol{\lambda}')\| \leq \frac{1}{\beta} \|\boldsymbol{\lambda} - \boldsymbol{\lambda}'\|.$$

**Proof** Suppose that there exist $\mathbf{x}$ and $\mathbf{x}' \in \mathcal{D}(\boldsymbol{\lambda})$ with $\mathbf{A}\mathbf{x} \neq \mathbf{A}\mathbf{x}'$. Then we have

$$d_\beta(\boldsymbol{\lambda}) = L_\beta(\mathbf{x}, \boldsymbol{\lambda}) = L_\beta(\mathbf{x}', \boldsymbol{\lambda}).$$

Due to the convexity of $L_\beta(\mathbf{x}, \boldsymbol{\lambda})$ with respect to $\mathbf{x}$, $\mathcal{D}(\boldsymbol{\lambda})$ must be convex, implying

$$\bar{\mathbf{x}} = (\mathbf{x} + \mathbf{x}')/2 \in \mathcal{D}(\boldsymbol{\lambda}).$$

By the convexity of $f$ and strict convexity of $\| \cdot \|^2$, we have

$$d_\beta(\boldsymbol{\lambda}) = \frac{1}{2}L_\beta(\mathbf{x}, \boldsymbol{\lambda}) + \frac{1}{2}L_\beta(\mathbf{x}', \boldsymbol{\lambda})$$

$$> f(\overline{\mathbf{x}}) + \langle \mathbf{A}\overline{\mathbf{x}} - \mathbf{b}, \boldsymbol{\lambda} \rangle + \frac{\beta}{2}\|\mathbf{A}\overline{\mathbf{x}} - \mathbf{b}\|^2 = L_\beta(\overline{\mathbf{x}}, \boldsymbol{\lambda}).$$

This contradicts the definition $d_\beta(\boldsymbol{\lambda}) = \min_{\mathbf{x}} L_\beta(\mathbf{x}, \boldsymbol{\lambda})$. Thus $\mathbf{A}\mathbf{x}$ is invariant over $\mathcal{D}(\boldsymbol{\lambda})$. So by Danskin's theorem (Theorem A.1), $\partial d_\beta(\mathbf{u})$ is a singleton $\{\mathbf{A}\mathbf{x}(\boldsymbol{\lambda}) - \mathbf{b}\}$, where $\mathbf{x}(\boldsymbol{\lambda}) \in \mathcal{D}(\boldsymbol{\lambda})$. By Proposition A.7, we know that $d_\beta(\boldsymbol{\lambda})$ is differentiable and

$$\nabla d_\beta(\boldsymbol{\lambda}) = \mathbf{A}\mathbf{x}(\boldsymbol{\lambda}) - \mathbf{b}.$$

Let

$$\mathbf{x} = \operatorname*{argmin}_{\mathbf{x}} L_\beta(\mathbf{x}, \boldsymbol{\lambda}) \quad \text{and} \quad \mathbf{x}' = \operatorname*{argmin}_{\mathbf{x}} L_\beta(\mathbf{x}, \boldsymbol{\lambda}').$$

Then we have

$$\mathbf{0} \in \partial f(\mathbf{x}) + \mathbf{A}^T\boldsymbol{\lambda} + \beta\mathbf{A}^T(\mathbf{A}\mathbf{x} - \mathbf{b}),$$

$$\mathbf{0} \in \partial f(\mathbf{x}') + \mathbf{A}^T\boldsymbol{\lambda}' + \beta\mathbf{A}^T(\mathbf{A}\mathbf{x}' - \mathbf{b}).$$

From the monotonicity of $\partial f$ (Proposition A.10), we have

$$\left\langle -(\mathbf{A}^T\boldsymbol{\lambda} + \beta\mathbf{A}^T(\mathbf{A}\mathbf{x} - \mathbf{b})) + (\mathbf{A}^T\boldsymbol{\lambda}' + \beta\mathbf{A}^T(\mathbf{A}\mathbf{x}' - \mathbf{b})), \mathbf{x} - \mathbf{x}' \right\rangle \geq 0$$

$$\Rightarrow \quad \langle \boldsymbol{\lambda} - \boldsymbol{\lambda}', \mathbf{A}\mathbf{x} - \mathbf{A}\mathbf{x}' \rangle + \beta\|\mathbf{A}\mathbf{x} - \mathbf{A}\mathbf{x}'\|^2 \leq 0$$

$$\Rightarrow \quad \beta\|\mathbf{A}\mathbf{x} - \mathbf{A}\mathbf{x}'\| \leq \|\boldsymbol{\lambda} - \boldsymbol{\lambda}'\|.$$

So we have

$$\|\nabla d_\beta(\boldsymbol{\lambda}) - \nabla d_\beta(\boldsymbol{\lambda}')\| = \|\mathbf{A}\mathbf{x} - \mathbf{A}\mathbf{x}'\| \leq \frac{1}{\beta}\|\boldsymbol{\lambda} - \boldsymbol{\lambda}'\|.$$

$\square$

Applying dual ascent to $d_\beta(\boldsymbol{\lambda})$, we have the following augmented Lagrangian method:

$$\mathbf{x}^{k+1} = \operatorname*{argmin}_{\mathbf{x}} L_\beta(\mathbf{x}, \boldsymbol{\lambda}^k), \tag{2.6}$$

$$\boldsymbol{\lambda}^{k+1} = \boldsymbol{\lambda}^k + \beta(\mathbf{A}\mathbf{x}^{k+1} - \mathbf{b}). \tag{2.7}$$

Note that the step size is chosen as $\beta$. One of the reasons will be shown immediately. Discussions on other choices of step sizes can be found in Sect. 7.1.2.

The augmented Lagrangian method can also be derived from the dual problem. Since the dual function of Problem (2.1) is (2.2), the dual problem is

$$\min_{\boldsymbol{\lambda}} \left( f^*(-\mathbf{A}^T\boldsymbol{\lambda}) + \langle \boldsymbol{\lambda}, \mathbf{b}\rangle \right). \tag{2.8}$$

We may use the proximal point method to solve (2.8):

$$\boldsymbol{\lambda}^{k+1} = \operatorname*{argmin}_{\boldsymbol{\lambda}} \left( f^*(-\mathbf{A}^T\boldsymbol{\lambda}) + \langle \boldsymbol{\lambda}, \mathbf{b}\rangle + \frac{1}{2\beta}\|\boldsymbol{\lambda} - \boldsymbol{\lambda}^k\|^2 \right). \tag{2.9}$$

The optimality condition of (2.9) is

$$\mathbf{0} \in -\mathbf{A}\partial f^* \left( -\mathbf{A}^T\boldsymbol{\lambda}^{k+1} \right) + \mathbf{b} + \frac{1}{\beta} \left( \boldsymbol{\lambda}^{k+1} - \boldsymbol{\lambda}^k \right).$$

So there exists

$$\mathbf{x}^{k+1} \in \partial f^* \left( -\mathbf{A}^T\boldsymbol{\lambda}^{k+1} \right), \tag{2.10}$$

such that

$$\mathbf{0} = -\mathbf{A}\mathbf{x}^{k+1} + \mathbf{b} + \frac{1}{\beta} \left( \boldsymbol{\lambda}^{k+1} - \boldsymbol{\lambda}^k \right),$$

which gives

$$\boldsymbol{\lambda}^{k+1} = \boldsymbol{\lambda}^k + \beta \left( \mathbf{A}\mathbf{x}^{k+1} - \mathbf{b} \right). \tag{2.11}$$

On the other hand, (2.10) and Point 5 of Proposition A.11 give $-\mathbf{A}^T\boldsymbol{\lambda}^{k+1} \in \partial f(\mathbf{x}^{k+1})$, i.e.,

$$\mathbf{0} \in \partial f(\mathbf{x}^{k+1}) + \mathbf{A}^T\boldsymbol{\lambda}^{k+1}$$
$$= \partial f(\mathbf{x}^{k+1}) + \mathbf{A}^T \left[ \boldsymbol{\lambda}^k + \beta(\mathbf{A}\mathbf{x}^{k+1} - \mathbf{b}) \right],$$

which gives

$$\mathbf{x}^{k+1} = \operatorname*{argmin}_{\mathbf{x}} \left( f(\mathbf{x}) + \left\langle \boldsymbol{\lambda}^k, \mathbf{A}\mathbf{x}\right\rangle + \frac{\beta}{2}\|\mathbf{A}\mathbf{x} - \mathbf{b}\|^2 \right), \tag{2.12}$$

supposing that the solution of (2.12) is unique. (2.12) and (2.11) constitute the augmented Lagrangian method.

### 2.1.3   Alternating Direction Method of Multipliers

Model (2.1) covers many problems in real applications. Consider a special case of Problem (2.1), which has the following separable structure and arises from diverse applications in machine learning, image processing, and computer vision:

$$\min_{\mathbf{x},\mathbf{y}} \ (f(\mathbf{x}) + g(\mathbf{y})), \quad s.t. \quad \mathbf{Ax} + \mathbf{By} = \mathbf{b}. \tag{2.13}$$

Introduce the augmented Lagrangian function

$$L_\beta(\mathbf{x}, \mathbf{y}, \boldsymbol{\lambda}) = f(\mathbf{x}) + g(\mathbf{y}) + \langle \mathbf{Ax} + \mathbf{By} - \mathbf{b}, \boldsymbol{\lambda} \rangle + \frac{\beta}{2} \|\mathbf{Ax} + \mathbf{By} - \mathbf{b}\|^2.$$

When we use the augmented Lagrangian method to solve Problem (2.13), we need to solve the following subproblem

$$(\mathbf{x}^{k+1}, \mathbf{y}^{k+1}) = \operatorname*{argmin}_{\mathbf{x},\mathbf{y}} \left( f(\mathbf{x}) + g(\mathbf{y}) + \left\langle \mathbf{Ax} + \mathbf{By} - \mathbf{b}, \boldsymbol{\lambda}^k \right\rangle \right.$$
$$\left. + \frac{\beta}{2} \|\mathbf{Ax} + \mathbf{By} - \mathbf{b}\|^2 \right), \tag{2.14}$$

which is minimized jointly with respect to $\mathbf{x}$ and $\mathbf{y}$. Sometimes, it is much simpler when we solve (2.14) for $\mathbf{x}$ and $\mathbf{y}$ separately, which motivates the ADMM [3, 4]. Different from the augmented Lagrangian method, ADMM updates $\mathbf{x}$ and $\mathbf{y}$ in an alternating (or called sequential) fashion. ADMM consists of the following iterations:

$$\mathbf{x}^{k+1} = \operatorname*{argmin}_{\mathbf{x}} \left( f(\mathbf{x}) + g(\mathbf{y}^k) + \left\langle \boldsymbol{\lambda}^k, \mathbf{Ax} + \mathbf{By}^k - \mathbf{b} \right\rangle \right.$$
$$\left. + \frac{\beta}{2} \|\mathbf{Ax} + \mathbf{By}^k - \mathbf{b}\|^2 \right), \tag{2.15a}$$

$$\mathbf{y}^{k+1} = \operatorname*{argmin}_{\mathbf{y}} \left( f(\mathbf{x}^{k+1}) + g(\mathbf{y}) + \left\langle \boldsymbol{\lambda}^k, \mathbf{Ax}^{k+1} + \mathbf{By} - \mathbf{b} \right\rangle \right.$$
$$\left. + \frac{\beta}{2} \|\mathbf{Ax}^{k+1} + \mathbf{By} - \mathbf{b}\|^2 \right), \tag{2.15b}$$

$$\boldsymbol{\lambda}^{k+1} = \boldsymbol{\lambda}^k + \beta(\mathbf{Ax}^{k+1} + \mathbf{By}^{k+1} - \mathbf{b}). \tag{2.15c}$$

ADMM is superior to the augmented Lagrangian method when the $\mathbf{x}$ and $\mathbf{y}$ subproblems, (2.15a) and (2.15b), can be more efficiently solved than the $(\mathbf{x}, \mathbf{y})$ subproblem in (2.14). We present ADMM in Algorithm 2.1 for the convenience of reference.

---

**Algorithm 2.1** Original ADMM

---

Initialize $\mathbf{x}^0$, $\mathbf{y}^0$, and $\boldsymbol{\lambda}^0$.
**for** $k = 0, 1, 2, 3, \cdots$ **do**
    Update $\mathbf{x}^{k+1}$, $\mathbf{y}^{k+1}$, and $\boldsymbol{\lambda}^{k+1}$ by (2.15a), (2.15b), and (2.15c), respectively.
    $k \leftarrow k + 1$.
**end for**

---

### 2.1.4 Relation to the Split Bregman Method

The split Bregman method was proposed by Goldstein and Osher [5] and has been widely used in image processing. Recall the Bregman distance (Definition A.15):

$$D_\phi^{\mathbf{v}}(\mathbf{y}, \mathbf{x}) = \phi(\mathbf{y}) - \phi(\mathbf{x}) - \langle \mathbf{v}, \mathbf{y} - \mathbf{x} \rangle,$$

where $\phi$ is convex but may not be differentiable and $\mathbf{v} \in \partial\phi(\mathbf{x})$. Consider Problem (2.1) and let

$$h(\mathbf{x}) = \frac{1}{2}\|\mathbf{A}\mathbf{x} - \mathbf{b}\|^2.$$

We can use the following Bregman method to solve Problem (2.1):

$$\mathbf{x}^{k+1} = \operatorname*{argmin}_{\mathbf{x}} \left( D_f^{\mathbf{v}^k}(\mathbf{x}, \mathbf{x}^k) + \beta h(\mathbf{x}) \right), \tag{2.16}$$

$$\mathbf{v}^{k+1} = \mathbf{v}^k - \beta\nabla h(\mathbf{x}^{k+1}), \tag{2.17}$$

where $\mathbf{v}^0 \in \partial f(\mathbf{x}^0) \cap \operatorname{Span}(\mathbf{A}^T)$.[2] From $\nabla h(\mathbf{x}) = \mathbf{A}^T(\mathbf{A}\mathbf{x} - \mathbf{b})$ and (2.17), we know that

$$\mathbf{v}^k \in \operatorname{Span}(\mathbf{A}^T), \quad \forall k \geq 0. \tag{2.18}$$

The optimality condition of step (2.16) ensures

$$0 \in \partial f(\mathbf{x}^{k+1}) - \mathbf{v}^k + \beta\nabla h(\mathbf{x}^{k+1}).$$

Combining with step (2.17), we have

$$\mathbf{v}^{k+1} \in \partial f(\mathbf{x}^{k+1}), \quad \forall k \geq 0.$$

---

[2] This can be realized by first choosing $\mathbf{v}^{-1} \in \operatorname{Span}(\mathbf{A}^T)$, then solving $\mathbf{x}^0 = \operatorname{argmin}_{\mathbf{x}} \left( f(\mathbf{x}) - \langle \mathbf{v}^{-1}, \mathbf{x} \rangle + \beta h(\mathbf{x}) \right)$, and finally setting $\mathbf{v}^0 = \mathbf{v}^{-1} - \beta\nabla h(\mathbf{x}^0)$.

So the Bregman distance $D_f^{\mathbf{v}^k}(\mathbf{x}, \mathbf{x}^k)$ is well defined for all $k \geq 0$. Recall that $\mathbf{x}^*$ is an optimal solution of problem (2.1) if and only if $\mathbf{A}\mathbf{x}^* = \mathbf{b}$ and there exists $\mathbf{v}^*$ such that $\mathbf{v}^* \in \partial f(\mathbf{x}^*) \cap \mathrm{Span}(\mathbf{A}^T)$. So we only need to ensure $\mathbf{A}\mathbf{x}^{k+1} = \mathbf{b}$ so that $\mathbf{x}^{k+1}$ is the optimal solution of problem (2.1). We know that $h(\mathbf{x}^{k+1})$ reduces to 0 from the convergence guarantee of the Bregman method.

On the other hand, by (2.18) we know that there exists $\boldsymbol{\lambda}^k$ such that $\mathbf{v}^k = -\mathbf{A}^T\boldsymbol{\lambda}^k$. So the above method reduces to the following iterations:

$$\mathbf{x}^{k+1} = \underset{\mathbf{x}}{\mathrm{argmin}} \left( f(\mathbf{x}) - \left\langle \mathbf{v}^k, \mathbf{x} \right\rangle + \beta h(\mathbf{x}) \right)$$

$$= \underset{\mathbf{x}}{\mathrm{argmin}} \left( f(\mathbf{x}) + \left\langle \boldsymbol{\lambda}^k, \mathbf{A}\mathbf{x} \right\rangle + \frac{\beta}{2} \|\mathbf{A}\mathbf{x} - \mathbf{b}\|^2 \right),$$

$$\mathbf{A}^T\boldsymbol{\lambda}^{k+1} = \mathbf{A}^T\boldsymbol{\lambda}^k + \beta\nabla h(\mathbf{x}^{k+1})$$

$$\Rightarrow \quad \boldsymbol{\lambda}^{k+1} = \boldsymbol{\lambda}^k + \beta(\mathbf{A}\mathbf{x}^{k+1} - \mathbf{b}),$$

which coincides with the augmented Lagrangian method given in (2.6)–(2.7). In the last step, we assume that $\mathbf{A}$ is of full row rank.

Now, we apply the above Bregman method to Problem (2.13) directly, leading to the following method:

$$\left(\mathbf{x}^{k+1}, \mathbf{y}^{k+1}\right) = \underset{\mathbf{x}, \mathbf{y}}{\mathrm{argmin}} \left( D_f^{\mathbf{v}_1^k}(\mathbf{x}, \mathbf{x}^k) + D_g^{\mathbf{v}_2^k}(\mathbf{y}, \mathbf{y}^k) + \beta h(\mathbf{x}, \mathbf{y}) \right),$$

$$\begin{pmatrix} \mathbf{v}_1^{k+1} \\ \mathbf{v}_2^{k+1} \end{pmatrix} = \begin{pmatrix} \mathbf{v}_1^k \\ \mathbf{v}_2^k \end{pmatrix} - \beta\nabla h(\mathbf{x}^{k+1}, \mathbf{y}^{k+1})$$

$$= \begin{pmatrix} \mathbf{v}_1^k \\ \mathbf{v}_2^k \end{pmatrix} - \beta \begin{pmatrix} \mathbf{A}^T \\ \mathbf{B}^T \end{pmatrix} \left(\mathbf{A}\mathbf{x}^{k+1} + \mathbf{B}\mathbf{y}^{k+1} - \mathbf{b}\right),$$

where we let

$$h(\mathbf{x}, \mathbf{y}) = \frac{1}{2}\|\mathbf{A}\mathbf{x} + \mathbf{B}\mathbf{y} - \mathbf{b}\|^2 \quad \text{and} \quad \mathbf{v}^0 = \begin{pmatrix} \mathbf{v}_1^0 \\ \mathbf{v}_2^0 \end{pmatrix} \in \begin{pmatrix} \partial f(\mathbf{x}^0) \\ \partial g(\mathbf{y}^0) \end{pmatrix} \cap \mathrm{Span}\left( \begin{pmatrix} \mathbf{A}^T \\ \mathbf{B}^T \end{pmatrix} \right).$$

Letting $\begin{pmatrix} \mathbf{v}_1^k \\ \mathbf{v}_2^k \end{pmatrix} = -\begin{pmatrix} \mathbf{A}^T \\ \mathbf{B}^T \end{pmatrix}\boldsymbol{\lambda}^k$, we eliminate $\mathbf{v}_1^k$ and $\mathbf{v}_2^k$ to get

$$(\mathbf{x}^{k+1}, \mathbf{y}^{k+1}) = \underset{\mathbf{x}, \mathbf{y}}{\mathrm{argmin}} \left( f(\mathbf{x}) + g(\mathbf{y}) + \left\langle \mathbf{A}^T\boldsymbol{\lambda}^k, \mathbf{x} \right\rangle + \left\langle \mathbf{B}^T\boldsymbol{\lambda}^k, \mathbf{y} \right\rangle + \beta h(\mathbf{x}, \mathbf{y}) \right)$$

$$= \underset{\mathbf{x}, \mathbf{y}}{\mathrm{argmin}} \left( f(\mathbf{x}) + g(\mathbf{y}) + \left\langle \boldsymbol{\lambda}^k, \mathbf{A}\mathbf{x} + \mathbf{B}\mathbf{y} - \mathbf{b} \right\rangle \right)$$

$$+\frac{\beta}{2}\|\mathbf{Ax} + \mathbf{By} - \mathbf{b}\|^2\bigg),$$

$$\boldsymbol{\lambda}^{k+1} = \boldsymbol{\lambda}^k + \beta\left(\mathbf{Ax}^{k+1} + \mathbf{By}^{k+1} - \mathbf{b}\right),$$

where we assume that $[\mathbf{A}, \mathbf{B}]$ is of full row rank. In the first step, we can compute $\mathbf{x}^{k+1}$ and $\mathbf{y}^{k+1}$ through alternating updates until convergence, and the corresponding method is called the split Bregman method [5]. When we only use one pass of updates, that is, firstly minimize with respect to $\mathbf{x}$, and then to $\mathbf{y}$, the split Bregman method coincides with ADMM.

## 2.2   Operator Splitting Viewpoint of ADMM

We first introduce the Douglas–Rachford splitting (DRS), a special operator splitting method, and then derive ADMM from DRS.

### 2.2.1   Douglas–Rachford splitting

We say that $\mathcal{T}$ is a set-valued operator $\mathcal{T} : \mathbb{R}^n \rightrightarrows \mathbb{R}^n$, if it maps a point in $\mathbb{R}^n$ to a (possibly empty) subset of $\mathbb{R}^n$. The inverse operator of $\mathcal{T}$ is denoted as $\mathcal{T}^{-1}$. Given a closed and convex function $f$ on $\mathbb{R}^n$, its subgradient $\partial f$ is a set-valued operator, actually a maximal monotone operator (Definition A.14, Proposition A.10), and $\mathrm{Argmin}_{\mathbf{x}\in\mathbb{R}^n} f(\mathbf{x}) = \{\mathbf{x}|\mathbf{0} \in \partial f(\mathbf{x})\}$. Define the resolvent of an operator $\mathcal{T}$ as $\mathcal{J}_{\mathcal{T}} = (I + \mathcal{T})^{-1}$, where $I$ is the identity mapping. $\mathcal{J}_{\mathcal{T}}$ is single-valued if $\mathcal{T}$ is maximal monotone (Proposition A.9). When $\mathcal{T} = \partial f$ and $\alpha > 0$, we have

$$\mathbf{x} = \mathcal{J}_{\alpha\mathcal{T}}(\mathbf{z}) = (I + \alpha\partial f)^{-1}(\mathbf{z})$$

$$\Leftrightarrow \mathbf{z} \in \mathbf{x} + \alpha\partial f(\mathbf{x})$$

$$\Leftrightarrow \mathbf{0} \in \partial\left(\alpha f(\mathbf{x}) + \frac{1}{2}\|\mathbf{x} - \mathbf{z}\|^2\right)$$

$$\Leftrightarrow \mathbf{x} = \mathrm{Prox}_{\alpha f}(\mathbf{z}) \equiv \underset{\mathbf{x}\in\mathbb{R}^n}{\mathrm{argmin}}\left(f(\mathbf{x}) + \frac{1}{2\alpha}\|\mathbf{x} - \mathbf{z}\|^2\right).$$

Consider the inclusion problem

$$\mathrm{find}_{\mathbf{x}\in\mathbb{R}^n} \ \mathbf{0} \in (\mathcal{A} + \mathcal{B})\mathbf{x},$$

where $\mathcal{A}$ and $\mathcal{B}$ are both maximal monotone. Douglas–Rachford splitting [1], a widely used operator splitting method, consists of the following steps:

$$\mathbf{v}^k = \mathcal{J}_{\alpha\mathcal{B}}(\mathbf{x}^k), \tag{2.19a}$$

$$\mathbf{u}^{k+1} = \mathcal{J}_{\alpha\mathcal{A}}(2\mathbf{v}^k - \mathbf{x}^k), \tag{2.19b}$$

$$\mathbf{x}^{k+1} = \mathbf{x}^k + \mathbf{u}^{k+1} - \mathbf{v}^k. \tag{2.19c}$$

(2.19a)–(2.19c) can be written as the following fixed-point iteration:

$$\mathbf{x}^{k+1} = \mathcal{T}(\mathbf{x}^k),$$

where

$$\mathcal{T}(\mathbf{x}) = \mathbf{x} + \mathcal{J}_{\alpha\mathcal{A}}(2\mathcal{J}_{\alpha\mathcal{B}}(\mathbf{x}) - \mathbf{x}) - \mathcal{J}_{\alpha\mathcal{B}}(\mathbf{x}).$$

We claim that $\mathbf{x}$ is a fixed point of $\mathcal{T}$ if and only if $\mathbf{0} \in (\mathcal{A}+\mathcal{B})\mathbf{v}$, where $\mathbf{v} = \mathcal{J}_{\alpha\mathcal{B}}(\mathbf{x})$. In fact, $\mathbf{v} = \mathcal{J}_{\alpha\mathcal{B}}(\mathbf{x})$ is equivalent to

$$\mathbf{x} \in \mathbf{v} + \alpha\mathcal{B}(\mathbf{v}) \Leftrightarrow \mathbf{x} - \mathbf{v} \in \alpha\mathcal{B}(\mathbf{v}).$$

On the other hand,

$$\mathbf{x} = \mathcal{T}(\mathbf{x}) \Leftrightarrow \mathbf{v} = \mathcal{J}_{\alpha\mathcal{A}}(2\mathbf{v} - \mathbf{x})$$

$$\Leftrightarrow 2\mathbf{v} - \mathbf{x} \in \mathbf{v} + \alpha\mathcal{A}(\mathbf{v})$$

$$\Leftrightarrow \mathbf{v} - \mathbf{x} \in \alpha\mathcal{A}(\mathbf{v})$$

$$\Leftrightarrow \mathbf{0} \in (\mathcal{A} + \mathcal{B})\mathbf{v}.$$

Next, we give an equivalent form of DRS. Switching $\mathbf{v}$-update and $\mathbf{u}$-update, (2.19a)–(2.19c) is equivalent to

$$\mathbf{u}^{k+1} = \mathcal{J}_{\alpha\mathcal{A}}(2\mathbf{v}^k - \mathbf{x}^k),$$

$$\mathbf{v}^{k+1} = \mathcal{J}_{\alpha\mathcal{B}}(\mathbf{u}^{k+1} + \mathbf{x}^k - \mathbf{v}^k),$$

$$\mathbf{x}^{k+1} = \mathbf{x}^k + \mathbf{u}^{k+1} - \mathbf{v}^k.$$

Letting $\mathbf{w}^k = \mathbf{v}^k - \mathbf{x}^k$, it is further equivalent to

$$\mathbf{u}^{k+1} = \mathcal{J}_{\alpha\mathcal{A}}(\mathbf{v}^k + \mathbf{w}^k), \tag{2.20a}$$

$$\mathbf{v}^{k+1} = \mathcal{J}_{\alpha\mathcal{B}}(\mathbf{u}^{k+1} - \mathbf{w}^k), \tag{2.20b}$$

$$\mathbf{w}^{k+1} = \mathbf{w}^k + \mathbf{v}^{k+1} - \mathbf{u}^{k+1}. \tag{2.20c}$$

### 2.2.2   From DRS to ADMM

ADMM can also be derived from DRS [2] and accordingly shares the theoretical properties of the operator splitting methods.

The dual function of Problem (2.13) is

$$\min_{\mathbf{x},\mathbf{y}} \left( f(\mathbf{x}) + g(\mathbf{y}) + \langle \mathbf{A}\mathbf{x} + \mathbf{B}\mathbf{y} - \mathbf{b}, \boldsymbol{\lambda} \rangle \right)$$

$$= -\max_{\mathbf{x}} \left( -\left\langle \mathbf{A}^T \boldsymbol{\lambda}, \mathbf{x} \right\rangle - f(\mathbf{x}) \right) - \max_{\mathbf{y}} \left( -\left\langle \mathbf{B}^T \boldsymbol{\lambda}, \mathbf{y} \right\rangle - g(\mathbf{y}) \right) - \langle \boldsymbol{\lambda}, \mathbf{b} \rangle$$

$$= -f^*(-\mathbf{A}^T \boldsymbol{\lambda}) - g^*(-\mathbf{B}^T \boldsymbol{\lambda}) - \langle \boldsymbol{\lambda}, \mathbf{b} \rangle .$$

So the dual problem of (2.13) can be written as

$$\min_{\boldsymbol{\lambda}} \left( f^*(-\mathbf{A}^T \boldsymbol{\lambda}) + g^*(-\mathbf{B}^T \boldsymbol{\lambda}) + \langle \boldsymbol{\lambda}, \mathbf{b} \rangle \right) .$$

Denote

$$\psi_1(\boldsymbol{\lambda}) = f^*(-\mathbf{A}^T \boldsymbol{\lambda}) + \mathbf{b}^T \boldsymbol{\lambda} \quad \text{and} \quad \psi_2(\boldsymbol{\lambda}) = g^*(-\mathbf{B}^T \boldsymbol{\lambda}). \tag{2.21}$$

We can use DRS (2.20a)–(2.20c) to solve Problem (2.21), which consists of the following iterations:

$$\mathbf{u}^{k+1} = \text{Prox}_{\beta\psi_1}(\mathbf{v}^k + \mathbf{w}^k),$$

$$\mathbf{v}^{k+1} = \text{Prox}_{\beta\psi_2}(\mathbf{u}^{k+1} - \mathbf{w}^k),$$

$$\mathbf{w}^{k+1} = \mathbf{w}^k + \mathbf{v}^{k+1} - \mathbf{u}^{k+1}.$$

From the optimality condition of the first step of DRS, we have

$$\mathbf{0} \in \partial \psi_1(\mathbf{u}^{k+1}) + \frac{1}{\beta}(\mathbf{u}^{k+1} - \mathbf{v}^k - \mathbf{w}^k)$$

$$= -\mathbf{A}\partial f^*(-\mathbf{A}^T \mathbf{u}^{k+1}) + \mathbf{b} + \frac{1}{\beta}(\mathbf{u}^{k+1} - \mathbf{v}^k - \mathbf{w}^k).$$

So there exists $\mathbf{x}^{k+1} \in \partial f^*(-\mathbf{A}^T \mathbf{u}^{k+1})$, such that

$$\mathbf{0} = -\mathbf{A}\mathbf{x}^{k+1} + \mathbf{b} + \frac{1}{\beta}(\mathbf{u}^{k+1} - \mathbf{v}^k - \mathbf{w}^k),$$

and by Point 5 of Proposition A.11 we have

$$-\mathbf{A}^T \mathbf{u}^{k+1} \in \partial f(\mathbf{x}^{k+1}).$$

Similarly, for the second step, we have

$$0 \in -\mathbf{B}\partial g^*(-\mathbf{B}^T\mathbf{v}^{k+1}) + \frac{1}{\beta}(\mathbf{v}^{k+1} - \mathbf{u}^{k+1} + \mathbf{w}^k).$$

There also exists $\mathbf{y}^{k+1} \in \partial g^*(-\mathbf{B}^T\mathbf{v}^{k+1})$, such that

$$0 = -\mathbf{B}\mathbf{y}^{k+1} + \frac{1}{\beta}(\mathbf{v}^{k+1} - \mathbf{u}^{k+1} + \mathbf{w}^k) = -\mathbf{B}\mathbf{y}^{k+1} + \frac{1}{\beta}\mathbf{w}^{k+1}$$

and

$$-\mathbf{B}^T\mathbf{v}^{k+1} \in \partial g(\mathbf{y}^{k+1}).$$

So we have

$$\mathbf{A}\mathbf{x}^{k+1} + \mathbf{B}\mathbf{y}^{k+1} - \mathbf{b} = \frac{1}{\beta}(\mathbf{v}^{k+1} - \mathbf{v}^k), \tag{2.22}$$

$$0 \in \partial g(\mathbf{y}^{k+1}) + \mathbf{B}^T\mathbf{v}^k + \beta\mathbf{B}^T(\mathbf{A}\mathbf{x}^{k+1} + \mathbf{B}\mathbf{y}^{k+1} - \mathbf{b}),$$

and

$$0 \in \partial f(\mathbf{x}^{k+1}) + \mathbf{A}^T\mathbf{v}^k + \beta\mathbf{A}^T\left(\mathbf{A}\mathbf{x}^{k+1} - \mathbf{b} + \frac{1}{\beta}\mathbf{w}^k\right)$$

$$= \partial f(\mathbf{x}^{k+1}) + \mathbf{A}^T\mathbf{v}^k + \beta\mathbf{A}^T(\mathbf{A}\mathbf{x}^{k+1} - \mathbf{b} + \mathbf{B}\mathbf{y}^k),$$

which further yield

$$\mathbf{x}^{k+1} = \underset{\mathbf{x}}{\mathrm{argmin}}\left(f(\mathbf{x}) + \left\langle \mathbf{A}\mathbf{x}, \mathbf{v}^k \right\rangle + \frac{\beta}{2}\|\mathbf{A}\mathbf{x} + \mathbf{B}\mathbf{y}^k - \mathbf{b}\|^2\right), \tag{2.23}$$

$$\mathbf{y}^{k+1} = \underset{\mathbf{y}}{\mathrm{argmin}}\left(g(\mathbf{y}) + \left\langle \mathbf{B}\mathbf{y}, \mathbf{v}^k \right\rangle + \frac{\beta}{2}\|\mathbf{A}\mathbf{x}^{k+1} + \mathbf{B}\mathbf{y} - \mathbf{b}\|^2\right). \tag{2.24}$$

(2.23), (2.24), and (2.22) constitute ADMM, where $\mathbf{v}$ acts as the Lagrange multiplier. Thus, ADMM is a special case of DRS.

# References

1. J. Douglas, H. Rachford, On the numerical solution of heat conduction problems in two and three space variables. Trans. Am. Math. Soc. **82**(2),421–439 (1956)
2. D. Gabay, Applications of the method of multipliers to variational inequalities. Math. Appl. **15**, 299–331 (1983)

3. D. Gabay, B. Mercier, A dual algorithm for the solution of nonlinear variational problems via finite element approximations. Comput. Math. **2**(1), 17–40 (1976)
4. R. Glowinski, A. Marrocco, Sur l'approximation par éléments finis d'ordre un, et la résolution, par pénalisation-dualité d'une classe de problèmes de Dirichlet non linéaires. Rev. fr. autom. inform. rech. opér., Anal. numér. **9**(R2), 41–76 (1975)
5. T. Goldstein, S.J. Osher, The split Bregman method for $\ell_1$-regularized problems. SIAM J. Imaging Sci. **2**(2), 323–343 (2009)
6. M.R. Hestenes, Multiplier and gradient methods. J. Optim. Theory Appl. **4**(5), 302–320 (1979)

# Chapter 3
# ADMM for Deterministic and Convex Optimization

In this chapter, we focus on the theoretical convergence and convergence rates for ADMM and its several variants for deterministic and convex optimization. We first introduce the convergence property of the original ADMM as well as its convergence rates, including the sublinear and the linear rates under different assumptions. Then we extend to two variants of ADMM, the linearized ADMM and the accelerated linearized ADMM, and give their sublinear and linear convergence rates, respectively. At last, we focus on how to use ADMM to solve multi-block separable problems and general problems with nonlinear (but still convex) constraints.

## 3.1 Original ADMM

In this section, we focus on the convergence and the convergence rate analysis of the original ADMM presented in Algorithm 2.1 (see Sect. 2.1.3).

### 3.1.1 Convergence Analysis

We first give several supporting lemmas, which are useful for both the convergence and the convergence rates analysis, and prove the convergence at the end of this section. The convergence of ADMM was studied in many references, e.g., [4, 5].

**Lemma 3.1** *Suppose that* $f(\mathbf{x})$ *and* $g(\mathbf{y})$ *are convex. Let* $(\mathbf{x}^*, \mathbf{y}^*, \boldsymbol{\lambda}^*)$ *be a KKT point (Definition A.21) of Problem (2.13), then we have*

$$f(\mathbf{x}) + g(\mathbf{y}) - f(\mathbf{x}^*) - g(\mathbf{y}^*) + \langle \boldsymbol{\lambda}^*, \mathbf{A}\mathbf{x} + \mathbf{B}\mathbf{y} - \mathbf{b} \rangle \geq 0, \forall \mathbf{x}, \mathbf{y}.$$

This lemma is actually a direct consequence of Point 1 of Proposition A.14.

**Lemma 3.2** *Suppose that* $f(\mathbf{x})$ *and* $g(\mathbf{y})$ *are convex. Let* $(\mathbf{x}^*, \mathbf{y}^*, \boldsymbol{\lambda}^*)$ *be a KKT point of Problem (2.13). If*

$$f(\mathbf{x}) + g(\mathbf{y}) - f(\mathbf{x}^*) - g(\mathbf{y}^*) + \langle \boldsymbol{\lambda}^*, \mathbf{A}\mathbf{x} + \mathbf{B}\mathbf{y} - \mathbf{b} \rangle \leq \alpha_1,$$

$$\|\mathbf{A}\mathbf{x} + \mathbf{B}\mathbf{y} - \mathbf{b}\| \leq \alpha_2,$$

*then we have*

$$-\|\boldsymbol{\lambda}^*\|\alpha_2 \leq f(\mathbf{x}) + g(\mathbf{y}) - f(\mathbf{x}^*) - g(\mathbf{y}^*) \leq \|\boldsymbol{\lambda}^*\|\alpha_2 + \alpha_1.$$

The next lemma measures the optimality conditions of ADMM and describes the KKT condition (Definition A.21) of Problem (2.13).

**Lemma 3.3** *For Algorithm 2.1, we have*

$$\mathbf{0} \in \partial f(\mathbf{x}^{k+1}) + \mathbf{A}^T \boldsymbol{\lambda}^k + \beta \mathbf{A}^T (\mathbf{A}\mathbf{x}^{k+1} + \mathbf{B}\mathbf{y}^k - \mathbf{b}), \tag{3.1}$$

$$\mathbf{0} \in \partial g(\mathbf{y}^{k+1}) + \mathbf{B}^T \boldsymbol{\lambda}^k + \beta \mathbf{B}^T (\mathbf{A}\mathbf{x}^{k+1} + \mathbf{B}\mathbf{y}^{k+1} - \mathbf{b}), \tag{3.2}$$

$$\boldsymbol{\lambda}^{k+1} - \boldsymbol{\lambda}^k = \beta(\mathbf{A}\mathbf{x}^{k+1} + \mathbf{B}\mathbf{y}^{k+1} - \mathbf{b}), \tag{3.3}$$

$$\mathbf{0} \in \partial f(\mathbf{x}^*) + \mathbf{A}^T \boldsymbol{\lambda}^*, \tag{3.4}$$

$$\mathbf{0} \in \partial g(\mathbf{y}^*) + \mathbf{B}^T \boldsymbol{\lambda}^*, \tag{3.5}$$

$$\mathbf{A}\mathbf{x}^* + \mathbf{B}\mathbf{y}^* = \mathbf{b}, \tag{3.6}$$

*where* $(\mathbf{x}^*, \mathbf{y}^*, \boldsymbol{\lambda}^*)$ *is any KKT point of Problem (2.13).*

Define two vectors:

$$\hat{\nabla} f(\mathbf{x}^{k+1}) = -\mathbf{A}^T \boldsymbol{\lambda}^k - \beta \mathbf{A}^T (\mathbf{A}\mathbf{x}^{k+1} + \mathbf{B}\mathbf{y}^k - \mathbf{b}),$$

$$\hat{\nabla} g(\mathbf{y}^{k+1}) = -\mathbf{B}^T \boldsymbol{\lambda}^k - \beta \mathbf{B}^T (\mathbf{A}\mathbf{x}^{k+1} + \mathbf{B}\mathbf{y}^{k+1} - \mathbf{b})$$

$$= -\mathbf{B}^T \boldsymbol{\lambda}^{k+1}. \tag{3.7}$$

Then we have

$$\hat{\nabla} f(\mathbf{x}^{k+1}) \in \partial f(\mathbf{x}^{k+1}) \quad \text{and} \quad \hat{\nabla} g(\mathbf{y}^{k+1}) \in \partial g(\mathbf{y}^{k+1}). \tag{3.8}$$

We further provide Lemmas 3.4–3.6.

**Lemma 3.4** *For Algorithm 2.1, we have*

$$\left\langle \hat{\nabla} g(\mathbf{y}^{k+1}), \mathbf{y}^{k+1} - \mathbf{y} \right\rangle = - \left\langle \boldsymbol{\lambda}^{k+1}, \mathbf{B}\mathbf{y}^{k+1} - \mathbf{B}\mathbf{y} \right\rangle, \quad \forall \mathbf{y}, \tag{3.9}$$

*and*

$$\left\langle \hat{\nabla} f(\mathbf{x}^{k+1}), \mathbf{x}^{k+1} - \mathbf{x} \right\rangle + \left\langle \hat{\nabla} g(\mathbf{y}^{k+1}), \mathbf{y}^{k+1} - \mathbf{y} \right\rangle$$

$$= - \left\langle \boldsymbol{\lambda}^{k+1}, \mathbf{A}\mathbf{x}^{k+1} + \mathbf{B}\mathbf{y}^{k+1} - \mathbf{A}\mathbf{x} - \mathbf{B}\mathbf{y} \right\rangle$$

$$+ \beta \left\langle \mathbf{B}\mathbf{y}^{k+1} - \mathbf{B}\mathbf{y}^k, \mathbf{A}\mathbf{x}^{k+1} - \mathbf{A}\mathbf{x} \right\rangle, \quad \forall \mathbf{x}, \mathbf{y}. \tag{3.10}$$

***Proof*** From the definitions of $\hat{\nabla} f(\mathbf{x}^{k+1})$ and $\hat{\nabla} g(\mathbf{y}^{k+1})$ and (3.3) we have

$$\left\langle \hat{\nabla} f(\mathbf{x}^{k+1}), \mathbf{x}^{k+1} - \mathbf{x} \right\rangle$$

$$= - \left\langle \mathbf{A}^T \boldsymbol{\lambda}^k + \beta \mathbf{A}^T (\mathbf{A}\mathbf{x}^{k+1} + \mathbf{B}\mathbf{y}^k - \mathbf{b}), \mathbf{x}^{k+1} - \mathbf{x} \right\rangle$$

$$= - \left\langle \boldsymbol{\lambda}^{k+1}, \mathbf{A}\mathbf{x}^{k+1} - \mathbf{A}\mathbf{x} \right\rangle + \beta \left\langle \mathbf{B}\mathbf{y}^{k+1} - \mathbf{B}\mathbf{y}^k, \mathbf{A}\mathbf{x}^{k+1} - \mathbf{A}\mathbf{x} \right\rangle$$

and

$$\left\langle \hat{\nabla} g(\mathbf{y}^{k+1}), \mathbf{y}^{k+1} - \mathbf{y} \right\rangle = - \left\langle \boldsymbol{\lambda}^{k+1}, \mathbf{B}\mathbf{y}^{k+1} - \mathbf{B}\mathbf{y} \right\rangle.$$

Adding them together, we have (3.10). □

**Lemma 3.5** *Suppose that $f(\mathbf{x})$ and $g(\mathbf{y})$ are convex. Then for Algorithm 2.1, we have*

$$\left\langle \hat{\nabla} f(\mathbf{x}^{k+1}), \mathbf{x}^{k+1} - \mathbf{x}^* \right\rangle + \left\langle \hat{\nabla} g(\mathbf{y}^{k+1}), \mathbf{y}^{k+1} - \mathbf{y}^* \right\rangle + \left\langle \boldsymbol{\lambda}^*, \mathbf{A}\mathbf{x}^{k+1} + \mathbf{B}\mathbf{y}^{k+1} - \mathbf{b} \right\rangle$$

$$\leq \frac{1}{2\beta} \|\boldsymbol{\lambda}^k - \boldsymbol{\lambda}^*\|^2 - \frac{1}{2\beta} \|\boldsymbol{\lambda}^{k+1} - \boldsymbol{\lambda}^*\|^2$$

$$+ \frac{\beta}{2} \|\mathbf{B}\mathbf{y}^k - \mathbf{B}\mathbf{y}^*\|^2 - \frac{\beta}{2} \|\mathbf{B}\mathbf{y}^{k+1} - \mathbf{B}\mathbf{y}^*\|^2$$

$$- \frac{1}{2\beta} \|\boldsymbol{\lambda}^{k+1} - \boldsymbol{\lambda}^k\|^2 - \frac{\beta}{2} \|\mathbf{B}\mathbf{y}^{k+1} - \mathbf{B}\mathbf{y}^k\|^2.$$

***Proof*** Letting $(\mathbf{x}, \mathbf{y}, \lambda) = (\mathbf{x}^*, \mathbf{y}^*, \lambda^*)$ in (3.10), adding $\langle \lambda^*, \mathbf{A}\mathbf{x}^{k+1} + \mathbf{B}\mathbf{y}^{k+1} - \mathbf{b} \rangle$ to both sides, and using (3.3) and (3.6), we have

$$\left\langle \hat{\nabla} f(\mathbf{x}^{k+1}), \mathbf{x}^{k+1} - \mathbf{x}^* \right\rangle + \left\langle \hat{\nabla} g(\mathbf{y}^{k+1}), \mathbf{y}^{k+1} - \mathbf{y}^* \right\rangle + \left\langle \lambda^*, \mathbf{A}\mathbf{x}^{k+1} + \mathbf{B}\mathbf{y}^{k+1} - \mathbf{b} \right\rangle$$

$$= -\left\langle \lambda^{k+1} - \lambda^*, \mathbf{A}\mathbf{x}^{k+1} + \mathbf{B}\mathbf{y}^{k+1} - \mathbf{b} \right\rangle + \beta \left\langle \mathbf{B}\mathbf{y}^{k+1} - \mathbf{B}\mathbf{y}^k, \mathbf{A}\mathbf{x}^{k+1} - \mathbf{A}\mathbf{x}^* \right\rangle$$

$$= -\frac{1}{\beta} \left\langle \lambda^{k+1} - \lambda^*, \lambda^{k+1} - \lambda^k \right\rangle + \left\langle \mathbf{B}\mathbf{y}^{k+1} - \mathbf{B}\mathbf{y}^k, \lambda^{k+1} - \lambda^k \right\rangle$$

$$- \beta \left\langle \mathbf{B}\mathbf{y}^{k+1} - \mathbf{B}\mathbf{y}^k, \mathbf{B}\mathbf{y}^{k+1} - \mathbf{B}\mathbf{y}^* \right\rangle \tag{3.11}$$

$$\overset{a}{=} \frac{1}{2\beta} \|\lambda^k - \lambda^*\|^2 - \frac{1}{2\beta} \|\lambda^{k+1} - \lambda^*\|^2 - \frac{1}{2\beta} \|\lambda^{k+1} - \lambda^k\|^2$$

$$+ \frac{\beta}{2} \|\mathbf{B}\mathbf{y}^k - \mathbf{B}\mathbf{y}^*\|^2 - \frac{\beta}{2} \|\mathbf{B}\mathbf{y}^{k+1} - \mathbf{B}\mathbf{y}^*\|^2 - \frac{\beta}{2} \|\mathbf{B}\mathbf{y}^{k+1} - \mathbf{B}\mathbf{y}^k\|^2$$

$$+ \left\langle \mathbf{B}\mathbf{y}^{k+1} - \mathbf{B}\mathbf{y}^k, \lambda^{k+1} - \lambda^k \right\rangle, \tag{3.12}$$

where $\overset{a}{=}$ uses (A.1). On the other hand, (3.9) gives

$$\left\langle \hat{\nabla} g(\mathbf{y}^k), \mathbf{y}^k - \mathbf{y} \right\rangle + \left\langle \lambda^k, \mathbf{B}\mathbf{y}^k - \mathbf{B}\mathbf{y} \right\rangle = 0. \tag{3.13}$$

Letting $\mathbf{y} = \mathbf{y}^k$ in (3.9) and $\mathbf{y} = \mathbf{y}^{k+1}$ in (3.13), and adding them together, we have

$$\left\langle \hat{\nabla} g(\mathbf{y}^{k+1}) - \hat{\nabla} g(\mathbf{y}^k), \mathbf{y}^{k+1} - \mathbf{y}^k \right\rangle + \left\langle \lambda^{k+1} - \lambda^k, \mathbf{B}\mathbf{y}^{k+1} - \mathbf{B}\mathbf{y}^k \right\rangle = 0.$$

The first term of the above equality is non-negative thanks to the monotonicity of $\partial g$. So we have

$$\left\langle \lambda^{k+1} - \lambda^k, \mathbf{B}\mathbf{y}^{k+1} - \mathbf{B}\mathbf{y}^k \right\rangle \leq 0. \tag{3.14}$$

Plugging it into (3.12), we have the conclusion. □

**Lemma 3.6** *Suppose that $f(\mathbf{x})$ and $g(\mathbf{y})$ are convex. Then for Algorithm 2.1, we have*

$$f(\mathbf{x}^{k+1}) + g(\mathbf{y}^{k+1}) - f(\mathbf{x}^*) - g(\mathbf{y}^*) + \left\langle \lambda^*, \mathbf{A}\mathbf{x}^{k+1} + \mathbf{B}\mathbf{y}^{k+1} - \mathbf{b} \right\rangle$$

$$\leq \frac{1}{2\beta} \|\lambda^k - \lambda^*\|^2 - \frac{1}{2\beta} \|\lambda^{k+1} - \lambda^*\|^2$$

$$+ \frac{\beta}{2} \|\mathbf{B}\mathbf{y}^k - \mathbf{B}\mathbf{y}^*\|^2 - \frac{\beta}{2} \|\mathbf{B}\mathbf{y}^{k+1} - \mathbf{B}\mathbf{y}^*\|^2$$

$$- \frac{1}{2\beta} \|\lambda^{k+1} - \lambda^k\|^2 - \frac{\beta}{2} \|\mathbf{B}\mathbf{y}^{k+1} - \mathbf{B}\mathbf{y}^k\|^2. \tag{3.15}$$

*If we further assume that $g(\mathbf{y})$ is $\mu$-strongly convex, then we have*

$$f(\mathbf{x}^{k+1}) + g(\mathbf{y}^{k+1}) - f(\mathbf{x}^*) - g(\mathbf{y}^*) + \left\langle \boldsymbol{\lambda}^*, \mathbf{A}\mathbf{x}^{k+1} + \mathbf{B}\mathbf{y}^{k+1} - \mathbf{b} \right\rangle$$

$$\leq \frac{1}{2\beta} \|\boldsymbol{\lambda}^k - \boldsymbol{\lambda}^*\|^2 - \frac{1}{2\beta} \|\boldsymbol{\lambda}^{k+1} - \boldsymbol{\lambda}^*\|^2$$

$$+ \frac{\beta}{2} \|\mathbf{B}\mathbf{y}^k - \mathbf{B}\mathbf{y}^*\|^2 - \frac{\beta}{2} \|\mathbf{B}\mathbf{y}^{k+1} - \mathbf{B}\mathbf{y}^*\|^2 - \frac{\mu}{2} \|\mathbf{y}^{k+1} - \mathbf{y}^*\|^2. \tag{3.16}$$

*If we further assume that $g(\mathbf{y})$ is $L$-smooth, then we have*

$$f(\mathbf{x}^{k+1}) + g(\mathbf{y}^{k+1}) - f(\mathbf{x}^*) - g(\mathbf{y}^*) + \left\langle \boldsymbol{\lambda}^*, \mathbf{A}\mathbf{x}^{k+1} + \mathbf{B}\mathbf{y}^{k+1} - \mathbf{b} \right\rangle$$

$$\leq \frac{1}{2\beta} \|\boldsymbol{\lambda}^k - \boldsymbol{\lambda}^*\|^2 - \frac{1}{2\beta} \|\boldsymbol{\lambda}^{k+1} - \boldsymbol{\lambda}^*\|^2$$

$$+ \frac{\beta}{2} \|\mathbf{B}\mathbf{y}^k - \mathbf{B}\mathbf{y}^*\|^2 - \frac{\beta}{2} \|\mathbf{B}\mathbf{y}^{k+1} - \mathbf{B}\mathbf{y}^*\|^2$$

$$- \frac{1}{2L} \|\nabla g(\mathbf{y}^{k+1}) - \nabla g(\mathbf{y}^*)\|^2. \tag{3.17}$$

***Proof*** We use Lemma 3.5 to prove these conclusions. From the convexity of $f(\mathbf{x})$ and $g(\mathbf{y})$ and (3.8), we have

$$f(\mathbf{x}^{k+1}) + g(\mathbf{y}^{k+1}) - f(\mathbf{x}^*) - g(\mathbf{y}^*) + \left\langle \boldsymbol{\lambda}^*, \mathbf{A}\mathbf{x}^{k+1} + \mathbf{B}\mathbf{y}^{k+1} - \mathbf{b} \right\rangle$$

$$\overset{a}{\leq} \left\langle \hat{\nabla} f(\mathbf{x}^{k+1}), \mathbf{x}^{k+1} - \mathbf{x}^* \right\rangle + \left\langle \hat{\nabla} g(\mathbf{y}^{k+1}), \mathbf{y}^{k+1} - \mathbf{y}^* \right\rangle$$

$$+ \left\langle \boldsymbol{\lambda}^*, \mathbf{A}\mathbf{x}^{k+1} + \mathbf{B}\mathbf{y}^{k+1} - \mathbf{b} \right\rangle$$

$$\leq \frac{1}{2\beta} \|\boldsymbol{\lambda}^k - \boldsymbol{\lambda}^*\|^2 - \frac{1}{2\beta} \|\boldsymbol{\lambda}^{k+1} - \boldsymbol{\lambda}^*\|^2$$

$$+ \frac{\beta}{2} \|\mathbf{B}\mathbf{y}^k - \mathbf{B}\mathbf{y}^*\|^2 - \frac{\beta}{2} \|\mathbf{B}\mathbf{y}^{k+1} - \mathbf{B}\mathbf{y}^*\|^2$$

$$- \frac{1}{2\beta} \|\boldsymbol{\lambda}^{k+1} - \boldsymbol{\lambda}^k\|^2 - \frac{\beta}{2} \|\mathbf{B}\mathbf{y}^{k+1} - \mathbf{B}\mathbf{y}^k\|^2.$$

When $g(\mathbf{y})$ is strongly convex, from (A.6) we will have an extra $\frac{\mu}{2} \|\mathbf{y}^{k+1} - \mathbf{y}^*\|^2$ on the left hand side of $\overset{a}{\leq}$, thus leading to (3.16). When $g(\mathbf{y})$ is $L$-smooth, from (A.5) we will have an extra $\frac{1}{2L} \|\nabla g(\mathbf{y}^{k+1}) - \nabla g(\mathbf{y}^*)\|^2$ on the left hand side of $\overset{a}{\leq}$, thus leading to (3.17). □

Now we are ready to prove the convergence of ADMM.

**Theorem 3.1** *Suppose that* $f(\mathbf{x})$ *and* $g(\mathbf{y})$ *are convex. Then for Algorithm 2.1, we have*

$$f(\mathbf{x}^{k+1}) - f(\mathbf{x}^*) + g(\mathbf{y}^{k+1}) - g(\mathbf{y}^*) \to 0,$$

$$\mathbf{A}\mathbf{x}^{k+1} + \mathbf{B}\mathbf{y}^{k+1} - \mathbf{b} \to \mathbf{0},$$

*as* $k \to \infty$.

**Proof** From Lemma 3.1 and (3.15), we have

$$\frac{1}{2\beta}\|\boldsymbol{\lambda}^{k+1} - \boldsymbol{\lambda}^k\|^2 + \frac{\beta}{2}\|\mathbf{B}\mathbf{y}^{k+1} - \mathbf{B}\mathbf{y}^k\|^2$$

$$\leq \frac{1}{2\beta}\|\boldsymbol{\lambda}^k - \boldsymbol{\lambda}^*\|^2 - \frac{1}{2\beta}\|\boldsymbol{\lambda}^{k+1} - \boldsymbol{\lambda}^*\|^2$$

$$+ \frac{\beta}{2}\|\mathbf{B}\mathbf{y}^k - \mathbf{B}\mathbf{y}^*\|^2 - \frac{\beta}{2}\|\mathbf{B}\mathbf{y}^{k+1} - \mathbf{B}\mathbf{y}^*\|^2. \tag{3.18}$$

Summing over $k = 0, \cdots, \infty$, we have

$$\sum_{k=0}^{\infty}\left(\frac{1}{2\beta}\|\boldsymbol{\lambda}^{k+1} - \boldsymbol{\lambda}^k\|^2 + \frac{\beta}{2}\|\mathbf{B}\mathbf{y}^{k+1} - \mathbf{B}\mathbf{y}^k\|^2\right)$$

$$\leq \frac{1}{2\beta}\|\boldsymbol{\lambda}^0 - \boldsymbol{\lambda}^*\|^2 + \frac{\beta}{2}\|\mathbf{B}\mathbf{y}^0 - \mathbf{B}\mathbf{y}^*\|^2.$$

Thus, we have

$$\boldsymbol{\lambda}^{k+1} - \boldsymbol{\lambda}^k \to \mathbf{0} \quad \text{and} \quad \mathbf{B}\mathbf{y}^{k+1} - \mathbf{B}\mathbf{y}^k \to \mathbf{0}.$$

Moreover, from (3.18) we obtain that $\frac{1}{2\beta}\|\boldsymbol{\lambda}^k - \boldsymbol{\lambda}^*\|^2 + \frac{\beta}{2}\|\mathbf{B}\mathbf{y}^k - \mathbf{B}\mathbf{y}^*\|^2$ is a non-increasing sequence. So $\|\boldsymbol{\lambda}^k - \boldsymbol{\lambda}^*\|^2$ and $\|\mathbf{B}\mathbf{y}^k - \mathbf{B}\mathbf{y}^*\|^2$ are bounded for all $k$. Then we have that $\|\boldsymbol{\lambda}^k\|$ is also bounded for all $k$. Since

$$\boldsymbol{\lambda}^{k+1} - \boldsymbol{\lambda}^k = \beta(\mathbf{A}\mathbf{x}^{k+1} + \mathbf{B}\mathbf{y}^{k+1} - \mathbf{b})$$

$$= \beta(\mathbf{A}\mathbf{x}^{k+1} - \mathbf{A}\mathbf{x}^*) + \beta(\mathbf{B}\mathbf{y}^{k+1} - \mathbf{B}\mathbf{y}^*),$$

we know that $\mathbf{A}\mathbf{x}^{k+1} + \mathbf{B}\mathbf{y}^{k+1} - \mathbf{b} \to \mathbf{0}$ and $\mathbf{A}\mathbf{x}^{k+1} - \mathbf{A}\mathbf{x}^*$ is also bounded.

From (3.10) and the convexity of $f$ and $g$, we have

$$f(\mathbf{x}^{k+1}) - f(\mathbf{x}^*) + g(\mathbf{y}^{k+1}) - g(\mathbf{y}^*)$$

$$\leq -\left\langle\boldsymbol{\lambda}^{k+1}, \mathbf{A}\mathbf{x}^{k+1} + \mathbf{B}\mathbf{y}^{k+1} - \mathbf{b}\right\rangle + \beta\left\langle\mathbf{B}\mathbf{y}^{k+1} - \mathbf{B}\mathbf{y}^k, \mathbf{A}\mathbf{x}^{k+1} - \mathbf{A}\mathbf{x}^*\right\rangle$$

$$\to 0.$$

On the other hand, from (3.4), (3.5), and (3.6), we have

$$f(\mathbf{x}^{k+1}) - f(\mathbf{x}^*) + g(\mathbf{y}^{k+1}) - g(\mathbf{y}^*)$$

$$\geq \left\langle -\mathbf{A}^T \boldsymbol{\lambda}^*, \mathbf{x}^{k+1} - \mathbf{x}^* \right\rangle + \left\langle -\mathbf{B}^T \boldsymbol{\lambda}^*, \mathbf{y}^{k+1} - \mathbf{y}^* \right\rangle$$

$$= - \left\langle \boldsymbol{\lambda}^*, \mathbf{A}\mathbf{x}^{k+1} + \mathbf{B}\mathbf{y}^{k+1} - \mathbf{b} \right\rangle$$

$$\to 0.$$

Thus, we have $f(\mathbf{x}^{k+1}) - f(\mathbf{x}^*) + g(\mathbf{y}^{k+1}) - g(\mathbf{y}^*) \to 0$. $\qquad\square$

### 3.1.2 Sublinear Convergence Rate

In this section, we introduce the sublinear rates of ADMM for generally convex problem (2.13), which are only based on the assumptions that both $f$ and $g$ are convex.

#### 3.1.2.1 Non-ergodic Convergence Rate

We first give the $O\left(\frac{1}{\sqrt{K}}\right)$ non-ergodic convergence rate of ADMM, where $K$ is the number of iterations and non-ergodic means the convergence rate is measured at the last iterate. Accordingly, ergodic means that the convergence rate is measured at the average of previous iterates. This result was first proved in [7] and then extended in [2]. Moreover, [2] proved that the $O\left(\frac{1}{\sqrt{K}}\right)$ non-ergodic convergence rate of ADMM is tight, meaning that it cannot be further improved.

**Lemma 3.7** *Suppose that $f(\mathbf{x})$ and $g(\mathbf{y})$ are convex. Then for Algorithm 2.1, we have*

$$\frac{1}{2\beta} \|\boldsymbol{\lambda}^{k+1} - \boldsymbol{\lambda}^k\|^2 + \frac{\beta}{2} \|\mathbf{B}\mathbf{y}^{k+1} - \mathbf{B}\mathbf{y}^k\|^2$$

$$\leq \frac{1}{2\beta} \|\boldsymbol{\lambda}^k - \boldsymbol{\lambda}^{k-1}\|^2 + \frac{\beta}{2} \|\mathbf{B}\mathbf{y}^k - \mathbf{B}\mathbf{y}^{k-1}\|^2.$$

**Proof** (3.10) gives

$$\left\langle \hat{\nabla} f(\mathbf{x}^k), \mathbf{x}^k - \mathbf{x} \right\rangle + \left\langle \hat{\nabla} g(\mathbf{y}^k), \mathbf{y}^k - \mathbf{y} \right\rangle$$

$$= - \left\langle \boldsymbol{\lambda}^k, \mathbf{A}\mathbf{x}^k + \mathbf{B}\mathbf{y}^k - \mathbf{A}\mathbf{x} - \mathbf{B}\mathbf{y} \right\rangle + \beta \left\langle \mathbf{B}\mathbf{y}^k - \mathbf{B}\mathbf{y}^{k-1}, \mathbf{A}\mathbf{x}^k - \mathbf{A}\mathbf{x} \right\rangle. \qquad (3.19)$$

Letting $(\mathbf{x}, \mathbf{y}, \boldsymbol{\lambda}) = (\mathbf{x}^k, \mathbf{y}^k, \boldsymbol{\lambda}^k)$ in (3.10) and $(\mathbf{x}, \mathbf{y}, \boldsymbol{\lambda}) = (\mathbf{x}^{k+1}, \mathbf{y}^{k+1}, \boldsymbol{\lambda}^{k+1})$ in (3.19), adding them together, and using (3.3), we have

$$
\left\langle \hat{\nabla} f(\mathbf{x}^{k+1}) - \hat{\nabla} f(\mathbf{x}^k), \mathbf{x}^{k+1} - \mathbf{x}^k \right\rangle + \left\langle \hat{\nabla} g(\mathbf{y}^{k+1}) - \hat{\nabla} g(\mathbf{y}^k), \mathbf{y}^{k+1} - \mathbf{y}^k \right\rangle
$$

$$
= -\left\langle \boldsymbol{\lambda}^{k+1} - \boldsymbol{\lambda}^k, \mathbf{A}\mathbf{x}^{k+1} + \mathbf{B}\mathbf{y}^{k+1} - \mathbf{A}\mathbf{x}^k - \mathbf{B}\mathbf{y}^k \right\rangle
$$

$$
\quad + \beta \left\langle \mathbf{B}\mathbf{y}^{k+1} - \mathbf{B}\mathbf{y}^k - (\mathbf{B}\mathbf{y}^k - \mathbf{B}\mathbf{y}^{k-1}), \mathbf{A}\mathbf{x}^{k+1} - \mathbf{A}\mathbf{x}^k \right\rangle
$$

$$
= -\frac{1}{\beta} \left\langle \boldsymbol{\lambda}^{k+1} - \boldsymbol{\lambda}^k, \boldsymbol{\lambda}^{k+1} - \boldsymbol{\lambda}^k - (\boldsymbol{\lambda}^k - \boldsymbol{\lambda}^{k-1}) \right\rangle
$$

$$
\quad + \Big\langle \mathbf{B}\mathbf{y}^{k+1} - \mathbf{B}\mathbf{y}^k - (\mathbf{B}\mathbf{y}^k - \mathbf{B}\mathbf{y}^{k-1}),
$$

$$
\quad\quad \boldsymbol{\lambda}^{k+1} - \boldsymbol{\lambda}^k - \beta\mathbf{B}\mathbf{y}^{k+1} - (\boldsymbol{\lambda}^k - \boldsymbol{\lambda}^{k-1} - \beta\mathbf{B}\mathbf{y}^k) \Big\rangle
$$

$$
\overset{a}{=} \frac{1}{2\beta}\left[ \|\boldsymbol{\lambda}^k - \boldsymbol{\lambda}^{k-1}\|^2 - \|\boldsymbol{\lambda}^{k+1} - \boldsymbol{\lambda}^k\|^2 - \|\boldsymbol{\lambda}^{k+1} - \boldsymbol{\lambda}^k - (\boldsymbol{\lambda}^k - \boldsymbol{\lambda}^{k-1})\|^2 \right]
$$

$$
\quad + \frac{\beta}{2}\Big[ \|\mathbf{B}\mathbf{y}^k - \mathbf{B}\mathbf{y}^{k-1}\|^2 - \|\mathbf{B}\mathbf{y}^{k+1} - \mathbf{B}\mathbf{y}^k\|^2
$$

$$
\quad\quad - \|\mathbf{B}\mathbf{y}^{k+1} - \mathbf{B}\mathbf{y}^k - (\mathbf{B}\mathbf{y}^k - \mathbf{B}\mathbf{y}^{k-1})\|^2 \Big]
$$

$$
\quad + \left\langle \mathbf{B}\mathbf{y}^{k+1} - \mathbf{B}\mathbf{y}^k - (\mathbf{B}\mathbf{y}^k - \mathbf{B}\mathbf{y}^{k-1}), \boldsymbol{\lambda}^{k+1} - \boldsymbol{\lambda}^k - (\boldsymbol{\lambda}^k - \boldsymbol{\lambda}^{k-1}) \right\rangle
$$

$$
= \frac{1}{2\beta}\left( \|\boldsymbol{\lambda}^k - \boldsymbol{\lambda}^{k-1}\|^2 - \|\boldsymbol{\lambda}^{k+1} - \boldsymbol{\lambda}^k\|^2 \right)
$$

$$
\quad + \frac{\beta}{2}\left( \|\mathbf{B}\mathbf{y}^k - \mathbf{B}\mathbf{y}^{k-1}\|^2 - \|\mathbf{B}\mathbf{y}^{k+1} - \mathbf{B}\mathbf{y}^k\|^2 \right)
$$

$$
\quad - \bigg[ \frac{1}{2\beta}\|\boldsymbol{\lambda}^{k+1} - \boldsymbol{\lambda}^k - (\boldsymbol{\lambda}^k - \boldsymbol{\lambda}^{k-1})\|^2
$$

$$
\quad\quad + \frac{\beta}{2}\|\mathbf{B}\mathbf{y}^{k+1} - \mathbf{B}\mathbf{y}^k - (\mathbf{B}\mathbf{y}^k - \mathbf{B}\mathbf{y}^{k-1})\|^2
$$

$$
\quad\quad - \left\langle \mathbf{B}\mathbf{y}^{k+1} - \mathbf{B}\mathbf{y}^k - (\mathbf{B}\mathbf{y}^k - \mathbf{B}\mathbf{y}^{k-1}), \boldsymbol{\lambda}^{k+1} - \boldsymbol{\lambda}^k - (\boldsymbol{\lambda}^k - \boldsymbol{\lambda}^{k-1}) \right\rangle \bigg]
$$

$$
\leq \frac{1}{2\beta}\left( \|\boldsymbol{\lambda}^k - \boldsymbol{\lambda}^{k-1}\|^2 - \|\boldsymbol{\lambda}^{k+1} - \boldsymbol{\lambda}^k\|^2 \right)
$$

$$
\quad + \frac{\beta}{2}\left( \|\mathbf{B}\mathbf{y}^k - \mathbf{B}\mathbf{y}^{k-1}\|^2 - \|\mathbf{B}\mathbf{y}^{k+1} - \mathbf{B}\mathbf{y}^k\|^2 \right),
$$

where $\overset{a}{=}$ uses (A.1). Using the monotonicity of $\partial f$ and $\partial g$, we have the conclusion.
$\square$

**Theorem 3.2** *Suppose that $f$ and $g$ are convex. Then for Algorithm 2.1, we have*

$$-\|\boldsymbol{\lambda}^*\|\sqrt{\frac{C}{\beta(K+1)}} \leq f(\mathbf{x}^{K+1}) + g(\mathbf{y}^{K+1}) - f(\mathbf{x}^*) - g(\mathbf{y}^*)$$

$$\leq \frac{C}{K+1} + \frac{2C}{\sqrt{K+1}} + \|\boldsymbol{\lambda}^*\|\sqrt{\frac{C}{\beta(K+1)}},$$

$$\|\mathbf{A}\mathbf{x}^{K+1} + \mathbf{B}\mathbf{y}^{K+1} - \mathbf{b}\| \leq \sqrt{\frac{C}{\beta(K+1)}},$$

*where $C = \frac{1}{\beta}\|\boldsymbol{\lambda}^0 - \boldsymbol{\lambda}^*\|^2 + \beta\|\mathbf{B}\mathbf{y}^0 - \mathbf{B}\mathbf{y}^*\|^2$.*

**Proof** Summing (3.18) over $k = 0, 1, \cdots, K$ and using the monotonicity of $\frac{1}{2\beta}\|\boldsymbol{\lambda}^{k+1} - \boldsymbol{\lambda}^k\|^2 + \frac{\beta}{2}\|\mathbf{B}\mathbf{y}^{k+1} - \mathbf{B}\mathbf{y}^k\|^2$ from Lemma 3.7, we have

$$\frac{1}{\beta}\|\boldsymbol{\lambda}^{K+1} - \boldsymbol{\lambda}^K\|^2 + \beta\|\mathbf{B}\mathbf{y}^{K+1} - \mathbf{B}\mathbf{y}^K\|^2$$

$$\leq \frac{1}{K+1}\left(\frac{1}{\beta}\|\boldsymbol{\lambda}^0 - \boldsymbol{\lambda}^*\|^2 + \beta\|\mathbf{B}\mathbf{y}^0 - \mathbf{B}\mathbf{y}^*\|^2\right).$$

Then we have

$$\beta\|\mathbf{A}\mathbf{x}^{K+1} + \mathbf{B}\mathbf{y}^{K+1} - \mathbf{b}\| = \|\boldsymbol{\lambda}^{K+1} - \boldsymbol{\lambda}^K\| \leq \sqrt{\frac{\beta C}{K+1}},$$

$$\|\mathbf{B}\mathbf{y}^{K+1} - \mathbf{B}\mathbf{y}^K\| \leq \sqrt{\frac{C}{\beta(K+1)}}.$$

On the other hand, (3.18) gives

$$\frac{1}{2\beta}\|\boldsymbol{\lambda}^{k+1} - \boldsymbol{\lambda}^*\|^2 + \frac{\beta}{2}\|\mathbf{B}\mathbf{y}^{k+1} - \mathbf{B}\mathbf{y}^*\|^2$$

$$\leq \frac{1}{2\beta}\|\boldsymbol{\lambda}^k - \boldsymbol{\lambda}^*\|^2 + \frac{\beta}{2}\|\mathbf{B}\mathbf{y}^k - \mathbf{B}\mathbf{y}^*\|^2$$

$$\leq \frac{1}{2\beta}\|\boldsymbol{\lambda}^0 - \boldsymbol{\lambda}^*\|^2 + \frac{\beta}{2}\|\mathbf{B}\mathbf{y}^0 - \mathbf{B}\mathbf{y}^*\|^2 = \frac{1}{2}C.$$

So we have

$$\|\boldsymbol{\lambda}^{K+1} - \boldsymbol{\lambda}^*\| \le \sqrt{\beta C}, \tag{3.20}$$

$$\|\mathbf{B}\mathbf{y}^{K+1} - \mathbf{B}\mathbf{y}^*\| \le \sqrt{\frac{C}{\beta}}.$$

Then from (3.11) and the convexity of $f$ and $g$, we have

$$f(\mathbf{x}^{K+1}) - f(\mathbf{x}^*) + g(\mathbf{y}^{K+1}) - g(\mathbf{y}^*) + \left\langle \boldsymbol{\lambda}^*, \mathbf{A}\mathbf{x}^{K+1} + \mathbf{B}\mathbf{y}^{K+1} - \mathbf{b} \right\rangle$$

$$\le \frac{1}{\beta}\|\boldsymbol{\lambda}^{K+1} - \boldsymbol{\lambda}^*\|\|\boldsymbol{\lambda}^{K+1} - \boldsymbol{\lambda}^K\| + \|\mathbf{B}\mathbf{y}^{K+1} - \mathbf{B}\mathbf{y}^K\|\|\boldsymbol{\lambda}^{K+1} - \boldsymbol{\lambda}^K\|$$

$$+ \beta\|\mathbf{B}\mathbf{y}^{K+1} - \mathbf{B}\mathbf{y}^K\|\|\mathbf{B}\mathbf{y}^{K+1} - \mathbf{B}\mathbf{y}^*\|$$

$$\le \frac{C}{K+1} + \frac{2C}{\sqrt{K+1}}.$$

From Lemma 3.2, we have the conclusion.                                      □

### 3.1.2.2   Ergodic Convergence Rate

Now we describe the ergodic $O\left(\frac{1}{K}\right)$ convergence rate, which was originally proved in [10]. Davis and Yin [2] also proved that the $O\left(\frac{1}{K}\right)$ ergodic convergence rate of ADMM is tight, thus it cannot be further improved.

**Theorem 3.3** *Suppose that $f(\mathbf{x})$ and $g(\mathbf{y})$ are convex. Then for Algorithm 2.1, we have*

$$|f(\hat{\mathbf{x}}^{K+1}) + g(\hat{\mathbf{y}}^{K+1}) - f(\mathbf{x}^*) - g(\mathbf{y}^*)| \le \frac{C}{2(K+1)} + \frac{2\sqrt{C}\|\boldsymbol{\lambda}^*\|}{\sqrt{\beta}(K+1)},$$

$$\|\mathbf{A}\hat{\mathbf{x}}^{K+1} + \mathbf{B}\hat{\mathbf{y}}^{K+1} - \mathbf{b}\| \le \frac{2\sqrt{C}}{\sqrt{\beta}(K+1)},$$

*where*

$$\hat{\mathbf{x}}^{K+1} = \frac{1}{K+1}\sum_{k=1}^{K+1}\mathbf{x}^k, \quad \hat{\mathbf{y}}^{K+1} = \frac{1}{K+1}\sum_{k=1}^{K+1}\mathbf{y}^k,$$

*and $C$ is defined in Theorem 3.2.*

**Proof** Summing (3.15) over $k = 0, 1, \cdots, K$, dividing both sides with $K + 1$, and using the definitions of $\hat{\mathbf{x}}^{K+1}$ and $\hat{\mathbf{y}}^{K+1}$ and the convexity of $f$ and $g$, we have

$$f(\hat{\mathbf{x}}^{K+1}) + g(\hat{\mathbf{y}}^{K+1}) - f(\mathbf{x}^*) - g(\mathbf{y}^*) + \left\langle \boldsymbol{\lambda}^*, \mathbf{A}\hat{\mathbf{x}}^{K+1} + \mathbf{B}\hat{\mathbf{y}}^{K+1} - \mathbf{b} \right\rangle$$

$$\leq \frac{C}{2(K + 1)}.$$

From (3.3) and (3.20), we have

$$\|\mathbf{A}\hat{\mathbf{x}}^{K+1} + \mathbf{B}\hat{\mathbf{y}}^{K+1} - \mathbf{b}\| = \frac{1}{\beta(K + 1)} \left\| \sum_{k=0}^{K} (\boldsymbol{\lambda}^{k+1} - \boldsymbol{\lambda}^k) \right\|$$

$$= \frac{1}{\beta(K + 1)} \|\boldsymbol{\lambda}^{K+1} - \boldsymbol{\lambda}^0\|$$

$$\leq \frac{1}{\beta(K + 1)} \left( \|\boldsymbol{\lambda}^0 - \boldsymbol{\lambda}^*\| + \|\boldsymbol{\lambda}^{K+1} - \boldsymbol{\lambda}^*\| \right)$$

$$\leq \frac{2\sqrt{\beta C}}{\beta(K + 1)}.$$

From Lemma 3.2, we have the conclusion. □

### 3.1.3 Linear Convergence Rate

We discuss the linear convergence in two scenarios. The first scenario is under the assumption that the objective is strongly convex and smooth. The second scenario is under the assumption that the objective is not necessarily strongly convex and smooth, but satisfies the error bound condition.

#### 3.1.3.1 Linear Convergence Under the Strong Convexity and Smoothness Assumption

We first discuss the scenario that $g$ is both strongly convex and smooth, and $\mathbf{B}$ is surjective [3].

**Theorem 3.4** *Suppose that $f(\mathbf{x})$ is convex and $g(\mathbf{y})$ is $\mu$-strongly convex and $L$-smooth. Assume that $\|\mathbf{B}^T\boldsymbol{\lambda}\| \geq \sigma\|\boldsymbol{\lambda}\|, \forall\boldsymbol{\lambda}$, where $\sigma > 0$. Let $\beta = \frac{\sqrt{\mu L}}{\sigma\|\mathbf{B}\|_2}$. Then we have*

$$\frac{1}{2\beta}\|\boldsymbol{\lambda}^{k+1} - \boldsymbol{\lambda}^*\|^2 + \frac{\beta}{2}\|\mathbf{B}\mathbf{y}^{k+1} - \mathbf{B}\mathbf{y}^*\|^2$$

$$\leq \left(1 + \frac{1}{2}\sqrt{\frac{\mu}{L}}\frac{\sigma}{\|\mathbf{B}\|_2}\right)^{-1}\left(\frac{1}{2\beta}\|\boldsymbol{\lambda}^k - \boldsymbol{\lambda}^*\|^2 + \frac{\beta}{2}\|\mathbf{B}\mathbf{y}^k - \mathbf{B}\mathbf{y}^*\|^2\right).$$

**Proof** Since $g$ is differentiable, $\partial g(\mathbf{y})$ is a singleton. By (3.7), (3.8), and (3.5), we have

$$\nabla g(y^{k+1}) = -\mathbf{B}^T\boldsymbol{\lambda}^{k+1} \quad \text{and} \quad \nabla g(y^*) = -\mathbf{B}^T\boldsymbol{\lambda}^*.$$

Then from $\|\mathbf{B}^T\boldsymbol{\lambda}\| \geq \sigma\|\boldsymbol{\lambda}\|$ we have

$$\frac{\sigma^2}{2L}\|\boldsymbol{\lambda}^{k+1} - \boldsymbol{\lambda}^*\|^2 \leq \frac{1}{2L}\|\mathbf{B}^T(\boldsymbol{\lambda}^{k+1} - \boldsymbol{\lambda}^*)\|^2$$

$$= \frac{1}{2L}\|\nabla g(\mathbf{y}^{k+1}) - \nabla g(\mathbf{y}^*)\|^2. \tag{3.21}$$

From (3.16), (3.17), (3.21), and Lemma 3.1, we have

$$\frac{\mu}{2\|\mathbf{B}\|_2^2}\|\mathbf{B}\mathbf{y}^{k+1} - \mathbf{B}\mathbf{y}^*\|^2 \leq \frac{1}{2\beta}\|\boldsymbol{\lambda}^k - \boldsymbol{\lambda}^*\|^2 - \frac{1}{2\beta}\|\boldsymbol{\lambda}^{k+1} - \boldsymbol{\lambda}^*\|^2$$

$$+ \frac{\beta}{2}\|\mathbf{B}\mathbf{y}^k - \mathbf{B}\mathbf{y}^*\|^2 - \frac{\beta}{2}\|\mathbf{B}\mathbf{y}^{k+1} - \mathbf{B}\mathbf{y}^*\|^2 \tag{3.22}$$

and

$$\frac{\sigma^2}{2L}\|\boldsymbol{\lambda}^{k+1} - \boldsymbol{\lambda}^*\|^2 \leq \frac{1}{2\beta}\|\boldsymbol{\lambda}^k - \boldsymbol{\lambda}^*\|^2 - \frac{1}{2\beta}\|\boldsymbol{\lambda}^{k+1} - \boldsymbol{\lambda}^*\|^2$$

$$+ \frac{\beta}{2}\|\mathbf{B}\mathbf{y}^k - \mathbf{B}\mathbf{y}^*\|^2 - \frac{\beta}{2}\|\mathbf{B}\mathbf{y}^{k+1} - \mathbf{B}\mathbf{y}^*\|^2. \tag{3.23}$$

Multiplying (3.23) by $t$, multiplying (3.22) by $1 - t$, adding them together, and rearranging the terms, we have

$$\left(\frac{\sigma^2 t}{2L} + \frac{1}{2\beta}\right)\|\boldsymbol{\lambda}^{k+1} - \boldsymbol{\lambda}^*\|^2 + \left[\frac{\beta}{2} + \frac{\mu(1-t)}{2\|\mathbf{B}\|_2^2}\right]\|\mathbf{B}\mathbf{y}^{k+1} - \mathbf{B}\mathbf{y}^*\|^2$$

$$\leq \frac{1}{2\beta}\|\boldsymbol{\lambda}^k - \boldsymbol{\lambda}^*\|^2 + \frac{\beta}{2}\|\mathbf{B}\mathbf{y}^k - \mathbf{B}\mathbf{y}^*\|^2. \tag{3.24}$$

Letting

$$\frac{\frac{\sigma^2 t}{2L} + \frac{1}{2\beta}}{\frac{1}{2\beta}} = \frac{\frac{\beta}{2} + \frac{\mu(1-t)}{2\|\mathbf{B}\|_2^2}}{\frac{\beta}{2}},$$

we have $t = \frac{\mu L}{\mu L + \|\mathbf{B}\|_2^2 \sigma^2 \beta^2}$ and (3.24) reduces to

$$\left( \frac{\mu\beta\sigma^2}{\mu L + \|\mathbf{B}\|_2^2 \sigma^2 \beta^2} + 1 \right) \left( \frac{1}{2\beta} \|\boldsymbol{\lambda}^{k+1} - \boldsymbol{\lambda}^*\|^2 + \frac{\beta}{2} \|\mathbf{B}\mathbf{y}^{k+1} - \mathbf{B}\mathbf{y}^*\|^2 \right)$$

$$\leq \frac{1}{2\beta} \|\boldsymbol{\lambda}^k - \boldsymbol{\lambda}^*\|^2 + \frac{\beta}{2} \|\mathbf{B}\mathbf{y}^k - \mathbf{B}\mathbf{y}^*\|^2.$$

Letting $\beta = \frac{\sqrt{\mu L}}{\sigma \|\mathbf{B}\|_2}$, which maximizes $\frac{\mu\beta\sigma^2}{\mu L + \|\mathbf{B}\|_2^2 \sigma^2 \beta^2} + 1$, we have

$$\frac{\mu\beta\sigma^2}{\mu L + \|\mathbf{B}\|_2^2 \sigma^2 \beta^2} + 1 = 1 + \frac{1}{2} \sqrt{\frac{\mu}{L}} \frac{\sigma}{\|\mathbf{B}\|_2}.$$

$\square$

### 3.1.3.2 Linear Convergence Under the Error Bound Condition

Now we move to the scenario under the error bound condition. There are many works studying the linear convergence of ADMM under the error bound condition, see [13, 21, 29, 30] for example. In this section, we introduce the results in [21] but with many simplifications. Denote

$$\phi(\mathbf{x}, \mathbf{y}, \boldsymbol{\lambda}) = \begin{pmatrix} \partial f(\mathbf{x}) + \mathbf{A}^T \boldsymbol{\lambda} \\ \partial g(\mathbf{y}) + \mathbf{B}^T \boldsymbol{\lambda} \\ \mathbf{A}\mathbf{x} + \mathbf{B}\mathbf{y} - \mathbf{b} \end{pmatrix} \quad \text{and} \quad \phi^{-1}(\mathbf{s}) = \{(\mathbf{x}, \mathbf{y}, \boldsymbol{\lambda}) | \mathbf{s} \in \phi(\mathbf{x}, \mathbf{y}, \boldsymbol{\lambda})\}.$$

From Lemma 3.3, we know that $(\mathbf{x}, \mathbf{y}, \boldsymbol{\lambda})$ is a KKT point if and only if $\mathbf{0} \in \phi(\mathbf{x}, \mathbf{y}, \boldsymbol{\lambda})$.

**Definition 3.1** The set-valued mapping $\phi(\mathbf{w})$ is called as satisfying the (global) error bound condition, if there exists constant $\kappa > 0$ such that

$$\text{dist}_{\mathbf{H}}(\mathbf{w}, \phi^{-1}(\mathbf{0})) \leq \kappa \, \text{dist}(\mathbf{0}, \phi(\mathbf{w})), \quad \forall \mathbf{w}, \tag{3.25}$$

where

$$\mathbf{H} = \begin{pmatrix} 0 & 0 & 0 \\ 0 & \beta \mathbf{B}^T \mathbf{B} & 0 \\ 0 & 0 & \frac{1}{\beta}\mathbf{I} \end{pmatrix} \quad \text{and} \quad \text{dist}_\mathbf{H}(\mathbf{w}, \phi^{-1}(\mathbf{0})) = \min_{\mathbf{w}^* \in \phi^{-1}(\mathbf{0})} \|\mathbf{w} - \mathbf{w}^*\|_\mathbf{H}.$$

See [30] for the examples in machine learning which satisfy (3.25). For example, consider the following problem:

$$\min_{\mathbf{x},\mathbf{y}} (f(\mathbf{x}) + g(\mathbf{y})), \quad s.t. \quad \mathbf{x} = \mathbf{y},$$

where $f(\mathbf{x}) = h(\mathbf{Lx}) + \langle \mathbf{q}, \mathbf{x} \rangle$ with strongly convex and smooth $h$, and $g(\mathbf{y})$ can be $\|\mathbf{y}\|_1$ or $\sum_J \|\mathbf{y}_J\|$, which corresponds to the sparse regularizer and the group sparse regularizer, respectively, in which $J$ is a subset of the indexes of $\mathbf{y}$ and $\mathbf{y}_J$ is a subvector of $\mathbf{y}$ by extracting the entries of $\mathbf{y}$ whose indexes are in $J$.

We give the linear convergence rate under the error bound condition (3.25) [21].

**Theorem 3.5** *Suppose that $f(\mathbf{x})$ and $g(\mathbf{y})$ are convex and $\phi(\mathbf{w})$ satisfies the error bound condition (3.25). Then for Algorithm 2.1, we have*

$$\text{dist}_\mathbf{H}^2 \left( \left( \mathbf{x}^{k+1}, \mathbf{y}^{k+1}, \boldsymbol{\lambda}^{k+1} \right), \phi^{-1}(\mathbf{0}) \right)$$

$$\leq \left[ 1 + \frac{1}{\kappa^2 \left( \beta \|\mathbf{A}\|_2^2 + \frac{1}{\beta} \right)} \right]^{-1} \text{dist}_\mathbf{H}^2((\mathbf{x}^k, \mathbf{y}^k, \boldsymbol{\lambda}^k), \phi^{-1}(\mathbf{0})).$$

***Proof*** From Lemma 3.3, we have

$$\begin{pmatrix} \beta \mathbf{A}^T \mathbf{B}(\mathbf{y}^{k+1} - \mathbf{y}^k) \\ \mathbf{0} \\ \frac{1}{\beta}(\boldsymbol{\lambda}^{k+1} - \boldsymbol{\lambda}^k) \end{pmatrix} \in \phi\left( \mathbf{x}^{k+1}, \mathbf{y}^{k+1}, \boldsymbol{\lambda}^{k+1} \right).$$

Thus we have

$$\frac{1}{\kappa^2} \text{dist}_\mathbf{H}^2 \left( \left( \mathbf{x}^{k+1}, \mathbf{y}^{k+1}, \boldsymbol{\lambda}^{k+1} \right), \phi^{-1}(\mathbf{0}) \right)$$

$$\overset{a}{\leq} \text{dist}^2 \left( \mathbf{0}, \phi\left( \mathbf{x}^{k+1}, \mathbf{y}^{k+1}, \boldsymbol{\lambda}^{k+1} \right) \right)$$

$$\leq \beta^2 \|\mathbf{A}^T \mathbf{B}(\mathbf{y}^{k+1} - \mathbf{y}^k)\|^2 + \frac{1}{\beta^2} \|\boldsymbol{\lambda}^{k+1} - \boldsymbol{\lambda}^k\|^2$$

$$\leq \beta^2 \|\mathbf{A}\|_2^2 \|\mathbf{B}(\mathbf{y}^{k+1} - \mathbf{y}^k)\|^2 + \frac{1}{\beta^2} \|\boldsymbol{\lambda}^{k+1} - \boldsymbol{\lambda}^k\|^2$$

$$\leq \left( \beta \|\mathbf{A}\|_2^2 + \frac{1}{\beta} \right) \left( \beta \|\mathbf{B}(\mathbf{y}^{k+1} - \mathbf{y}^k)\|^2 + \frac{1}{\beta} \|\boldsymbol{\lambda}^{k+1} - \boldsymbol{\lambda}^k\|^2 \right), \qquad (3.26)$$

where we use the error bound condition in $\overset{a}{\leq}$.

Next, we choose

$$(\mathbf{x}^*, \mathbf{y}^*, \boldsymbol{\lambda}^*) \in \underset{(\mathbf{x}, \mathbf{y}, \boldsymbol{\lambda}) \in \phi^{-1}(\mathbf{0})}{\text{Argmin}} \left( \beta \|\mathbf{B}\mathbf{y}^k - \mathbf{B}\mathbf{y}\|^2 + \frac{1}{\beta} \|\boldsymbol{\lambda}^k - \boldsymbol{\lambda}\|^2 \right).$$

From the definition of $\mathbf{H}$, we have

$$\text{dist}_{\mathbf{H}}^2 \left( \left( \mathbf{x}^{k+1}, \mathbf{y}^{k+1}, \boldsymbol{\lambda}^{k+1} \right), \phi^{-1}(\mathbf{0}) \right)$$

$$= \underset{(\mathbf{x}, \mathbf{y}, \boldsymbol{\lambda}) \in \phi^{-1}(\mathbf{0})}{\min} \left( \frac{1}{\beta} \|\boldsymbol{\lambda}^{k+1} - \boldsymbol{\lambda}\|^2 + \beta \|\mathbf{B}\mathbf{y}^{k+1} - \mathbf{B}\mathbf{y}\|^2 \right)$$

$$\leq \frac{1}{\beta} \|\boldsymbol{\lambda}^{k+1} - \boldsymbol{\lambda}^*\|^2 + \beta \|\mathbf{B}\mathbf{y}^{k+1} - \mathbf{B}\mathbf{y}^*\|^2$$

$$\overset{a}{\leq} \frac{1}{\beta} \|\boldsymbol{\lambda}^k - \boldsymbol{\lambda}^*\|^2 + \beta \|\mathbf{B}\mathbf{y}^k - \mathbf{B}\mathbf{y}^*\|^2$$

$$- \frac{1}{\beta} \|\boldsymbol{\lambda}^{k+1} - \boldsymbol{\lambda}^k\|^2 - \beta \|\mathbf{B}\mathbf{y}^{k+1} - \mathbf{B}\mathbf{y}^k\|^2$$

$$\overset{b}{=} \text{dist}_{\mathbf{H}}^2 \left( \left( \mathbf{x}^k, \mathbf{y}^k, \boldsymbol{\lambda}^k \right), \phi^{-1}(\mathbf{0}) \right)$$

$$- \frac{1}{\beta} \|\boldsymbol{\lambda}^{k+1} - \boldsymbol{\lambda}^k\|^2 - \beta \|\mathbf{B}\mathbf{y}^{k+1} - \mathbf{B}\mathbf{y}^k\|^2$$

$$\overset{c}{\leq} \text{dist}_{\mathbf{H}}^2 \left( \left( \mathbf{x}^k, \mathbf{y}^k, \boldsymbol{\lambda}^k \right), \phi^{-1}(\mathbf{0}) \right)$$

$$- \frac{1}{\kappa^2 \left( \beta \|\mathbf{A}\|_2^2 + \frac{1}{\beta} \right)} \text{dist}_{\mathbf{H}}^2 \left( \left( \mathbf{x}^{k+1}, \mathbf{y}^{k+1}, \boldsymbol{\lambda}^{k+1} \right), \phi^{-1}(\mathbf{0}) \right),$$

where $\overset{a}{\leq}$ uses (3.15) and Lemma 3.1, $\overset{b}{=}$ is by the choice of $(\mathbf{x}^*, \mathbf{y}^*, \boldsymbol{\lambda}^*)$, and $\overset{c}{\leq}$ uses (3.26). $\qquad \square$

## 3.2  Bregman ADMM

ADMM needs to solve two subproblems to update $\mathbf{x}$ and $\mathbf{y}$, respectively, which are time-consuming when they have no closed-form solutions. To address this issue, we can use the linearization technique to make the subproblems computationally efficient. See, for example, [8, 17, 25, 27, 28], and Algorithms 3.2 and 3.3. We introduce a more general method, called Bregman ADMM, in this section.

Recall the Bregman distance (Definition A.15):

$$D_\phi(\mathbf{y}, \mathbf{x}) = \phi(\mathbf{y}) - \phi(\mathbf{x}) - \langle \nabla\phi(\mathbf{x}), \mathbf{y} - \mathbf{x} \rangle,$$

where $\phi$ is a convex and differentiable function. We consider the general Bregman ADMM, which consists of the following iterations:

$$\mathbf{x}^{k+1} = \underset{\mathbf{x}}{\operatorname{argmin}} \left( f(\mathbf{x}) + g(\mathbf{y}^k) + \left\langle \boldsymbol{\lambda}^k, \mathbf{A}\mathbf{x} + \mathbf{B}\mathbf{y}^k - \mathbf{b} \right\rangle \right.$$
$$\left. + \frac{\beta}{2} \|\mathbf{A}\mathbf{x} + \mathbf{B}\mathbf{y}^k - \mathbf{b}\|^2 + D_\phi(\mathbf{x}, \mathbf{x}^k) \right), \tag{3.27a}$$

$$\mathbf{y}^{k+1} = \underset{\mathbf{y}}{\operatorname{argmin}} \left( f(\mathbf{x}^{k+1}) + g(\mathbf{y}) + \left\langle \boldsymbol{\lambda}^k, \mathbf{A}\mathbf{x}^{k+1} + \mathbf{B}\mathbf{y} - \mathbf{b} \right\rangle \right.$$
$$\left. + \frac{\beta}{2} \|\mathbf{A}\mathbf{x}^{k+1} + \mathbf{B}\mathbf{y} - \mathbf{b}\|^2 + D_\psi(\mathbf{y}, \mathbf{y}^k) \right), \tag{3.27b}$$

$$\boldsymbol{\lambda}^{k+1} = \boldsymbol{\lambda}^k + \beta(\mathbf{A}\mathbf{x}^{k+1} + \mathbf{B}\mathbf{y}^{k+1} - \mathbf{b}). \tag{3.27c}$$

We present the general Bregman ADMM in Algorithm 3.1.

---

**Algorithm 3.1** Bregman ADMM

---

Initialize $\mathbf{x}^0, \mathbf{y}^0, \boldsymbol{\lambda}^0$.
**for** $k = 0, 1, 2, 3, \cdots$ **do**
    Update $\mathbf{x}^{k+1}, \mathbf{y}^{k+1}$, and $\boldsymbol{\lambda}^{k+1}$ by (3.27a), (3.27b), and (3.27c), respectively.
**end for**

---

We can choose different $\phi$ and $\psi$ to give different variants of Bregman ADMM. Specifically, when

$$\phi(\mathbf{x}) = \frac{\beta\|\mathbf{A}\|_2^2}{2} \|\mathbf{x} - \mathbf{u}_1\|^2 - \frac{\beta}{2} \|\mathbf{A}\mathbf{x} - \mathbf{u}_2\|^2, \tag{3.28}$$

$$\psi(\mathbf{y}) = \frac{\beta\|\mathbf{B}\|_2^2}{2} \|\mathbf{y} - \mathbf{v}_1\|^2 - \frac{\beta}{2} \|\mathbf{B}\mathbf{y} - \mathbf{v}_2\|^2, \tag{3.29}$$

where $\mathbf{u}_i$ and $\mathbf{v}_i$ $(i = 1, 2)$ are any constant vectors, we have

$$D_\phi(\mathbf{x}, \mathbf{x}') = \frac{\beta \|\mathbf{A}\|_2^2}{2} \|\mathbf{x} - \mathbf{x}'\|^2 - \frac{\beta}{2} \|\mathbf{A}(\mathbf{x} - \mathbf{x}')\|^2,$$

$$D_\psi(\mathbf{y}, \mathbf{y}') = \frac{\beta \|\mathbf{B}\|_2^2}{2} \|\mathbf{y} - \mathbf{y}'\|^2 - \frac{\beta}{2} \|\mathbf{B}(\mathbf{y} - \mathbf{y}')\|^2,$$

which are independent of $\mathbf{u}_i$ and $\mathbf{v}_i$ $(i = 1, 2)$. Then steps (3.27a) and (3.27b) reduce to

$$\mathbf{x}^{k+1} = \underset{\mathbf{x}}{\operatorname{argmin}} \left( f(\mathbf{x}) + g(\mathbf{y}^k) + \left\langle \boldsymbol{\lambda}^k, \mathbf{A}\mathbf{x} + \mathbf{B}\mathbf{y}^k - \mathbf{b} \right\rangle \right.$$

$$+ \frac{\beta}{2} \|\mathbf{A}\mathbf{x}^k + \mathbf{B}\mathbf{y}^k - \mathbf{b}\|^2 + \beta \left\langle \mathbf{A}^T(\mathbf{A}\mathbf{x}^k + \mathbf{B}\mathbf{y}^k - \mathbf{b}), \mathbf{x} - \mathbf{x}^k \right\rangle$$

$$\left. + \frac{\beta \|\mathbf{A}\|_2^2}{2} \|\mathbf{x} - \mathbf{x}^k\|^2 \right)$$

$$= \operatorname{Prox}_{(\beta \|\mathbf{A}\|_2^2)^{-1} f} \left( \mathbf{x}^k - \left( \beta \|\mathbf{A}\|_2^2 \right)^{-1} \mathbf{A}^T \left[ \boldsymbol{\lambda}^k + \beta (\mathbf{A}\mathbf{x}^k + \mathbf{B}\mathbf{y}^k - \mathbf{b}) \right] \right),$$

$$\tag{3.30}$$

$$\mathbf{y}^{k+1} = \underset{\mathbf{y}}{\operatorname{argmin}} \left( f(\mathbf{x}^{k+1}) + g(\mathbf{y}) + \left\langle \boldsymbol{\lambda}^k, \mathbf{A}\mathbf{x}^{k+1} + \mathbf{B}\mathbf{y} - \mathbf{b} \right\rangle \right.$$

$$+ \frac{\beta}{2} \|\mathbf{A}\mathbf{x}^{k+1} + \mathbf{B}\mathbf{y}^k - \mathbf{b}\|^2 + \beta \left\langle \mathbf{B}^T(\mathbf{A}\mathbf{x}^{k+1} + \mathbf{B}\mathbf{y}^k - \mathbf{b}), \mathbf{y} - \mathbf{y}^k \right\rangle$$

$$\left. + \frac{\beta \|\mathbf{B}\|_2^2}{2} \|\mathbf{y} - \mathbf{y}^k\|^2 \right)$$

$$= \operatorname{Prox}_{(\beta \|\mathbf{B}\|_2^2)^{-1} g} \left( \mathbf{y}^k - \left( \beta \|\mathbf{B}\|_2^2 \right)^{-1} \mathbf{B}^T \left[ \boldsymbol{\lambda}^k + \beta (\mathbf{A}\mathbf{x}^{k+1} + \mathbf{B}\mathbf{y}^k - \mathbf{b}) \right] \right).$$

$$\tag{3.31}$$

It is equivalent to approximating $\frac{\beta}{2} \|\mathbf{A}\mathbf{x} + \mathbf{B}\mathbf{y}^k - \mathbf{b}\|^2$ in (2.15a) and $\frac{\beta}{2} \|\mathbf{A}\mathbf{x}^{k+1} + \mathbf{B}\mathbf{y} - \mathbf{b}\|^2$ in (2.15b) by their quadratic upper bounds at $\mathbf{x}^k$ and $\mathbf{y}^k$, respectively. Note that

$$\frac{\beta}{2} \|\mathbf{A}\mathbf{x}^k + \mathbf{B}\mathbf{y}^k - \mathbf{b}\|^2 + \beta \left\langle \mathbf{A}^T(\mathbf{A}\mathbf{x}^k + \mathbf{B}\mathbf{y}^k - \mathbf{b}), \mathbf{x} - \mathbf{x}^k \right\rangle$$

and

$$\frac{\beta}{2} \|\mathbf{A}\mathbf{x}^{k+1} + \mathbf{B}\mathbf{y}^k - \mathbf{b}\|^2 + \beta \left\langle \mathbf{B}^T(\mathbf{A}\mathbf{x}^{k+1} + \mathbf{B}\mathbf{y}^k - \mathbf{b}), \mathbf{y} - \mathbf{y}^k \right\rangle$$

are the linear approximations of

$$\frac{\beta}{2}\|\mathbf{A}\mathbf{x} + \mathbf{B}\mathbf{y}^k - \mathbf{b}\|^2 \quad \text{and} \quad \frac{\beta}{2}\|\mathbf{A}\mathbf{x}^{k+1} + \mathbf{B}\mathbf{y} - \mathbf{b}\|^2$$

at $\mathbf{x}^k$ and $\mathbf{y}^k$, respectively. So we call iterations (3.30), (3.31), and (3.27c) linearized ADMM, which involves the proximal mappings of $f$ and $g$. In many cases, the proximal mappings of $f$ and $g$ are easily computable. For example, the proximal mappings of $\ell_1$-norm, $\ell_2$-norm, and matrix operator norm and nuclear norm all have closed-form solutions [16, 19].

We summarize the linearized ADMM for the case of the proximal mappings of $f$ and $g$ being easily computable in Algorithm 3.2.

---

**Algorithm 3.2** Linearized ADMM for the case of the proximal mappings of $f$ and $g$ being easily computable (LADMM-1)

---

Initialize $\mathbf{x}^0, \mathbf{y}^0, \boldsymbol{\lambda}^0$.
**for** $k = 0, 1, 2, 3, \cdots$ **do**
    Update $\mathbf{x}^{k+1}, \mathbf{y}^{k+1}$, and $\boldsymbol{\lambda}^{k+1}$ by (3.30), (3.31), and (3.27c), respectively.
**end for**

---

When the proximal mappings of $f$ and $g$ are not easily computable, but $f$ and $g$ are $L_f$-smooth and $L_g$-smooth, respectively, we may choose

$$\phi(\mathbf{x}) = \frac{L_f + \beta\|\mathbf{A}\|_2^2}{2}\|\mathbf{x} - \mathbf{u}_1\|^2 - f(\mathbf{x}) - \frac{\beta}{2}\|\mathbf{A}\mathbf{x} - \mathbf{u}_2\|^2, \qquad (3.32)$$

$$\psi(\mathbf{y}) = \frac{L_g + \beta\|\mathbf{B}\|_2^2}{2}\|\mathbf{y} - \mathbf{v}_1\|^2 - g(\mathbf{y}) - \frac{\beta}{2}\|\mathbf{B}\mathbf{y} - \mathbf{v}_2\|^2, \qquad (3.33)$$

where $\mathbf{u}_i$ and $\mathbf{v}_i$ ($i = 1, 2$) are any constant vectors. Then

$$D_\phi(\mathbf{x}, \mathbf{x}') = \frac{L_f + \beta\|\mathbf{A}\|_2^2}{2}\|\mathbf{x} - \mathbf{x}'\|^2 - f(\mathbf{x}) + f(\mathbf{x}') + \langle \nabla f(\mathbf{x}'), \mathbf{x} - \mathbf{x}'\rangle$$

$$- \frac{\beta}{2}\|\mathbf{A}(\mathbf{x} - \mathbf{x}')\|^2,$$

$$D_\psi(\mathbf{y}, \mathbf{y}') = \frac{L_g + \beta\|\mathbf{B}\|_2^2}{2}\|\mathbf{y} - \mathbf{y}'\|^2 - g(\mathbf{y}) + g(\mathbf{y}') + \langle \nabla g(\mathbf{y}'), \mathbf{y} - \mathbf{y}'\rangle$$

$$- \frac{\beta}{2}\|\mathbf{B}(\mathbf{y} - \mathbf{y}')\|^2, \qquad (3.34)$$

which are also independent of $\mathbf{u}_i$ and $\mathbf{v}_i$ ($i = 1, 2$), and steps (3.27a) and (3.27b) reduce to

$$
\begin{aligned}
\mathbf{x}^{k+1} = \operatorname*{argmin}_{\mathbf{x}} & \left( f(\mathbf{x}^k) + \left\langle \nabla f(\mathbf{x}^k), \mathbf{x} - \mathbf{x}^k \right\rangle + g(\mathbf{y}^k) + \left\langle \boldsymbol{\lambda}^k, \mathbf{A}\mathbf{x} + \mathbf{B}\mathbf{y}^k - \mathbf{b} \right\rangle \right. \\
& + \frac{\beta}{2} \|\mathbf{A}\mathbf{x}^k + \mathbf{B}\mathbf{y}^k - \mathbf{b}\|^2 + \beta \left\langle \mathbf{A}^T (\mathbf{A}\mathbf{x}^k + \mathbf{B}\mathbf{y}^k - \mathbf{b}), \mathbf{x} - \mathbf{x}^k \right\rangle \\
& \left. + \frac{L_f + \beta \|\mathbf{A}\|_2^2}{2} \|\mathbf{x} - \mathbf{x}^k\|^2 \right) \\
= \mathbf{x}^k - & \left( L_f + \beta \|\mathbf{A}\|_2^2 \right)^{-1} \left\{ \nabla f(\mathbf{x}^k) + \mathbf{A}^T \left[ \boldsymbol{\lambda}^k + \beta(\mathbf{A}\mathbf{x}^k + \mathbf{B}\mathbf{y}^k - \mathbf{b}) \right] \right\},
\end{aligned}
$$
$$(3.35)$$

$$
\begin{aligned}
\mathbf{y}^{k+1} = \operatorname*{argmin}_{\mathbf{y}} & \left( f(\mathbf{x}^{k+1}) + g(\mathbf{y}^k) + \left\langle \nabla g(\mathbf{y}^k), \mathbf{y} - \mathbf{y}^k \right\rangle + \left\langle \boldsymbol{\lambda}^k, \mathbf{A}\mathbf{x}^{k+1} + \mathbf{B}\mathbf{y} - \mathbf{b} \right\rangle \right. \\
& + \frac{\beta}{2} \|\mathbf{A}\mathbf{x}^{k+1} + \mathbf{B}\mathbf{y}^k - \mathbf{b}\|^2 + \beta \left\langle \mathbf{B}^T (\mathbf{A}\mathbf{x}^{k+1} + \mathbf{B}\mathbf{y}^k - \mathbf{b}), \mathbf{y} - \mathbf{y}^k \right\rangle \\
& \left. + \frac{L_g + \beta \|\mathbf{B}\|_2^2}{2} \|\mathbf{y} - \mathbf{y}^k\|^2 \right) \\
= \mathbf{y}^k - & \left( L_g + \beta \|\mathbf{B}\|_2^2 \right)^{-1} \left\{ \nabla g(\mathbf{y}^k) + \mathbf{B}^T \left[ \boldsymbol{\lambda}^k + \beta(\mathbf{A}\mathbf{x}^{k+1} + \mathbf{B}\mathbf{y}^k - \mathbf{b}) \right] \right\}.
\end{aligned}
$$
$$(3.36)$$

Similarly, it is equivalent to approximating $f(\mathbf{x}) + \frac{\beta}{2}\|\mathbf{A}\mathbf{x} + \mathbf{B}\mathbf{y}^k - \mathbf{b}\|^2$ in (2.15a) and $g(\mathbf{y}) + \frac{\beta}{2}\|\mathbf{A}\mathbf{x}^{k+1} + \mathbf{B}\mathbf{y} - \mathbf{b}\|^2$ in (2.15b) by their quadratic upper bounds at $\mathbf{x}^k$ and $\mathbf{y}^k$, respectively. Also note that

$$
f(\mathbf{x}^k) + \left\langle \nabla f(\mathbf{x}^k), \mathbf{x} - \mathbf{x}^k \right\rangle
$$
$$
+ \frac{\beta}{2} \|\mathbf{A}\mathbf{x}^k + \mathbf{B}\mathbf{y}^k - \mathbf{b}\|^2 + \beta \left\langle \mathbf{A}^T (\mathbf{A}\mathbf{x}^k + \mathbf{B}\mathbf{y}^k - \mathbf{b}), \mathbf{x} - \mathbf{x}^k \right\rangle
$$

and

$$
g(\mathbf{y}^k) + \left\langle \nabla g(\mathbf{y}^k), \mathbf{y} - \mathbf{y}^k \right\rangle
$$
$$
+ \frac{\beta}{2} \|\mathbf{A}\mathbf{x}^{k+1} + \mathbf{B}\mathbf{y}^k - \mathbf{b}\|^2 + \beta \left\langle \mathbf{B}^T (\mathbf{A}\mathbf{x}^{k+1} + \mathbf{B}\mathbf{y}^k - \mathbf{b}), \mathbf{y} - \mathbf{y}^k \right\rangle
$$

are the linear approximation of

$$f(\mathbf{x}) + \frac{\beta}{2}\|\mathbf{Ax} + \mathbf{By}^k - \mathbf{b}\|^2 \quad \text{and} \quad g(\mathbf{y}) + \frac{\beta}{2}\|\mathbf{Ax}^{k+1} + \mathbf{By} - \mathbf{b}\|^2,$$

at $\mathbf{x}^k$ and $\mathbf{y}^k$, respectively. So we also call the iterations (3.35), (3.36), and (3.27c) linearized ADMM.

We summarize the linearized ADMM for the case of the proximal mappings of $f$ and $g$ not being easily computable but the gradients of $f$ and $g$ being Lipschitz continuous in Algorithm 3.3.

---

**Algorithm 3.3** Linearized ADMM for the case of the proximal mappings of $f$ and $g$ not being easily computable but the gradients of $f$ and $g$ being Lipschitz continuous (LADMM-2)

Initialize $\mathbf{x}^0, \mathbf{y}^0, \lambda^0$.
**for** $k = 0, 1, 2, 3, \cdots$ **do**
        Update $\mathbf{x}^{k+1}, \mathbf{y}^{k+1}$, and $\lambda^{k+1}$ by (3.35), (3.36), and (3.27c), respectively.
**end for**

---

### 3.2.1   Sublinear Convergence

The analysis of sublinear convergence rate of Bregman ADMM is similar to that of the original ADMM. The proofs in this section are adapted from those in Sect. 3.1.2. For Algorithm 3.1, Lemma 3.3 still holds with (3.1), (3.2), and (3.8) being replaced by

$$\begin{aligned}
\mathbf{0} \in{}& \partial f(\mathbf{x}^{k+1}) + \mathbf{A}^T\lambda^k + \beta\mathbf{A}^T(\mathbf{Ax}^{k+1} + \mathbf{By}^k - \mathbf{b}) \\
& + \nabla\phi(\mathbf{x}^{k+1}) - \nabla\phi(\mathbf{x}^k), \\
\mathbf{0} \in{}& \partial g(\mathbf{y}^{k+1}) + \mathbf{B}^T\lambda^k + \beta\mathbf{B}^T(\mathbf{Ax}^{k+1} + \mathbf{By}^{k+1} - \mathbf{b}) \\
& + \nabla\psi(\mathbf{y}^{k+1}) - \nabla\psi(\mathbf{y}^k),
\end{aligned} \tag{3.37}$$

and

$$\begin{aligned}
\hat{\nabla} f(\mathbf{x}^{k+1}) - \nabla\phi(\mathbf{x}^{k+1}) + \nabla\phi(\mathbf{x}^k) \in \partial f(\mathbf{x}^{k+1}), \\
\hat{\nabla} g(\mathbf{y}^{k+1}) - \nabla\psi(\mathbf{y}^{k+1}) + \nabla\psi(\mathbf{y}^k) \in \partial g(\mathbf{y}^{k+1}),
\end{aligned}$$

respectively. Lemma 3.4 still holds without any modification. However, Lemma 3.7 does not hold any more and Lemma 3.5 should be replaced by the following lemma.

**Lemma 3.8** *Suppose that $f(\mathbf{x})$ and $g(\mathbf{y})$ are convex. Then for Algorithm 3.1, we have*

$$\left\langle \hat{\nabla} f(\mathbf{x}^{k+1}), \mathbf{x}^{k+1} - \mathbf{x}^* \right\rangle + \left\langle \hat{\nabla} g(\mathbf{y}^{k+1}), \mathbf{y}^{k+1} - \mathbf{y}^* \right\rangle + \left\langle \boldsymbol{\lambda}^*, \mathbf{A}\mathbf{x}^{k+1} + \mathbf{B}\mathbf{y}^{k+1} - \mathbf{b} \right\rangle$$

$$\leq \frac{1}{2\beta} \|\boldsymbol{\lambda}^k - \boldsymbol{\lambda}^*\|^2 - \frac{1}{2\beta} \|\boldsymbol{\lambda}^{k+1} - \boldsymbol{\lambda}^*\|^2$$

$$+ \frac{\beta}{2} \|\mathbf{B}\mathbf{y}^k - \mathbf{B}\mathbf{y}^*\|^2 - \frac{\beta}{2} \|\mathbf{B}\mathbf{y}^{k+1} - \mathbf{B}\mathbf{y}^*\|^2.$$

**Proof** From the proof of Lemma 3.5, we have

$$\left\langle \hat{\nabla} f(\mathbf{x}^{k+1}), \mathbf{x}^{k+1} - \mathbf{x}^* \right\rangle + \left\langle \hat{\nabla} g(\mathbf{y}^{k+1}), \mathbf{y}^{k+1} - \mathbf{y}^* \right\rangle + \left\langle \boldsymbol{\lambda}^*, \mathbf{A}\mathbf{x}^{k+1} + \mathbf{B}\mathbf{y}^{k+1} - \mathbf{b} \right\rangle$$

$$= \frac{1}{2\beta} \|\boldsymbol{\lambda}^k - \boldsymbol{\lambda}^*\|^2 - \frac{1}{2\beta} \|\boldsymbol{\lambda}^{k+1} - \boldsymbol{\lambda}^*\|^2 - \frac{1}{2\beta} \|\boldsymbol{\lambda}^{k+1} - \boldsymbol{\lambda}^k\|^2$$

$$+ \frac{\beta}{2} \|\mathbf{B}\mathbf{y}^k - \mathbf{B}\mathbf{y}^*\|^2 - \frac{\beta}{2} \|\mathbf{B}\mathbf{y}^{k+1} - \mathbf{B}\mathbf{y}^*\|^2 - \frac{\beta}{2} \|\mathbf{B}\mathbf{y}^{k+1} - \mathbf{B}\mathbf{y}^k\|^2$$

$$+ \left\langle \mathbf{B}\mathbf{y}^{k+1} - \mathbf{B}\mathbf{y}^k, \boldsymbol{\lambda}^{k+1} - \boldsymbol{\lambda}^k \right\rangle$$

$$\overset{a}{\leq} \frac{1}{2\beta} \|\boldsymbol{\lambda}^k - \boldsymbol{\lambda}^*\|^2 - \frac{1}{2\beta} \|\boldsymbol{\lambda}^{k+1} - \boldsymbol{\lambda}^*\|^2$$

$$+ \frac{\beta}{2} \|\mathbf{B}\mathbf{y}^k - \mathbf{B}\mathbf{y}^*\|^2 - \frac{\beta}{2} \|\mathbf{B}\mathbf{y}^{k+1} - \mathbf{B}\mathbf{y}^*\|^2,$$

where $\overset{a}{\leq}$ uses $\frac{1}{2\beta}\|\mathbf{u}\|^2 + \frac{\beta}{2}\|\mathbf{v}\|^2 \geq \langle \mathbf{u}, \mathbf{v} \rangle$. □

Accordingly, Lemma 3.6 should be replaced by the following lemma.

**Lemma 3.9** *Suppose that $f(\mathbf{x})$ and $g(\mathbf{y})$ are convex. Then for Algorithm 3.1, we have*

$$f(\mathbf{x}^{k+1}) + g(\mathbf{y}^{k+1}) - f(\mathbf{x}^*) - g(\mathbf{y}^*) + \left\langle \boldsymbol{\lambda}^*, \mathbf{A}\mathbf{x}^{k+1} + \mathbf{B}\mathbf{y}^{k+1} - \mathbf{b} \right\rangle$$

$$\leq \frac{1}{2\beta} \|\boldsymbol{\lambda}^k - \boldsymbol{\lambda}^*\|^2 - \frac{1}{2\beta} \|\boldsymbol{\lambda}^{k+1} - \boldsymbol{\lambda}^*\|^2$$

$$+ \frac{\beta}{2} \|\mathbf{B}\mathbf{y}^k - \mathbf{B}\mathbf{y}^*\|^2 - \frac{\beta}{2} \|\mathbf{B}\mathbf{y}^{k+1} - \mathbf{B}\mathbf{y}^*\|^2$$

$$+ D_\phi(\mathbf{x}^*, \mathbf{x}^k) - D_\phi(\mathbf{x}^*, \mathbf{x}^{k+1}) - D_\phi(\mathbf{x}^{k+1}, \mathbf{x}^k)$$

$$+ D_\psi(\mathbf{y}^*, \mathbf{y}^k) - D_\psi(\mathbf{y}^*, \mathbf{y}^{k+1}) - D_\psi(\mathbf{y}^{k+1}, \mathbf{y}^k). \tag{3.38}$$

*If we further assume that $g(\mathbf{y})$ is $\mu$-strongly convex, then we have*

$$f(\mathbf{x}^{k+1}) + g(\mathbf{y}^{k+1}) - f(\mathbf{x}^*) - g(\mathbf{y}^*) + \left\langle \boldsymbol{\lambda}^*, \mathbf{A}\mathbf{x}^{k+1} + \mathbf{B}\mathbf{y}^{k+1} - \mathbf{b} \right\rangle$$

$$\leq \frac{1}{2\beta} \|\boldsymbol{\lambda}^k - \boldsymbol{\lambda}^*\|^2 - \frac{1}{2\beta} \|\boldsymbol{\lambda}^{k+1} - \boldsymbol{\lambda}^*\|^2$$

$$+ \frac{\beta}{2} \|\mathbf{B}\mathbf{y}^k - \mathbf{B}\mathbf{y}^*\|^2 - \frac{\beta}{2} \|\mathbf{B}\mathbf{y}^{k+1} - \mathbf{B}\mathbf{y}^*\|^2 - \frac{\mu}{2} \|\mathbf{y}^{k+1} - \mathbf{y}^*\|^2$$

$$+ D_\phi(\mathbf{x}^*, \mathbf{x}^k) - D_\phi(\mathbf{x}^*, \mathbf{x}^{k+1}) - D_\phi(\mathbf{x}^{k+1}, \mathbf{x}^k)$$

$$+ D_\psi(\mathbf{y}^*, \mathbf{y}^k) - D_\psi(\mathbf{y}^*, \mathbf{y}^{k+1}) - D_\psi(\mathbf{y}^{k+1}, \mathbf{y}^k). \tag{3.39}$$

*If we further assume that $g(\mathbf{y})$ is $L$-smooth, then we have*

$$f(\mathbf{x}^{k+1}) + g(\mathbf{y}^{k+1}) - f(\mathbf{x}^*) - g(\mathbf{y}^*) + \left\langle \boldsymbol{\lambda}^*, \mathbf{A}\mathbf{x}^{k+1} + \mathbf{B}\mathbf{y}^{k+1} - \mathbf{b} \right\rangle$$

$$\leq \frac{1}{2\beta} \|\boldsymbol{\lambda}^k - \boldsymbol{\lambda}^*\|^2 - \frac{1}{2\beta} \|\boldsymbol{\lambda}^{k+1} - \boldsymbol{\lambda}^*\|^2$$

$$+ \frac{\beta}{2} \|\mathbf{B}\mathbf{y}^k - \mathbf{B}\mathbf{y}^*\|^2 - \frac{\beta}{2} \|\mathbf{B}\mathbf{y}^{k+1} - \mathbf{B}\mathbf{y}^*\|^2 - \frac{1}{2L} \|\nabla g(\mathbf{y}^{k+1}) - \nabla g(\mathbf{y}^*)\|^2$$

$$+ D_\phi(\mathbf{x}^*, \mathbf{x}^k) - D_\phi(\mathbf{x}^*, \mathbf{x}^{k+1}) - D_\phi(\mathbf{x}^{k+1}, \mathbf{x}^k)$$

$$+ D_\psi(\mathbf{y}^*, \mathbf{y}^k) - D_\psi(\mathbf{y}^*, \mathbf{y}^{k+1}) - D_\psi(\mathbf{y}^{k+1}, \mathbf{y}^k). \tag{3.40}$$

**Proof** We use Lemma 3.8 and Point 2 of Lemma A.2 to prove these conclusions. From the convexity of $f(\mathbf{x})$ and $g(\mathbf{y})$, we have

$$f(\mathbf{x}^{k+1}) + g(\mathbf{y}^{k+1}) - f(\mathbf{x}^*) - g(\mathbf{y}^*) + \left\langle \boldsymbol{\lambda}^*, \mathbf{A}\mathbf{x}^{k+1} + \mathbf{B}\mathbf{y}^{k+1} - \mathbf{b} \right\rangle$$

$$\leq \left\langle \hat{\nabla} f(\mathbf{x}^{k+1}) - \left(\nabla\phi(\mathbf{x}^{k+1}) - \nabla\phi(\mathbf{x}^k)\right), \mathbf{x}^{k+1} - \mathbf{x}^* \right\rangle$$

$$+ \left\langle \hat{\nabla} g(\mathbf{y}^{k+1}) - \left(\nabla\psi(\mathbf{y}^{k+1}) - \nabla\psi(\mathbf{y}^k)\right), \mathbf{y}^{k+1} - \mathbf{y}^* \right\rangle$$

$$+ \left\langle \boldsymbol{\lambda}^*, \mathbf{A}\mathbf{x}^{k+1} + \mathbf{B}\mathbf{y}^{k+1} - \mathbf{b} \right\rangle$$

$$\leq \frac{1}{2\beta} \|\boldsymbol{\lambda}^k - \boldsymbol{\lambda}^*\|^2 - \frac{1}{2\beta} \|\boldsymbol{\lambda}^{k+1} - \boldsymbol{\lambda}^*\|^2$$

$$+ \frac{\beta}{2} \|\mathbf{B}\mathbf{y}^k - \mathbf{B}\mathbf{y}^*\|^2 - \frac{\beta}{2} \|\mathbf{B}\mathbf{y}^{k+1} - \mathbf{B}\mathbf{y}^*\|^2$$

$$+ D_\phi(\mathbf{x}^*, \mathbf{x}^k) - D_\phi(\mathbf{x}^*, \mathbf{x}^{k+1}) - D_\phi(\mathbf{x}^{k+1}, \mathbf{x}^k)$$

$$+ D_\psi(\mathbf{y}^*, \mathbf{y}^k) - D_\psi(\mathbf{y}^*, \mathbf{y}^{k+1}) - D_\psi(\mathbf{y}^{k+1}, \mathbf{y}^k).$$

(3.39) and (3.40) can be obtained in the same way as in the proof of Lemma 3.6.  $\square$

### 3.2.1.1 Ergodic Convergence Rate

Considering the ergodic convergence rate of Bregman ADMM, Theorem 3.3 still holds with a little modification, resulting in the following theorem. The proof is quite similar to that of Theorem 3.3, by summing (3.38) instead.

**Theorem 3.6** *Suppose that $f(\mathbf{x})$ and $g(\mathbf{y})$ are convex. Then for Algorithm 3.1, we have*

$$|f(\hat{\mathbf{x}}^{K+1}) + g(\hat{\mathbf{y}}^{K+1}) - f(\mathbf{x}^*) - g(\mathbf{y}^*)| \leq \frac{D}{2(K+1)} + \frac{2\sqrt{D}\|\boldsymbol{\lambda}^*\|}{\sqrt{\beta}(K+1)},$$

$$\|\mathbf{A}\hat{\mathbf{x}}^{K+1} + \mathbf{B}\hat{\mathbf{y}}^{K+1} - \mathbf{b}\| \leq \frac{2\sqrt{D}}{\sqrt{\beta}(K+1)},$$

*where*

$$\hat{\mathbf{x}}^{K+1} = \frac{1}{K+1}\sum_{k=1}^{K+1}\mathbf{x}^k, \quad \hat{\mathbf{y}}^{K+1} = \frac{1}{K+1}\sum_{k=1}^{K+1}\mathbf{y}^k, \text{ and}$$

$$D = \frac{1}{\beta}\|\boldsymbol{\lambda}^0 - \boldsymbol{\lambda}^*\|^2 + \beta\|\mathbf{B}\mathbf{y}^0 - \mathbf{B}\mathbf{y}^*\|^2 + 2D_\phi(\mathbf{x}^*, \mathbf{x}^0) + 2D_\psi(\mathbf{y}^*, \mathbf{y}^0).$$

### 3.2.1.2 Non-ergodic Convergence Rate

When we restrict

$$\psi = 0 \quad \text{and} \quad D_\phi(\mathbf{x}, \mathbf{y}) = \frac{1}{2}\|\mathbf{x} - \mathbf{y}\|_{\mathbf{M}}^2 \equiv \frac{1}{2}\langle \mathbf{M}(\mathbf{x} - \mathbf{y}), \mathbf{x} - \mathbf{y}\rangle$$

for any fixed symmetric and positive semidefinite matrix $\mathbf{M}$ (that is, $\phi(\mathbf{x}) = \frac{1}{2}\mathbf{x}^T\mathbf{M}\mathbf{x}$), the $O\left(\frac{1}{\sqrt{K}}\right)$ non-ergodic convergence rate still holds. In this case, Lemmas 3.7 and 3.5 should be replaced by the following two lemmas, respectively.

**Lemma 3.10** *Suppose that $f(\mathbf{x})$ and $g(\mathbf{y})$ are convex. Let*

$$\psi = 0 \quad \text{and} \quad D_\phi(\mathbf{x}, \mathbf{y}) = \frac{1}{2}\|\mathbf{x} - \mathbf{y}\|_{\mathbf{M}}^2$$

*for some symmetric and positive semidefinite matrix $\mathbf{M}$. Then for Algorithm 3.1, we have*

$$\frac{1}{2\beta}\|\boldsymbol{\lambda}^{k+1} - \boldsymbol{\lambda}^k\|^2 + \frac{\beta}{2}\|\mathbf{B}\mathbf{y}^{k+1} - \mathbf{B}\mathbf{y}^k\|^2 + \frac{1}{2}\|\mathbf{x}^{k+1} - \mathbf{x}^k\|_{\mathbf{M}}^2$$

$$\leq \frac{1}{2\beta}\|\boldsymbol{\lambda}^k - \boldsymbol{\lambda}^{k-1}\|^2 + \frac{\beta}{2}\|\mathbf{B}\mathbf{y}^k - \mathbf{B}\mathbf{y}^{k-1}\|^2 + \frac{1}{2}\|\mathbf{x}^k - \mathbf{x}^{k-1}\|_{\mathbf{M}}^2.$$

***Proof*** From the proof of Lemma 3.7, we have

$$\left\langle \hat{\nabla} f(\mathbf{x}^{k+1}) - \hat{\nabla} f(\mathbf{x}^k), \mathbf{x}^{k+1} - \mathbf{x}^k \right\rangle + \left\langle \hat{\nabla} g(\mathbf{y}^{k+1}) - \hat{\nabla} g(\mathbf{y}^k), \mathbf{y}^{k+1} - \mathbf{y}^k \right\rangle$$

$$\leq \frac{1}{2\beta} \left( \|\boldsymbol{\lambda}^k - \boldsymbol{\lambda}^{k-1}\|^2 - \|\boldsymbol{\lambda}^{k+1} - \boldsymbol{\lambda}^k\|^2 \right)$$

$$+ \frac{\beta}{2} \left( \|\mathbf{B}\mathbf{y}^k - \mathbf{B}\mathbf{y}^{k-1}\|^2 - \|\mathbf{B}\mathbf{y}^{k+1} - \mathbf{B}\mathbf{y}^k\|^2 \right).$$

On the other hand, we have

$$\left\langle -\mathbf{M}(\mathbf{x}^{k+1} - \mathbf{x}^k) + \mathbf{M}(\mathbf{x}^k - \mathbf{x}^{k-1}), \mathbf{x}^{k+1} - \mathbf{x}^k \right\rangle$$

$$= \frac{1}{2} \left( \|\mathbf{x}^k - \mathbf{x}^{k-1}\|_{\mathbf{M}}^2 - \|\mathbf{x}^{k+1} - \mathbf{x}^k\|_{\mathbf{M}}^2 - \|(\mathbf{x}^{k+1} - \mathbf{x}^k) - (\mathbf{x}^k - \mathbf{x}^{k-1})\|_{\mathbf{M}}^2 \right)$$

$$\leq \frac{1}{2} \left( \|\mathbf{x}^k - \mathbf{x}^{k-1}\|_{\mathbf{M}}^2 - \|\mathbf{x}^{k+1} - \mathbf{x}^k\|_{\mathbf{M}}^2 \right). \tag{3.41}$$

Adding the above two inequalities together, we have

$$\left\langle \hat{\nabla} f(\mathbf{x}^{k+1}) - \mathbf{M}(\mathbf{x}^{k+1} - \mathbf{x}^k) - \left( \hat{\nabla} f(\mathbf{x}^k) - \mathbf{M}(\mathbf{x}^k - \mathbf{x}^{k-1}) \right), \mathbf{x}^{k+1} - \mathbf{x}^k \right\rangle$$

$$+ \left\langle \hat{\nabla} g(\mathbf{y}^{k+1}) - \hat{\nabla} g(\mathbf{y}^k), \mathbf{y}^{k+1} - \mathbf{y}^k \right\rangle$$

$$\leq \frac{1}{2\beta} \left( \|\boldsymbol{\lambda}^k - \boldsymbol{\lambda}^{k-1}\|^2 - \|\boldsymbol{\lambda}^{k+1} - \boldsymbol{\lambda}^k\|^2 \right)$$

$$+ \frac{\beta}{2} \left( \|\mathbf{B}\mathbf{y}^k - \mathbf{B}\mathbf{y}^{k-1}\|^2 - \|\mathbf{B}\mathbf{y}^{k+1} - \mathbf{B}\mathbf{y}^k\|^2 \right)$$

$$+ \frac{1}{2} \left( \|\mathbf{x}^k - \mathbf{x}^{k-1}\|_{\mathbf{M}}^2 - \|\mathbf{x}^{k+1} - \mathbf{x}^k\|_{\mathbf{M}}^2 \right).$$

Using the monotonicity of $\partial f$ and $\partial g$, we have the conclusion. □

**Lemma 3.11** *Suppose that $f(\mathbf{x})$ and $g(\mathbf{y})$ are convex. Let*

$$\psi = 0 \quad and \quad D_\phi(\mathbf{x}, \mathbf{y}) = \frac{1}{2} \|\mathbf{x} - \mathbf{y}\|_{\mathbf{M}}^2$$

*for some symmetric and positive semidefinite matrix $\mathbf{M}$. Then for Algorithm 3.1, we have*

$$\left\langle \hat{\nabla} f(\mathbf{x}^{k+1}), \mathbf{x}^{k+1} - \mathbf{x}^* \right\rangle + \left\langle \hat{\nabla} g(\mathbf{y}^{k+1}), \mathbf{y}^{k+1} - \mathbf{y}^* \right\rangle + \left\langle \boldsymbol{\lambda}^*, \mathbf{A}\mathbf{x}^{k+1} + \mathbf{B}\mathbf{y}^{k+1} - \mathbf{b} \right\rangle$$

$$\leq \frac{1}{2\beta} \|\boldsymbol{\lambda}^k - \boldsymbol{\lambda}^*\|^2 - \frac{1}{2\beta} \|\boldsymbol{\lambda}^{k+1} - \boldsymbol{\lambda}^*\|^2$$

$$+ \frac{\beta}{2}\|\mathbf{By}^k - \mathbf{By}^*\|^2 - \frac{\beta}{2}\|\mathbf{By}^{k+1} - \mathbf{By}^*\|^2$$

$$- \frac{1}{2\beta}\|\boldsymbol{\lambda}^{k+1} - \boldsymbol{\lambda}^k\|^2 - \frac{\beta}{2}\|\mathbf{By}^{k+1} - \mathbf{By}^k\|^2.$$

The difference from Lemma 3.8 is that (3.14) holds since $\psi = 0$. Similar to (3.38), we have

$$f(\mathbf{x}^{k+1}) + g(\mathbf{y}^{k+1}) - f(\mathbf{x}^*) - g(\mathbf{y}^*) + \left\langle \boldsymbol{\lambda}^*, \mathbf{Ax}^{k+1} + \mathbf{By}^{k+1} - \mathbf{b} \right\rangle$$

$$\leq \frac{1}{2\beta}\|\boldsymbol{\lambda}^k - \boldsymbol{\lambda}^*\|^2 + \frac{\beta}{2}\|\mathbf{By}^k - \mathbf{By}^*\|^2 + \frac{1}{2}\|\mathbf{x}^k - \mathbf{x}^*\|_{\mathbf{M}}^2$$

$$- \left( \frac{1}{2\beta}\|\boldsymbol{\lambda}^{k+1} - \boldsymbol{\lambda}^*\|^2 + \frac{\beta}{2}\|\mathbf{By}^{k+1} - \mathbf{By}^*\|^2 + \frac{1}{2}\|\mathbf{x}^{k+1} - \mathbf{x}^*\|_{\mathbf{M}}^2 \right)$$

$$- \left( \frac{1}{2\beta}\|\boldsymbol{\lambda}^{k+1} - \boldsymbol{\lambda}^k\|^2 + \frac{\beta}{2}\|\mathbf{By}^{k+1} - \mathbf{By}^k\|^2 + \frac{1}{2}\|\mathbf{x}^{k+1} - \mathbf{x}^k\|_{\mathbf{M}}^2 \right).$$

Similar to Theorem 3.2, we have the following theorem.

**Theorem 3.7** *Suppose that $f$ and $g$ are both generally convex. Let*

$$\psi = 0 \quad and \quad D_\phi(\mathbf{x}, \mathbf{y}) = \frac{1}{2}\|\mathbf{x} - \mathbf{y}\|_{\mathbf{M}}^2$$

*for some symmetric and positive semidefinite matrix $\mathbf{M}$. Then for Algorithm 3.1, we have*

$$-\|\boldsymbol{\lambda}^*\|\sqrt{\frac{C}{\beta(K+1)}} \leq f(\mathbf{x}^{K+1}) + g(\mathbf{y}^{K+1}) - f(\mathbf{x}^*) - g(\mathbf{y}^*)$$

$$\leq \frac{C}{K+1} + \frac{3C}{\sqrt{K+1}} + \|\boldsymbol{\lambda}^*\|\sqrt{\frac{C}{\beta(K+1)}},$$

$$\|\mathbf{Ax}^{K+1} + \mathbf{By}^{K+1} - \mathbf{b}\| \leq \sqrt{\frac{C}{\beta(K+1)}},$$

*where $C = \frac{1}{\beta}\|\boldsymbol{\lambda}^0 - \boldsymbol{\lambda}^*\|^2 + \beta\|\mathbf{By}^0 - \mathbf{By}^*\|^2 + \|\mathbf{x}^0 - \mathbf{x}^*\|_{\mathbf{M}}^2.$*

**Proof** Note that (3.11) also holds for Algorithm 3.1. From

$$\hat{\nabla} f(\mathbf{x}^{k+1}) - \mathbf{M}(\mathbf{x}^{k+1} - \mathbf{x}^k) \in \partial f(\mathbf{x}^{k+1}) \quad and \quad \hat{\nabla} g(\mathbf{y}^{k+1}) \in \partial g(\mathbf{y}^{k+1}),$$

we have

$$f(\mathbf{x}^{k+1}) + g(\mathbf{y}^{k+1}) - f(\mathbf{x}^*) - g(\mathbf{y}^*) + \left\langle \boldsymbol{\lambda}^*, \mathbf{A}\mathbf{x}^{k+1} + \mathbf{B}\mathbf{y}^{k+1} - \mathbf{b} \right\rangle$$

$$\leq -\frac{1}{\beta} \left\langle \boldsymbol{\lambda}^{k+1} - \boldsymbol{\lambda}^*, \boldsymbol{\lambda}^{k+1} - \boldsymbol{\lambda}^k \right\rangle + \left\langle \mathbf{B}\mathbf{y}^{k+1} - \mathbf{B}\mathbf{y}^k, \boldsymbol{\lambda}^{k+1} - \boldsymbol{\lambda}^k \right\rangle$$

$$- \beta \left\langle \mathbf{B}\mathbf{y}^{k+1} - \mathbf{B}\mathbf{y}^k, \mathbf{B}\mathbf{y}^{k+1} - \mathbf{B}\mathbf{y}^* \right\rangle - \left\langle \mathbf{M}(\mathbf{x}^{k+1} - \mathbf{x}^k), \mathbf{x}^{k+1} - \mathbf{x}^* \right\rangle.$$

Following the proof of Theorem 3.2, we have the conclusion. Note that since there is an additional term $-\left\langle \mathbf{M}(\mathbf{x}^{k+1} - \mathbf{x}^k), \mathbf{x}^{k+1} - \mathbf{x}^* \right\rangle$, we get $\frac{3C}{\sqrt{K+1}}$ here, rather than $\frac{2C}{\sqrt{K+1}}$ in Theorem 3.2.                                                                          □

*Remark 3.1* We explain why we consider the scenario of

$$\psi = 0 \quad \text{and} \quad \phi(\mathbf{x}) = \frac{1}{2}\mathbf{x}^T \mathbf{M}\mathbf{x}.$$

When $\psi \neq 0$, (3.14) does not hold since

$$\hat{\nabla} g(\mathbf{y}^{k+1}) - \nabla \psi(\mathbf{y}^{k+1}) + \nabla \psi(\mathbf{y}^k) \in \partial g(\mathbf{y}^{k+1}),$$

rather than $\hat{\nabla} g(\mathbf{y}^{k+1}) \in \partial g(\mathbf{y}^{k+1})$. So we can only get Lemma 3.8, rather than Lemma 3.11. From the proof of Theorem 3.2, we know that it is crucial to keep the term

$$\frac{1}{2\beta} \|\boldsymbol{\lambda}^{k+1} - \boldsymbol{\lambda}^k\|^2 + \frac{\beta}{2} \|\mathbf{B}\mathbf{y}^{k+1} - \mathbf{B}\mathbf{y}^k\|^2.$$

On the other hand, when $\phi$ is a general smooth convex function, instead of (3.41), we should upper bound

$$\left\langle -(\nabla\phi(\mathbf{x}^{k+1}) - \nabla\phi(\mathbf{x}^k)) + (\nabla\phi(\mathbf{x}^k) - \nabla\phi(\mathbf{x}^{k-1})), \mathbf{x}^{k+1} - \mathbf{x}^k \right\rangle,$$

for example, in the following form, which does not always hold,

$$\left\langle -(\nabla\phi(\mathbf{x}^{k+1}) - \nabla\phi(\mathbf{x}^k)) + (\nabla\phi(\mathbf{x}^k) - \nabla\phi(\mathbf{x}^{k-1})), \mathbf{x}^{k+1} - \mathbf{x}^k \right\rangle$$

$$\leq D_\phi(\mathbf{x}^k, \mathbf{x}^{k-1}) - D_\phi(\mathbf{x}^{k+1}, \mathbf{x}^k).$$

### 3.2.2 Linear Convergence

Now, we focus on the linear convergence of the Bregman ADMM. We consider two scenarios. The first scenario is under the assumption that $g(\mathbf{y})$ is $\mu_g$-strongly convex and $L_g$-smooth. The second scenario is under the assumption that $g(\mathbf{y})$ is $\mu_g$-strongly convex and $L_g$-smooth, and $f(\mathbf{x})$ is $\mu_f$-strongly convex. For the first scenario, we only linearize the second subproblem, i.e., let $\phi = 0$. For the second scenario, we linearize both subproblems. The proofs in this section are adapted from those in Sect. 3.1.3.1, and we try to make the convergence rates as sharp as possible.

**Theorem 3.8** *Assume that $f(\mathbf{x})$ is convex, $g(\mathbf{y})$ is $\mu_g$-strongly convex and $L_g$-smooth, $\phi = 0$, and $\psi(\mathbf{y})$ is convex and $L_\psi$-smooth. Assume that $\|\mathbf{B}^T\boldsymbol{\lambda}\| \geq \sigma\|\boldsymbol{\lambda}\|, \forall\boldsymbol{\lambda}$, where $\sigma > 0$. Then for Algorithm 3.1 we have*

$$\frac{1}{2\beta}\|\boldsymbol{\lambda}^{k+1} - \boldsymbol{\lambda}^*\|^2 + \frac{\beta}{2}\|\mathbf{B}\mathbf{y}^{k+1} - \mathbf{B}\mathbf{y}^*\|^2 + D_\psi(\mathbf{y}^*, \mathbf{y}^{k+1})$$

$$\leq \left(1 + \frac{1}{3}\min\left\{\frac{\beta\sigma^2}{L_g + L_\psi}, \frac{\mu_g}{\beta\|\mathbf{B}\|_2^2}, \frac{\mu_g}{L_\psi}\right\}\right)^{-1}$$

$$\times \left(\frac{1}{2\beta}\|\boldsymbol{\lambda}^k - \boldsymbol{\lambda}^*\|^2 + \frac{\beta}{2}\|\mathbf{B}\mathbf{y}^k - \mathbf{B}\mathbf{y}^*\|^2 + D_\psi(\mathbf{y}^*, \mathbf{y}^k)\right).$$

**Proof** Similar to the induction in (3.21) and using (A.5) and (3.37), we have

$$\frac{\sigma^2}{2(L_g + L_\psi)}\|\boldsymbol{\lambda}^{k+1} - \boldsymbol{\lambda}^*\|^2 - D_\psi(\mathbf{y}^{k+1}, \mathbf{y}^k)$$

$$\leq \frac{1}{2(L_g + L_\psi)}\|\mathbf{B}^T(\boldsymbol{\lambda}^{k+1} - \boldsymbol{\lambda}^*)\|^2 - \frac{1}{2L_\psi}\|\nabla\psi(\mathbf{y}^{k+1}) - \nabla\psi(\mathbf{y}^k)\|^2$$

$$\overset{a}{\leq} \frac{1}{2L_g}\|\mathbf{B}^T(\boldsymbol{\lambda}^{k+1} - \boldsymbol{\lambda}^*) + \nabla\psi(\mathbf{y}^{k+1}) - \nabla\psi(\mathbf{y}^k)\|^2$$

$$= \frac{1}{2L_g}\|\nabla g(\mathbf{y}^{k+1}) - \nabla g(\mathbf{y}^*)\|^2, \tag{3.42}$$

where $\overset{a}{\leq}$ uses $\|\mathbf{u} + \mathbf{v}\|^2 \geq (1 - \nu)\|\mathbf{u}\|^2 - (\frac{1}{\nu} - 1)\|\mathbf{v}\|^2$ for $\nu = \frac{L_\psi}{L_g + L_\psi}$. From (3.39), (3.40), (3.42) and Lemma 3.1 and using (A.4), we have

$$\frac{\mu_g}{2\|\mathbf{B}\|_2^2}\|\mathbf{B}\mathbf{y}^{k+1} - \mathbf{B}\mathbf{y}^*\|^2 \leq \frac{\mu_g}{2}\|\mathbf{y}^{k+1} - \mathbf{y}^*\|^2$$

$$\leq \frac{1}{2\beta}\|\boldsymbol{\lambda}^k - \boldsymbol{\lambda}^*\|^2 - \frac{1}{2\beta}\|\boldsymbol{\lambda}^{k+1} - \boldsymbol{\lambda}^*\|^2$$

$$+\frac{\beta}{2}\|\mathbf{By}^k - \mathbf{By}^*\|^2 - \frac{\beta}{2}\|\mathbf{By}^{k+1} - \mathbf{By}^*\|^2$$
$$+D_\psi(\mathbf{y}^*, \mathbf{y}^k) - D_\psi(\mathbf{y}^*, \mathbf{y}^{k+1}),$$

$$\frac{\mu_g}{L_\psi} D_\psi(\mathbf{y}^*, \mathbf{y}^{k+1}) \leq \frac{\mu_g}{2}\|\mathbf{y}^{k+1} - \mathbf{y}^*\|^2$$

$$\leq \frac{1}{2\beta}\|\boldsymbol{\lambda}^k - \boldsymbol{\lambda}^*\|^2 - \frac{1}{2\beta}\|\boldsymbol{\lambda}^{k+1} - \boldsymbol{\lambda}^*\|^2$$
$$+\frac{\beta}{2}\|\mathbf{By}^k - \mathbf{By}^*\|^2 - \frac{\beta}{2}\|\mathbf{By}^{k+1} - \mathbf{By}^*\|^2$$
$$+D_\psi(\mathbf{y}^*, \mathbf{y}^k) - D_\psi(\mathbf{y}^*, \mathbf{y}^{k+1}),$$

and

$$\frac{\sigma^2}{2(L_g + L_\psi)}\|\boldsymbol{\lambda}^{k+1} - \boldsymbol{\lambda}^*\|^2 \leq \frac{1}{2\beta}\|\boldsymbol{\lambda}^k - \boldsymbol{\lambda}^*\|^2 - \frac{1}{2\beta}\|\boldsymbol{\lambda}^{k+1} - \boldsymbol{\lambda}^*\|^2$$
$$+\frac{\beta}{2}\|\mathbf{By}^k - \mathbf{By}^*\|^2 - \frac{\beta}{2}\|\mathbf{By}^{k+1} - \mathbf{By}^*\|^2$$
$$+D_\psi(\mathbf{y}^*, \mathbf{y}^k) - D_\psi(\mathbf{y}^*, \mathbf{y}^{k+1}).$$

Adding them together, we have

$$\left(1 + \frac{1}{3}\min\left\{\frac{\beta\sigma^2}{L_g + L_\psi}, \frac{\mu_g}{\beta\|\mathbf{B}\|_2^2}, \frac{\mu_g}{L_\psi}\right\}\right)$$
$$\times\left(\frac{1}{2\beta}\|\boldsymbol{\lambda}^{k+1} - \boldsymbol{\lambda}^*\|^2 + \frac{\beta}{2}\|\mathbf{By}^{k+1} - \mathbf{By}^*\|^2 + D_\psi(\mathbf{y}^*, \mathbf{y}^{k+1})\right)$$
$$\leq \frac{1}{2\beta}\|\boldsymbol{\lambda}^k - \boldsymbol{\lambda}^*\|^2 + \frac{\beta}{2}\|\mathbf{By}^k - \mathbf{By}^*\|^2 + D_\psi(\mathbf{y}^*, \mathbf{y}^k).$$

So we have the conclusion.                                                      □

*Remark 3.2* From Theorem 3.8, we see that the complexity for achieving an $\epsilon$-approximate solution is

$$O\left(\left(\frac{L_g + L_\psi}{\beta\sigma^2} + \frac{\beta\|\mathbf{B}\|_2^2}{\mu_g} + \frac{L_\psi}{\mu_g}\right)\log\frac{1}{\epsilon}\right).$$

When

$$\psi(\mathbf{y}) = \frac{L_g + \beta\|\mathbf{B}\|_2^2}{2}\|\mathbf{y}\|^2 - g(\mathbf{y}) - \frac{\beta}{2}\|\mathbf{By}\|^2,$$

**Table 3.1** Complexity comparisons between ADMM and two variants of linearized ADMM (LADMM)

| Method | Rates | Linearization |
|--------|-------|---------------|
| ADMM | $O\left(\sqrt{\frac{L_g}{\mu_g}} \frac{\|\mathbf{B}\|_2}{\sigma} \log \frac{1}{\epsilon}\right)$ | None |
| LADMM-1 | $O\left(\left(\sqrt{\frac{L_g}{\mu_g}} \frac{\|\mathbf{B}\|_2}{\sigma} + \frac{\|\mathbf{B}\|_2^2}{\sigma^2}\right) \log \frac{1}{\epsilon}\right)$ | Only on the augmented term |
| LADMM-2 | $O\left(\left(\frac{\|\mathbf{B}\|_2^2}{\sigma^2} + \frac{L_g}{\mu_g}\right) \log \frac{1}{\epsilon}\right)$ | On $g$ and the augmented term |

we have $L_\psi = L_g + \beta \|\mathbf{B}\|_2^2$ and

$$
O\left(\left(\frac{L_g + L_\psi}{\beta \sigma^2} + \frac{\beta \|\mathbf{B}\|_2^2}{\mu_g} + \frac{L_\psi}{\mu_g}\right) \log \frac{1}{\epsilon}\right)
$$

$$
= O\left(\left(\frac{L_g}{\beta \sigma^2} + \frac{\|\mathbf{B}\|_2^2}{\sigma^2} + \frac{\beta \|\mathbf{B}\|_2^2}{\mu_g} + \frac{L_g}{\mu_g}\right) \log \frac{1}{\epsilon}\right).
$$

Letting $\beta = \frac{\sqrt{\mu_g L_g}}{\sigma \|\mathbf{B}\|_2}$ and using $ab \leq \frac{1}{2}(a^2 + b^2)$ for all $a, b \geq 0$, the complexity becomes

$$
O\left(\left(\frac{\|\mathbf{B}\|_2}{\sigma} \sqrt{\frac{L_g}{\mu_g}} + \frac{\|\mathbf{B}\|_2^2}{\sigma^2} + \frac{L_g}{\mu_g}\right) \log \frac{1}{\epsilon}\right) = O\left(\left(\frac{\|\mathbf{B}\|_2^2}{\sigma^2} + \frac{L_g}{\mu_g}\right) \log \frac{1}{\epsilon}\right).
$$

$$\tag{3.43}$$

Similarly, when

$$
\psi(\mathbf{y}) = \frac{\beta \|\mathbf{B}\|_2^2}{2} \|\mathbf{y}\|^2 - \frac{\beta}{2} \|\mathbf{B}\mathbf{y}\|^2,
$$

we have $L_\psi = \beta \|\mathbf{B}\|_2^2$. Accordingly, also letting $\beta = \frac{\sqrt{\mu_g L_g}}{\sigma \|\mathbf{B}\|_2}$ the complexity becomes

$$
O\left(\left(\frac{\|\mathbf{B}\|_2}{\sigma} \sqrt{\frac{L_g}{\mu_g}} + \frac{\|\mathbf{B}\|_2^2}{\sigma^2}\right) \log \frac{1}{\epsilon}\right),
$$

which equals the left hand side of (3.43) without the $\frac{L_g}{\mu_g}$ term. We list the comparisons in Table 3.1.

**Theorem 3.9** *Assume that $f(\mathbf{x})$ is $\mu_f$-strongly convex, $g(\mathbf{y})$ is $\mu_g$-strongly convex and $L_g$-smooth, $\phi$ is convex and $L_\phi$-smooth, and $\psi(\mathbf{y})$ is convex and $L_\psi$-smooth. Assume that $\|\mathbf{B}^T\boldsymbol{\lambda}\| \geq \sigma\|\boldsymbol{\lambda}\|, \forall\boldsymbol{\lambda}$, where $\sigma > 0$. Then for Algorithm 3.1 we have*

$$\frac{1}{2\beta}\|\boldsymbol{\lambda}^{k+1} - \boldsymbol{\lambda}^*\|^2 + \frac{\beta}{2}\|\mathbf{B}\mathbf{y}^{k+1} - \mathbf{B}\mathbf{y}^*\|^2 + D_\phi(\mathbf{x}^*, \mathbf{x}^{k+1}) + D_\psi(\mathbf{y}^*, \mathbf{y}^{k+1})$$

$$\leq \left(1 + \frac{1}{4}\min\left\{\frac{\beta\sigma^2}{L_g + L_\psi}, \frac{\mu_g}{\beta\|\mathbf{B}\|_2^2}, \frac{\mu_g}{L_\psi}, \frac{\mu_f}{L_\phi}\right\}\right)^{-1}$$

$$\times \left(\frac{1}{2\beta}\|\boldsymbol{\lambda}^k - \boldsymbol{\lambda}^*\|^2 + \frac{\beta}{2}\|\mathbf{B}\mathbf{y}^k - \mathbf{B}\mathbf{y}^*\|^2 + D_\phi(\mathbf{x}^*, \mathbf{x}^k) + D_\psi(\mathbf{y}^*, \mathbf{y}^k)\right).$$

**Proof** From the strong convexity of $f$ we can obtain an inequality similar to (3.39), then we have

$$\frac{\mu_f}{L_\phi}D_\phi(\mathbf{x}^*, \mathbf{x}^{k+1}) \leq \frac{\mu_f}{2}\|\mathbf{x}^{k+1} - \mathbf{x}^*\|^2$$

$$\leq \frac{1}{2\beta}\|\boldsymbol{\lambda}^k - \boldsymbol{\lambda}^*\|^2 - \frac{1}{2\beta}\|\boldsymbol{\lambda}^{k+1} - \boldsymbol{\lambda}^*\|^2$$

$$+ \frac{\beta}{2}\|\mathbf{B}\mathbf{y}^k - \mathbf{B}\mathbf{y}^*\|^2 - \frac{\beta}{2}\|\mathbf{B}\mathbf{y}^{k+1} - \mathbf{B}\mathbf{y}^*\|^2$$

$$+ D_\phi(\mathbf{x}^*, \mathbf{x}^k) - D_\phi(\mathbf{x}^*, \mathbf{x}^{k+1})$$

$$+ D_\psi(\mathbf{y}^*, \mathbf{y}^k) - D_\psi(\mathbf{y}^*, \mathbf{y}^{k+1}).$$

Following the same proof of Theorem 3.8, we have

$$\left(1 + \frac{1}{4}\min\left\{\frac{\beta\sigma^2}{L_g + L_\psi}, \frac{\mu_g}{\beta\|\mathbf{B}\|_2^2}, \frac{\mu_g}{L_\psi}, \frac{\mu_f}{L_\phi}\right\}\right)$$

$$\times \left(\frac{1}{2\beta}\|\boldsymbol{\lambda}^{k+1} - \boldsymbol{\lambda}^*\|^2 + \frac{\beta}{2}\|\mathbf{B}\mathbf{y}^{k+1} - \mathbf{B}\mathbf{y}^*\|^2 + D_\phi(\mathbf{x}^*, \mathbf{x}^{k+1}) + D_\psi(\mathbf{y}^*, \mathbf{y}^{k+1})\right)$$

$$\leq \frac{1}{2\beta}\|\boldsymbol{\lambda}^k - \boldsymbol{\lambda}^*\|^2 + \frac{\beta}{2}\|\mathbf{B}\mathbf{y}^k - \mathbf{B}\mathbf{y}^*\|^2 + D_\phi(\mathbf{x}^*, \mathbf{x}^k) + D_\psi(\mathbf{y}^*, \mathbf{y}^k)$$

and obtain the conclusion.                                                                     □

## 3.3   Accelerated Linearized ADMM

In this section, we introduce how to combine ADMM with Nesterov's acceleration techniques.

### 3.3.1   Sublinear Convergence Rate

We first consider the case that both $f$ and $g$ are generally convex and $g$ is $L_g$-smooth. In this case, we introduce the following accelerated linearized ADMM, originally proposed in [24], and then extended in [22], where we linearize $g$ at the auxiliary variable $\mathbf{v}^k$ in the $\mathbf{y}$ update step. The algorithm consists of steps (3.44a)–(3.44f), and we present it in Algorithm 3.4.

$$\mathbf{v}^k = \theta_k \mathbf{y}^k + (1 - \theta_k)\widetilde{\mathbf{y}}^k, \tag{3.44a}$$

$$\mathbf{x}^{k+1} = \underset{\mathbf{x}}{\operatorname{argmin}} \left( f(\mathbf{x}) + \left\langle \boldsymbol{\lambda}^k, \mathbf{A}\mathbf{x} + \mathbf{B}\mathbf{y}^k - \mathbf{b} \right\rangle + \frac{\beta}{2} \|\mathbf{A}\mathbf{x} + \mathbf{B}\mathbf{y}^k - \mathbf{b}\|^2 \right), \tag{3.44b}$$

$$\mathbf{y}^{k+1} = \underset{\mathbf{y}}{\operatorname{argmin}} \left( g(\mathbf{v}^k) + \left\langle \nabla g(\mathbf{v}^k), \mathbf{y} - \mathbf{v}^k \right\rangle + \left\langle \boldsymbol{\lambda}^k, \mathbf{A}\mathbf{x}^{k+1} + \mathbf{B}\mathbf{y} - \mathbf{b} \right\rangle \right.$$
$$+ \beta \left\langle \mathbf{B}^T (\mathbf{A}\mathbf{x}^{k+1} + \mathbf{B}\mathbf{y}^k - \mathbf{b}), \mathbf{y} - \mathbf{y}^k \right\rangle$$
$$\left. + \frac{L_g \theta_k + \beta \|\mathbf{B}\|_2^2}{2} \|\mathbf{y} - \mathbf{y}^k\|^2 \right), \tag{3.44c}$$

$$\widetilde{\mathbf{x}}^{k+1} = \theta_k \mathbf{x}^{k+1} + (1 - \theta_k)\widetilde{\mathbf{x}}^k, \tag{3.44d}$$

$$\widetilde{\mathbf{y}}^{k+1} = \theta_k \mathbf{y}^{k+1} + (1 - \theta_k)\widetilde{\mathbf{y}}^k, \tag{3.44e}$$

$$\boldsymbol{\lambda}^{k+1} = \boldsymbol{\lambda}^k + \beta(\mathbf{A}\mathbf{x}^{k+1} + \mathbf{B}\mathbf{y}^{k+1} - \mathbf{b}). \tag{3.44f}$$

---

**Algorithm 3.4** The first accelerated linearized ADMM for non-strongly convex problems

---

Initialize $\mathbf{x}^0 = \widetilde{\mathbf{x}}^0$, $\mathbf{y}^0 = \widetilde{\mathbf{y}}^0$, $\boldsymbol{\lambda}^0$.
**for** $k = 0, 1, 2, 3, \cdots$ **do**
  Update the variables by (3.44a)–(3.44f), respectively.
**end for**

---

The accelerated linearized ADMM has a convergence rate of $O\left(\frac{1}{K} + \frac{L_g}{K^2}\right)$, which is faster than the $O\left(\frac{L_g}{K}\right)$ one of the linearized ADMM (see Theorem 3.6, where we omit the other constants and only keep $L_g$ from $\psi(\mathbf{y}^*, \mathbf{y}^0)$ defined in (3.34)) when $L_g$ is very large. Below we give the analysis.

Denote

$$\ell_k = f(\widetilde{\mathbf{x}}^k) - f(\mathbf{x}^*) + g(\widetilde{\mathbf{y}}^k) - g(\mathbf{y}^*) + \left\langle \boldsymbol{\lambda}^*, \mathbf{A}\widetilde{\mathbf{x}}^k + \mathbf{B}\widetilde{\mathbf{y}}^k - \mathbf{b} \right\rangle.$$

By Lemma 3.1, $\ell_k \geq 0$.

**Lemma 3.12** *Suppose that $f(\mathbf{x})$ and $g(\mathbf{y})$ are convex and $g(\mathbf{y})$ is $L_g$-smooth. Let $\theta_k \in (0, 1], k \geq 0$, satisfy*

$$\frac{1 - \theta_k}{\theta_k^2} = \frac{1}{\theta_{k-1}^2} \; for \; k \geq 1, \quad \theta_0 = 1, \quad and \quad \theta_{-1} = \infty.$$

*Then for Algorithm 3.4, we have*

$$\frac{\ell_{k+1}}{\theta_k^2} \leq \frac{\ell_k}{\theta_{k-1}^2} + \frac{1}{2\beta\theta_k} \left( \|\boldsymbol{\lambda}^k - \boldsymbol{\lambda}^*\|^2 - \|\boldsymbol{\lambda}^{k+1} - \boldsymbol{\lambda}^*\|^2 \right)$$
$$+ \left( \frac{L_g}{2} + \frac{\beta\|\mathbf{B}\|_2^2}{2\theta_k} \right) \left( \|\mathbf{y}^k - \mathbf{y}^*\|^2 - \|\mathbf{y}^{k+1} - \mathbf{y}^*\|^2 \right) \quad (3.45)$$

*and*

$$\frac{\ell_{k+1}}{\theta_k} \leq \frac{\ell_k}{\theta_{k-1}} + \frac{1}{2\beta} \left( \|\boldsymbol{\lambda}^k - \boldsymbol{\lambda}^*\|^2 - \|\boldsymbol{\lambda}^{k+1} - \boldsymbol{\lambda}^*\|^2 \right)$$
$$+ \left( \frac{L_g\theta_k}{2} + \frac{\beta\|\mathbf{B}\|_2^2}{2} \right) \|\mathbf{y}^k - \mathbf{y}^*\|^2$$
$$- \left( \frac{L_g\theta_{k+1}}{2} + \frac{\beta\|\mathbf{B}\|_2^2}{2} \right) \|\mathbf{y}^{k+1} - \mathbf{y}^*\|^2. \quad (3.46)$$

***Proof*** We denote $L_g$ as $L$ in this proof for notation simplicity. From the optimality conditions of (3.44b) and (3.44c), we have

$$\mathbf{0} \in \partial f(\mathbf{x}^{k+1}) + \mathbf{A}^T\boldsymbol{\lambda}^k + \beta\mathbf{A}^T(\mathbf{A}\mathbf{x}^{k+1} + \mathbf{B}\mathbf{y}^k - \mathbf{b})$$
$$= \partial f(\mathbf{x}^{k+1}) + \mathbf{A}^T\boldsymbol{\lambda}^{k+1} - \beta\mathbf{A}^T(\mathbf{B}\mathbf{y}^{k+1} - \mathbf{B}\mathbf{y}^k) \quad (3.47)$$

and

$$\mathbf{0} = \nabla g(\mathbf{v}^k) + \mathbf{B}^T\boldsymbol{\lambda}^k + \beta\mathbf{B}^T(\mathbf{A}\mathbf{x}^{k+1} + \mathbf{B}\mathbf{y}^k - \mathbf{b})$$
$$+ (L\theta_k + \beta\|\mathbf{B}\|_2^2)(\mathbf{y}^{k+1} - \mathbf{y}^k)$$
$$= \nabla g(\mathbf{v}^k) + \mathbf{B}^T\boldsymbol{\lambda}^{k+1} - \beta\mathbf{B}^T\mathbf{B}(\mathbf{y}^{k+1} - \mathbf{y}^k)$$
$$+ (L\theta_k + \beta\|\mathbf{B}\|_2^2)(\mathbf{y}^{k+1} - \mathbf{y}^k). \quad (3.48)$$

(3.47) gives

$$f(\mathbf{x}^*) - f(\mathbf{x}^{k+1})$$
$$\geq \left\langle \boldsymbol{\lambda}^{k+1}, \mathbf{A}\mathbf{x}^{k+1} - \mathbf{A}\mathbf{x}^* \right\rangle - \beta \left\langle \mathbf{B}\mathbf{y}^{k+1} - \mathbf{B}\mathbf{y}^k, \mathbf{A}\mathbf{x}^{k+1} - \mathbf{A}\mathbf{x}^* \right\rangle.$$

Thus, from (3.44d) and the convexity of $f$, we have

$$f(\widetilde{\mathbf{x}}^{k+1}) - f(\mathbf{x}^*)$$
$$\leq \theta_k \left( f(\mathbf{x}^{k+1}) - f(\mathbf{x}^*) \right) + (1 - \theta_k) \left( f(\widetilde{\mathbf{x}}^k) - f(\mathbf{x}^*) \right)$$
$$\leq \theta_k \left\langle \boldsymbol{\lambda}^{k+1}, \mathbf{A}\mathbf{x}^* - \mathbf{A}\mathbf{x}^{k+1} \right\rangle + \beta \theta_k \left\langle \mathbf{B}\mathbf{y}^{k+1} - \mathbf{B}\mathbf{y}^k, \mathbf{A}\mathbf{x}^{k+1} - \mathbf{A}\mathbf{x}^* \right\rangle$$
$$+ (1 - \theta_k) \left( f(\widetilde{\mathbf{x}}^k) - f(\mathbf{x}^*) \right). \tag{3.49}$$

From the smoothness and the convexity of $g$, we have

$$g(\widetilde{\mathbf{y}}^{k+1})$$
$$\leq g(\mathbf{v}^k) + \left\langle \nabla g(\mathbf{v}^k), \widetilde{\mathbf{y}}^{k+1} - \mathbf{v}^k \right\rangle + \frac{L}{2} \|\widetilde{\mathbf{y}}^{k+1} - \mathbf{v}^k\|^2$$
$$\overset{a}{=} g(\mathbf{v}^k) + \left\langle \nabla g(\mathbf{v}^k), (1 - \theta_k)\widetilde{\mathbf{y}}^k + \theta_k \mathbf{y}^{k+1} - \mathbf{v}^k \right\rangle + \frac{L\theta_k^2}{2} \|\mathbf{y}^{k+1} - \mathbf{y}^k\|^2$$
$$= (1 - \theta_k) \left( g(\mathbf{v}^k) + \left\langle \nabla g(\mathbf{v}^k), \widetilde{\mathbf{y}}^k - \mathbf{v}^k \right\rangle \right)$$
$$+ \theta_k \left( g(\mathbf{v}^k) + \left\langle \nabla g(\mathbf{v}^k), \mathbf{y}^{k+1} - \mathbf{v}^k \right\rangle \right) + \frac{L\theta_k^2}{2} \|\mathbf{y}^{k+1} - \mathbf{y}^k\|^2$$
$$\leq (1 - \theta_k) g(\widetilde{\mathbf{y}}^k)$$
$$+ \theta_k \left( g(\mathbf{v}^k) + \left\langle \nabla g(\mathbf{v}^k), \mathbf{y}^* - \mathbf{v}^k \right\rangle + \left\langle \nabla g(\mathbf{v}^k), \mathbf{y}^{k+1} - \mathbf{y}^* \right\rangle \right)$$
$$+ \frac{L\theta_k^2}{2} \|\mathbf{y}^{k+1} - \mathbf{y}^k\|^2 \tag{3.50}$$
$$\overset{b}{\leq} (1 - \theta_k) g(\widetilde{\mathbf{y}}^k) + \theta_k g(\mathbf{y}^*) + \frac{L\theta_k^2}{2} \|\mathbf{y}^{k+1} - \mathbf{y}^k\|^2$$
$$+ \theta_k \left\langle \mathbf{B}^T \boldsymbol{\lambda}^{k+1} - \beta \mathbf{B}^T \mathbf{B}(\mathbf{y}^{k+1} - \mathbf{y}^k) \right.$$
$$+ (L\theta_k + \beta \|\mathbf{B}\|_2^2)(\mathbf{y}^{k+1} - \mathbf{y}^k), \mathbf{y}^* - \mathbf{y}^{k+1} \Big\rangle$$
$$= (1 - \theta_k) g(\widetilde{\mathbf{y}}^k) + \theta_k g(\mathbf{y}^*) + \frac{L\theta_k^2}{2} \|\mathbf{y}^{k+1} - \mathbf{y}^k\|^2$$
$$+ \theta_k \left\langle \boldsymbol{\lambda}^{k+1}, \mathbf{B}\mathbf{y}^* - \mathbf{B}\mathbf{y}^{k+1} \right\rangle - \beta \theta_k \left\langle \mathbf{B}(\mathbf{y}^{k+1} - \mathbf{y}^k), \mathbf{B}(\mathbf{y}^* - \mathbf{y}^{k+1}) \right\rangle$$
$$+ \left( L\theta_k^2 + \beta \theta_k \|\mathbf{B}\|_2^2 \right) \left\langle \mathbf{y}^{k+1} - \mathbf{y}^k, \mathbf{y}^* - \mathbf{y}^{k+1} \right\rangle,$$

where $\overset{a}{=}$ uses (3.44a) and (3.44e), and we plug (3.48) into $\overset{b}{\leq}$. Adding it with (3.49), we have

$$f(\widetilde{\mathbf{x}}^{k+1}) - f(\mathbf{x}^*) + g(\widetilde{\mathbf{y}}^{k+1}) - g(\mathbf{y}^*)$$

$$\leq (1 - \theta_k)\left( f(\widetilde{\mathbf{x}}^k) - f(\mathbf{x}^*) + g(\widetilde{\mathbf{y}}^k) - g(\mathbf{y}^*) \right) + \frac{L\theta_k^2}{2}\|\mathbf{y}^{k+1} - \mathbf{y}^k\|^2$$

$$+ \theta_k \left\langle \boldsymbol{\lambda}^{k+1}, \mathbf{b} - \mathbf{A}\mathbf{x}^{k+1} - \mathbf{B}\mathbf{y}^{k+1} \right\rangle$$

$$+ \beta\theta_k \left\langle \mathbf{B}(\mathbf{y}^{k+1} - \mathbf{y}^k), \mathbf{A}\mathbf{x}^{k+1} + \mathbf{B}\mathbf{y}^{k+1} - \mathbf{b} \right\rangle$$

$$+ \left( L\theta_k^2 + \beta\theta_k\|\mathbf{B}\|_2^2 \right)\left\langle \mathbf{y}^{k+1} - \mathbf{y}^k, \mathbf{y}^* - \mathbf{y}^{k+1} \right\rangle.$$

Adding $\left\langle \boldsymbol{\lambda}^*, \mathbf{A}\widetilde{\mathbf{x}}^{k+1} + \mathbf{B}\widetilde{\mathbf{y}}^{k+1} - \mathbf{b} \right\rangle$ to both sides and using Lemma A.1, we have

$$f(\widetilde{\mathbf{x}}^{k+1}) - f(\mathbf{x}^*) + g(\widetilde{\mathbf{y}}^{k+1}) - g(\mathbf{y}^*) + \left\langle \boldsymbol{\lambda}^*, \mathbf{A}\widetilde{\mathbf{x}}^{k+1} + \mathbf{B}\widetilde{\mathbf{y}}^{k+1} - \mathbf{b} \right\rangle$$

$$\overset{a}{\leq} (1 - \theta_k)\left( f(\widetilde{\mathbf{x}}^k) - f(\mathbf{x}^*) + g(\widetilde{\mathbf{y}}^k) - g(\mathbf{y}^*) + \left\langle \boldsymbol{\lambda}^*, \mathbf{A}\widetilde{\mathbf{x}}^k + \mathbf{B}\widetilde{\mathbf{y}}^k - \mathbf{b} \right\rangle \right)$$

$$- \frac{\theta_k}{\beta} \left\langle \boldsymbol{\lambda}^{k+1} - \boldsymbol{\lambda}^*, \boldsymbol{\lambda}^{k+1} - \boldsymbol{\lambda}^k \right\rangle$$

$$+ \beta\theta_k \left\langle \mathbf{B}(\mathbf{y}^{k+1} - \mathbf{y}^k), \mathbf{A}\mathbf{x}^{k+1} + \mathbf{B}\mathbf{y}^{k+1} - \mathbf{b} \right\rangle$$

$$+ \left( L\theta_k^2 + \beta\theta_k\|\mathbf{B}\|_2^2 \right)\left\langle \mathbf{y}^{k+1} - \mathbf{y}^k, \mathbf{y}^* - \mathbf{y}^{k+1} \right\rangle + \frac{L\theta_k^2}{2}\|\mathbf{y}^{k+1} - \mathbf{y}^k\|^2$$

$$= (1 - \theta_k)\left( f(\widetilde{\mathbf{x}}^k) - f(\mathbf{x}^*) + g(\widetilde{\mathbf{y}}^k) - g(\mathbf{y}^*) + \left\langle \boldsymbol{\lambda}^*, \mathbf{A}\widetilde{\mathbf{x}}^k + \mathbf{B}\widetilde{\mathbf{y}}^k - \mathbf{b} \right\rangle \right)$$

$$+ \frac{\theta_k}{2\beta}\left( \|\boldsymbol{\lambda}^k - \boldsymbol{\lambda}^*\|^2 - \|\boldsymbol{\lambda}^{k+1} - \boldsymbol{\lambda}^*\|^2 - \|\boldsymbol{\lambda}^{k+1} - \boldsymbol{\lambda}^k\|^2 \right)$$

$$+ \frac{\beta\theta_k}{2}\left( \|\mathbf{A}\mathbf{x}^{k+1} + \mathbf{B}\mathbf{y}^{k+1} - \mathbf{b}\|^2 + \|\mathbf{B}(\mathbf{y}^{k+1} - \mathbf{y}^k)\|^2 \right.$$

$$\left. - \|\mathbf{A}\mathbf{x}^{k+1} + \mathbf{B}\mathbf{y}^k - \mathbf{b}\|^2 \right)$$

$$+ \frac{L\theta_k^2 + \beta\theta_k\|\mathbf{B}\|_2^2}{2}\left( \|\mathbf{y}^k - \mathbf{y}^*\|^2 - \|\mathbf{y}^{k+1} - \mathbf{y}^*\|^2 \right)$$

$$- \frac{\beta\theta_k\|\mathbf{B}\|_2^2}{2}\|\mathbf{y}^{k+1} - \mathbf{y}^k\|^2$$

$$\overset{b}{\leq} (1 - \theta_k)\left( f(\widetilde{\mathbf{x}}^k) - f(\mathbf{x}^*) + g(\widetilde{\mathbf{y}}^k) - g(\mathbf{y}^*) + \left\langle \boldsymbol{\lambda}^*, \mathbf{A}\widetilde{\mathbf{x}}^k + \mathbf{B}\widetilde{\mathbf{y}}^k - \mathbf{b} \right\rangle \right)$$

$$+ \frac{\theta_k}{2\beta} \left( \|\boldsymbol{\lambda}^k - \boldsymbol{\lambda}^*\|^2 - \|\boldsymbol{\lambda}^{k+1} - \boldsymbol{\lambda}^*\|^2 \right)$$

$$+ \frac{L\theta_k^2 + \beta\theta_k\|\mathbf{B}\|_2^2}{2} \left( \|\mathbf{y}^k - \mathbf{y}^*\|^2 - \|\mathbf{y}^{k+1} - \mathbf{y}^*\|^2 \right),$$

where $\overset{a}{\leq}$ uses (3.44d) and (3.44e), and $\overset{b}{\leq}$ uses (3.44f). Dividing both sides by $\theta_k^2$ and using $\frac{1-\theta_k}{\theta_k^2} = \frac{1}{\theta_{k-1}^2}$, we have the first conclusion. We can easily check that $\theta_k$ is decreasing. Thus, we have $\frac{1-\theta_k}{\theta_k} = \frac{1}{\theta_{k-1}} \frac{\theta_k}{\theta_{k-1}} \leq \frac{1}{\theta_{k-1}}$. Dividing both sides by $\theta_k$, we have the second conclusion. □

**Theorem 3.10** *Suppose that $f(\mathbf{x})$ and $g(\mathbf{y})$ are convex and $g(\mathbf{y})$ is $L_g$-smooth. Let $\theta_k \in (0, 1]$, $k \geq 0$, satisfy*

$$\frac{1-\theta_k}{\theta_k^2} = \frac{1}{\theta_{k-1}^2} \text{ for } k \geq 1, \quad \theta_0 = 1, \quad \text{and} \quad \theta_{-1} = \infty.$$

*Assume that*

$$\|\boldsymbol{\lambda}^k - \boldsymbol{\lambda}^*\|^2 \leq D_{\boldsymbol{\lambda}} \quad \text{and} \quad \|\mathbf{y}^k - \mathbf{y}^*\|^2 \leq D_{\mathbf{y}}, \quad \forall k.$$

*Then for Algorithm 3.4, we have*

$$|f(\widetilde{\mathbf{x}}^{K+1}) + g(\widetilde{\mathbf{y}}^{K+1}) - f(\mathbf{x}^*) - g(\mathbf{y}^*)| \leq O\left( \frac{D_{\mathbf{y}} + D_{\boldsymbol{\lambda}} + \|\boldsymbol{\lambda}^*\|\sqrt{D_{\boldsymbol{\lambda}}}}{K} + \frac{L_g}{K^2} \right),$$

$$\|\mathbf{A}\widetilde{\mathbf{x}}^{K+1} + \mathbf{B}\widetilde{\mathbf{y}}^{K+1} - \mathbf{b}\| \leq O\left( \frac{\sqrt{D_{\boldsymbol{\lambda}}}}{K} \right).$$

**Proof** Summing (3.45) over $k = 0, \cdots, K$, we have

$$\frac{\ell_{K+1}}{\theta_K^2} \leq \frac{1}{2\beta} \sum_{k=1}^{K} \|\boldsymbol{\lambda}^k - \boldsymbol{\lambda}^*\|^2 \left( \frac{1}{\theta_k} - \frac{1}{\theta_{k-1}} \right) + \frac{1}{2\beta} \|\boldsymbol{\lambda}^0 - \boldsymbol{\lambda}^*\|^2$$

$$+ \frac{\beta\|\mathbf{B}\|_2^2}{2} \sum_{k=1}^{K} \|\mathbf{y}^k - \mathbf{y}^*\|^2 \left( \frac{1}{\theta_k} - \frac{1}{\theta_{k-1}} \right) + \left( \frac{L_g}{2} + \frac{\beta\|\mathbf{B}\|_2^2}{2} \right) \|\mathbf{y}^0 - \mathbf{y}^*\|^2$$

$$\leq \frac{D_{\boldsymbol{\lambda}}}{2\beta\theta_K} + \frac{1}{2\beta} \|\boldsymbol{\lambda}^0 - \boldsymbol{\lambda}^*\|^2 + \frac{\beta D_{\mathbf{y}}\|\mathbf{B}\|_2^2}{2\theta_K} + \left( \frac{L_g}{2} + \frac{\beta\|\mathbf{B}\|_2^2}{2} \right) \|\mathbf{y}^0 - \mathbf{y}^*\|^2.$$

So we have

$$\ell_{K+1} \le \theta_K \left( \frac{D_\lambda}{2\beta} + \frac{\beta D_\mathbf{y} \|\mathbf{B}\|_2^2}{2} \right)$$

$$+\theta_K^2 \left[ \frac{1}{2\beta} \|\boldsymbol{\lambda}^0 - \boldsymbol{\lambda}^*\|^2 + \left( \frac{L_g}{2} + \frac{\beta\|\mathbf{B}\|_2^2}{2} \right) \|\mathbf{y}^0 - \mathbf{y}^*\|^2 \right]$$

$$\overset{a}{=} O \left( \frac{D_\mathbf{y} + D_\lambda}{K} + \frac{L_g}{K^2} \right),$$

where $\overset{a}{=}$ uses

$$\theta_k \le \frac{2}{k+1}, \quad k \ge 0, \tag{3.51}$$

from

$$\left( \frac{1}{\theta_k} - \frac{1}{2} \right)^2 = \frac{1}{\theta_{k-1}^2} + \frac{1}{4} \ge \frac{1}{\theta_{k-1}^2} \quad \text{and} \quad \theta_0 = 1.$$

On the other hand, note that

$$\frac{1}{\theta_k^2} \left( A\widetilde{\mathbf{x}}^{k+1} + B\widetilde{\mathbf{y}}^{k+1} - \mathbf{b} \right)$$

$$= \frac{1}{\theta_k} \left( A\mathbf{x}^{k+1} + B\mathbf{y}^{k+1} - \mathbf{b} \right) + \frac{1 - \theta_k}{\theta_k^2} \left( A\widetilde{\mathbf{x}}^k + B\widetilde{\mathbf{y}}^k - \mathbf{b} \right)$$

$$= \frac{1}{\theta_k} \left( A\mathbf{x}^{k+1} + B\mathbf{y}^{k+1} - \mathbf{b} \right) + \frac{1}{\theta_{k-1}^2} \left( A\widetilde{\mathbf{x}}^k + B\widetilde{\mathbf{y}}^k - \mathbf{b} \right), \quad k \ge 0.$$

Thus, we have

$$\frac{1}{\theta_K^2} \left( A\widetilde{\mathbf{x}}^{K+1} + B\widetilde{\mathbf{y}}^{K+1} - \mathbf{b} \right)$$

$$\overset{a}{=} \sum_{k=0}^{K} \frac{1}{\theta_k} \left( A\mathbf{x}^{k+1} + B\mathbf{y}^{k+1} - \mathbf{b} \right)$$

$$= \frac{1}{\beta} \sum_{k=0}^{K} \frac{1}{\theta_k} \left[ (\boldsymbol{\lambda}^{k+1} - \boldsymbol{\lambda}^*) - (\boldsymbol{\lambda}^k - \boldsymbol{\lambda}^*) \right]$$

$$= \frac{1}{\beta} \sum_{k=0}^{K} \left( \frac{1}{\theta_{k-1}} - \frac{1}{\theta_k} \right) (\boldsymbol{\lambda}^k - \boldsymbol{\lambda}^*) + \frac{1}{\beta\theta_K} \left( \boldsymbol{\lambda}^{K+1} - \boldsymbol{\lambda}^* \right),$$

where $\stackrel{a}{=}$ uses $\frac{1}{\theta_{-1}} = 0$. Hence

$$\frac{1}{\theta_K^2} \left\| A\widetilde{x}^{K+1} + B\widetilde{y}^{K+1} - b \right\|$$

$$\leq \frac{1}{\beta} \sum_{k=0}^{K} \left( \frac{1}{\theta_k} - \frac{1}{\theta_{k-1}} \right) \left\| \lambda^k - \lambda^* \right\| + \frac{1}{\beta\theta_K} \left\| \lambda^{K+1} - \lambda^* \right\|$$

$$\stackrel{a}{\leq} \frac{2\sqrt{D_\lambda}}{\beta\theta_K},$$

where $\frac{1}{\theta_{-1}} = 0$ is used again in $\stackrel{a}{\leq}$. From Lemma 3.2 and (3.51), we have the conclusion. $\qquad\square$

*Remark 3.3* When $D_y$ and $D_\lambda$ are small constants independent of $L_g$, the accelerated linearized ADMM in Algorithm 3.4 has a faster convergence rate when $L_g$ is very large. However, this is a strong assumption and in general, we cannot prove it. In fact, from (3.46) we know

$$\frac{\ell_{k+1}}{\theta_k} + \frac{1}{2\beta} \|\lambda^{k+1} - \lambda^*\|^2 + \left( \frac{L_g\theta_{k+1}}{2} + \frac{\beta\|B\|_2^2}{2} \right) \|y^{k+1} - y^*\|^2$$

$$\leq \frac{1}{2\beta} \|\lambda^0 - \lambda^*\|^2 + \left( \frac{L_g}{2} + \frac{\beta\|B\|_2^2}{2} \right) \|y^0 - y^*\|^2 \equiv C.$$

That is, we have

$$\|\lambda^{k+1} - \lambda^*\|^2 \leq 2\beta C \quad \text{and} \quad \|y^{k+1} - y^*\|^2 \leq \frac{2C}{\beta\|B\|_2^2},$$

where the bound $C$ depends on $L_g$.

Next, we give another accelerated linearized ADMM [14], which consists of steps (3.52a)–(3.52e) and is presented in Algorithm 3.5.

$$u^k = x^k + \frac{\theta_k(1 - \theta_{k-1})}{\theta_{k-1}} (x^k - x^{k-1}), \tag{3.52a}$$

$$v^k = y^k + \frac{\theta_k(1 - \theta_{k-1})}{\theta_{k-1}} (y^k - y^{k-1}), \tag{3.52b}$$

$$x^{k+1} = \operatorname*{argmin}_{x} \left( f_1(x) + \left\langle \nabla f_2(u^k), x \right\rangle + \frac{L}{2} \|x - u^k\|^2 + \left\langle \lambda^k, Ax \right\rangle \right.$$

$$\left. + \frac{\beta}{\theta_k} \left\langle A^T (Au^k + Bv^k - b), x \right\rangle + \frac{\beta\|A\|_2^2}{2\theta_k} \|x - u^k\|^2 \right), \tag{3.52c}$$

$$\mathbf{y}^{k+1} = \underset{\mathbf{y}}{\operatorname{argmin}} \left( g_1(\mathbf{y}) + \left\langle \nabla g_2(\mathbf{v}^k), \mathbf{y} \right\rangle + \frac{L}{2} \|\mathbf{y} - \mathbf{v}^k\|^2 + \left\langle \boldsymbol{\lambda}^k, \mathbf{B}\mathbf{y} \right\rangle \right.$$

$$\left. + \frac{\beta}{\theta_k} \left\langle \mathbf{B}^T(\mathbf{A}\mathbf{x}^{k+1} + \mathbf{B}\mathbf{v}^k - \mathbf{b}), \mathbf{y} \right\rangle + \frac{\beta\|\mathbf{B}\|_2^2}{2\theta_k} \|\mathbf{y} - \mathbf{v}^k\|^2 \right), \qquad (3.52d)$$

$$\boldsymbol{\lambda}^{k+1} = \boldsymbol{\lambda}^k + \beta\tau(\mathbf{A}\mathbf{x}^{k+1} + \mathbf{B}\mathbf{y}^{k+1} - \mathbf{b}). \qquad (3.52e)$$

---

**Algorithm 3.5** The second accelerated linearized ADMM for non-strongly convex problems

---

Initialize $\mathbf{x}^0 = \mathbf{x}^{-1}$, $\mathbf{y}^0 = \mathbf{y}^{-1}$, $\boldsymbol{\lambda}^0$.
**for** $k = 0, 1, 2, 3, \cdots$ **do**
    Update the variables by (3.52a)–(3.52e), respectively.
**end for**

---

The second accelerated linearized ADMM has three differences from the first one (Algorithm 3.4). Firstly, Algorithm 3.5 linearizes both subproblems, while Algorithm 3.4 only linearizes the second subproblem. Secondly, Algorithm 3.5 can be used to solve composite problems, that is,

$$f(\mathbf{x}) = f_1(\mathbf{x}) + f_2(\mathbf{x}) \quad \text{and} \quad g(\mathbf{y}) = g_1(\mathbf{y}) + g_2(\mathbf{y})$$

with nonsmooth $f_1$ and $g_1$ and $L$-smooth $f_2$ and $g_2$. Thirdly, Algorithm 3.4 has the convergence rate measured at the averaged points $(\tilde{\mathbf{x}}^k, \tilde{\mathbf{y}}^k)$. Thus, the convergence rate is in the ergodic sense. As a comparison, the convergence rate of Algorithm 3.5 is in the non-ergodic sense. Note that when $\theta_k = 1$ for all $k$ and $\tau = 1$, Algorithm 3.5 reduces to the non-accelerated linearized ADMM (Algorithm 3.2 when $f_2 = g_2 = 0$ (in which case $L = 0$) and Algorithm 3.3 when $f_1 = g_1 = 0$). Below we give the analysis on Algorithm 3.5.

Define several auxiliary variables

$$\bar{\boldsymbol{\lambda}}_1^{k+1} = \boldsymbol{\lambda}^k + \frac{\beta}{\theta_k} \left( \mathbf{A}\mathbf{u}^k + \mathbf{B}\mathbf{v}^k - \mathbf{b} \right),$$

$$\bar{\boldsymbol{\lambda}}_2^{k+1} = \boldsymbol{\lambda}^k + \frac{\beta}{\theta_k} \left( \mathbf{A}\mathbf{x}^{k+1} + \mathbf{B}\mathbf{v}^k - \mathbf{b} \right),$$

$$\hat{\boldsymbol{\lambda}}^k = \boldsymbol{\lambda}^k + \frac{\beta(1 - \theta_k)}{\theta_k} \left( \mathbf{A}\mathbf{x}^k + \mathbf{B}\mathbf{y}^k - \mathbf{b} \right), \qquad (3.53)$$

$$\mathbf{r}^{k+1} = \frac{1}{\theta_k}\mathbf{x}^{k+1} - \frac{1 - \theta_k}{\theta_k}\mathbf{x}^k,$$

$$\mathbf{s}^{k+1} = \frac{1}{\theta_k}\mathbf{y}^{k+1} - \frac{1 - \theta_k}{\theta_k}\mathbf{y}^k,$$

and let sequence $\{\theta_k\}$ satisfy

$$\frac{1 - \theta_{k+1}}{\theta_{k+1}} = \frac{1}{\theta_k} - \tau, \quad \theta_0 = 1, \quad \text{and} \quad \theta_{-1} = 1/\tau, \tag{3.54}$$

where $0 < \tau < 1$. We first give the following lemma.

**Lemma 3.13** *For the above definitions in (3.53), we have*

$$\hat{\lambda}^{k+1} - \hat{\lambda}^k = \frac{\beta}{\theta_k}\left[ \mathbf{A}\mathbf{x}^{k+1} + \mathbf{B}\mathbf{y}^{k+1} - \mathbf{b} - (1 - \theta_k)(\mathbf{A}\mathbf{x}^k + \mathbf{B}\mathbf{y}^k - \mathbf{b}) \right],$$

$$\left\| \hat{\lambda}^{k+1} - \overline{\lambda}_2^{k+1} \right\| = \frac{\beta}{\theta_k}\left\| \mathbf{B}\mathbf{y}^{k+1} - \mathbf{B}\mathbf{v}^k \right\|,$$

$$\hat{\lambda}^{K+1} - \hat{\lambda}^0 = \frac{\beta}{\theta_K}(\mathbf{A}\mathbf{x}^{K+1} + \mathbf{B}\mathbf{y}^{K+1} - \mathbf{b}) + \beta\tau \sum_{k=1}^{K}\left( \mathbf{A}\mathbf{x}^k + \mathbf{B}\mathbf{y}^k - \mathbf{b} \right),$$

$$\mathbf{u}^k - (1 - \theta_k)\mathbf{x}^k = \theta_k \mathbf{r}^k,$$

$$\mathbf{v}^k - (1 - \theta_k)\mathbf{y}^k = \theta_k \mathbf{s}^k.$$

*Proof* From the definitions of $\hat{\lambda}^k$ and $\lambda^{k+1}$ and $\frac{1 - \theta_{k+1}}{\theta_{k+1}} = \frac{1}{\theta_k} - \tau$, we have

$$\hat{\lambda}^{k+1} = \lambda^{k+1} + \beta\frac{1 - \theta_{k+1}}{\theta_{k+1}}\left( \mathbf{A}\mathbf{x}^{k+1} + \mathbf{B}\mathbf{y}^{k+1} - \mathbf{b} \right)$$

$$= \lambda^{k+1} + \beta\left( \frac{1}{\theta_k} - \tau \right)\left( \mathbf{A}\mathbf{x}^{k+1} + \mathbf{B}\mathbf{y}^{k+1} - \mathbf{b} \right)$$

$$= \lambda^k + \beta\tau\left( \mathbf{A}\mathbf{x}^{k+1} + \mathbf{B}\mathbf{y}^{k+1} - \mathbf{b} \right) + \beta\left( \frac{1}{\theta_k} - \tau \right)\left( \mathbf{A}\mathbf{x}^{k+1} + \mathbf{B}\mathbf{y}^{k+1} - \mathbf{b} \right)$$

$$= \lambda^k + \frac{\beta}{\theta_k}\left( \mathbf{A}\mathbf{x}^{k+1} + \mathbf{B}\mathbf{y}^{k+1} - \mathbf{b} \right) \tag{3.55}$$

$$= \hat{\lambda}^k - \beta\frac{1 - \theta_k}{\theta_k}\left( \mathbf{A}\mathbf{x}^k + \mathbf{B}\mathbf{y}^k - \mathbf{b} \right) + \frac{\beta}{\theta_k}\left( \mathbf{A}\mathbf{x}^{k+1} + \mathbf{B}\mathbf{y}^{k+1} - \mathbf{b} \right) \tag{3.56}$$

$$= \hat{\lambda}^k + \frac{\beta}{\theta_k}\left[ \mathbf{A}\mathbf{x}^{k+1} + \mathbf{B}\mathbf{y}^{k+1} - \mathbf{b} - (1 - \theta_k)(\mathbf{A}\mathbf{x}^k + \mathbf{B}\mathbf{y}^k - \mathbf{b}) \right].$$

On the other hand, from (3.55) and the definition of $\overline{\lambda}_2^{k+1}$ we have

$$\left\| \hat{\lambda}^{k+1} - \overline{\lambda}_2^{k+1} \right\| = \frac{\beta}{\theta_k}\left\| \mathbf{B}(\mathbf{y}^{k+1} - \mathbf{v}^k) \right\|.$$

From (3.56), $\frac{1-\theta_k}{\theta_k} = \frac{1}{\theta_{k-1}} - \tau$, and $\frac{1}{\theta_{-1}} = \tau$, we have

$$\hat{\boldsymbol{\lambda}}^{K+1} - \hat{\boldsymbol{\lambda}}^0 = \sum_{k=0}^{K} \left( \hat{\boldsymbol{\lambda}}^{k+1} - \hat{\boldsymbol{\lambda}}^k \right)$$

$$= \beta \sum_{k=0}^{K} \left[ \frac{1}{\theta_k} (\mathbf{A}\mathbf{x}^{k+1} + \mathbf{B}\mathbf{y}^{k+1} - \mathbf{b}) - \frac{1-\theta_k}{\theta_k} \left( \mathbf{A}\mathbf{x}^k + \mathbf{B}\mathbf{y}^k - \mathbf{b} \right) \right]$$

$$= \beta \sum_{k=0}^{K} \left[ \frac{1}{\theta_k} (\mathbf{A}\mathbf{x}^{k+1} + \mathbf{B}\mathbf{y}^{k+1} - \mathbf{b}) - \frac{1}{\theta_{k-1}} (\mathbf{A}\mathbf{x}^k + \mathbf{B}\mathbf{y}^k - \mathbf{b}) \right.$$

$$\left. + \tau \left( \mathbf{A}\mathbf{x}^k + \mathbf{B}\mathbf{y}^k - \mathbf{b} \right) \right]$$

$$= \frac{\beta}{\theta_K} (\mathbf{A}\mathbf{x}^{K+1} + \mathbf{B}\mathbf{y}^{K+1} - \mathbf{b}) + \beta\tau \sum_{k=1}^{K} \left( \mathbf{A}\mathbf{x}^k + \mathbf{B}\mathbf{y}^k - \mathbf{b} \right).$$

For the fourth identity, we have

$$(1 - \theta_k)\mathbf{x}^k + \theta_k \mathbf{r}^k = (1 - \theta_k)\mathbf{x}^k + \frac{\theta_k}{\theta_{k-1}} \left[ \mathbf{x}^k - (1 - \theta_{k-1})\mathbf{x}^{k-1} \right]$$

$$= \mathbf{x}^k + \frac{\theta_k(1 - \theta_{k-1})}{\theta_{k-1}} (\mathbf{x}^k - \mathbf{x}^{k-1}).$$

The right-hand side is the definition of $\mathbf{u}^k$. Similarly, we can also have the last identity $\mathbf{v}^k - (1 - \theta_k)\mathbf{y}^k = \theta_k \mathbf{s}^k$.                                     □

**Lemma 3.14** *Suppose that $f_1$, $f_2$, $g_1$, and $g_2$ are convex, and $f_2$ and $g_2$ are L-smooth. With the definitions in (3.53) and (3.54), for Algorithm 3.5 we have*

$$\frac{1}{\theta_k} \left( f(\mathbf{x}^{k+1}) + g(\mathbf{y}^{k+1}) - f(\mathbf{x}^*) - g(\mathbf{y}^*) + \left\langle \boldsymbol{\lambda}^*, \mathbf{A}\mathbf{x}^{k+1} + \mathbf{B}\mathbf{y}^{k+1} - \mathbf{b} \right\rangle \right)$$

$$- \frac{1}{\theta_{k-1}} \left( f(\mathbf{x}^k) + g(\mathbf{y}^k) - f(\mathbf{x}^*) - g(\mathbf{y}^*) + \left\langle \boldsymbol{\lambda}^*, \mathbf{A}\mathbf{x}^k + \mathbf{B}\mathbf{y}^k - \mathbf{b} \right\rangle \right)$$

$$+ \tau \left( f(\mathbf{x}^k) + g(\mathbf{y}^k) - f(\mathbf{x}^*) - g(\mathbf{y}^*) + \left\langle \boldsymbol{\lambda}^*, \mathbf{A}\mathbf{x}^k + \mathbf{B}\mathbf{y}^k - \mathbf{b} \right\rangle \right)$$

$$\leq \frac{\beta}{2} \left( \|\mathbf{A}\mathbf{x}^* - \mathbf{A}\mathbf{r}^{k+1}\|^2 - \|\mathbf{A}\mathbf{x}^* - \mathbf{A}\mathbf{r}^k\|^2 \right)$$

$$+ \frac{1}{2\beta} \left( \|\hat{\boldsymbol{\lambda}}^k - \boldsymbol{\lambda}^*\|^2 - \|\hat{\boldsymbol{\lambda}}^{k+1} - \boldsymbol{\lambda}^*\|^2 \right)$$

$$+ \frac{1}{2} \left( L\theta_k + \beta \|\mathbf{A}\|_2^2 \right) \|\mathbf{x}^* - \mathbf{r}^k\|^2 - \frac{1}{2} \left( L\theta_{k+1} + \beta \|\mathbf{A}\|_2^2 \right) \|\mathbf{x}^* - \mathbf{r}^{k+1}\|^2$$

$$+ \frac{1}{2} \left( L\theta_k + \beta \|\mathbf{B}\|_2^2 \right) \|\mathbf{y}^* - \mathbf{s}^k\|^2 - \frac{1}{2} \left( L\theta_{k+1} + \beta \|\mathbf{B}\|_2^2 \right) \|\mathbf{y}^* - \mathbf{s}^{k+1}\|^2.$$

$$(3.57)$$

**Proof** From the optimality conditions of steps (3.52c) and (3.52d) and the definitions of $\bar{\boldsymbol{\lambda}}_1^{k+1}$ and $\bar{\boldsymbol{\lambda}}_2^{k+1}$, we have

$$\mathbf{0} \in \partial f_1(\mathbf{x}^{k+1}) + \nabla f_2(\mathbf{u}^k) + \mathbf{A}^T \bar{\boldsymbol{\lambda}}_1^{k+1} + \left( L + \frac{\beta \|\mathbf{A}\|_2^2}{\theta_k} \right) (\mathbf{x}^{k+1} - \mathbf{u}^k),$$

$$\mathbf{0} \in \partial g_1(\mathbf{y}^{k+1}) + \nabla g_2(\mathbf{v}^k) + \mathbf{B}^T \bar{\boldsymbol{\lambda}}_2^{k+1} + \left( L + \frac{\beta \|\mathbf{B}\|_2^2}{\theta_k} \right) (\mathbf{y}^{k+1} - \mathbf{v}^k).$$

From the convexity of $f_1$ and $g_1$, we have

$$f_1(\mathbf{x}) - f_1(\mathbf{x}^{k+1})$$

$$\geq - \left\langle \nabla f_2(\mathbf{u}^k) + \mathbf{A}^T \bar{\boldsymbol{\lambda}}_1^{k+1} + \left( L + \frac{\beta \|\mathbf{A}\|_2^2}{\theta_k} \right) (\mathbf{x}^{k+1} - \mathbf{u}^k), \mathbf{x} - \mathbf{x}^{k+1} \right\rangle,$$

$$g_1(\mathbf{y}) - g_1(\mathbf{y}^{k+1})$$

$$\geq - \left\langle \nabla g_2(\mathbf{v}^k) + \mathbf{B}^T \bar{\boldsymbol{\lambda}}_2^{k+1} + \left( L + \frac{\beta \|\mathbf{B}\|_2^2}{\theta_k} \right) (\mathbf{y}^{k+1} - \mathbf{v}^k), \mathbf{y} - \mathbf{y}^{k+1} \right\rangle.$$

On the other hand, from the smoothness and convexity of $f_2$ and $g_2$, we have

$$f_2(\mathbf{x}^{k+1})$$

$$\leq f_2(\mathbf{u}^k) + \left\langle \nabla f_2(\mathbf{u}^k), \mathbf{x}^{k+1} - \mathbf{u}^k \right\rangle + \frac{L}{2} \|\mathbf{x}^{k+1} - \mathbf{u}^k\|^2$$

$$= f_2(\mathbf{u}^k) + \left\langle \nabla f_2(\mathbf{u}^k), \mathbf{x} - \mathbf{u}^k \right\rangle + \left\langle \nabla f_2(\mathbf{u}^k), \mathbf{x}^{k+1} - \mathbf{x} \right\rangle + \frac{L}{2} \|\mathbf{x}^{k+1} - \mathbf{u}^k\|^2$$

$$\leq f_2(\mathbf{x}) + \left\langle \nabla f_2(\mathbf{u}^k), \mathbf{x}^{k+1} - \mathbf{x} \right\rangle + \frac{L}{2} \|\mathbf{x}^{k+1} - \mathbf{u}^k\|^2$$

and

$$g_2(\mathbf{y}^{k+1}) \leq g_2(\mathbf{y}) + \left\langle \nabla g_2(\mathbf{v}^k), \mathbf{y}^{k+1} - \mathbf{y} \right\rangle + \frac{L}{2} \|\mathbf{y}^{k+1} - \mathbf{v}^k\|^2.$$

So we have

$$
\begin{aligned}
f(\mathbf{x}) &- f(\mathbf{x}^{k+1}) \\
&= f_1(\mathbf{x}) + f_2(\mathbf{x}) - f_1(\mathbf{x}^{k+1}) - f_2(\mathbf{x}^{k+1}) \\
&\geq -\left\langle \mathbf{A}^T \overline{\boldsymbol{\lambda}}_1^{k+1} + \left( L + \frac{\beta \|\mathbf{A}\|_2^2}{\theta_k} \right)(\mathbf{x}^{k+1} - \mathbf{u}^k), \mathbf{x} - \mathbf{x}^{k+1} \right\rangle - \frac{L}{2} \|\mathbf{x}^{k+1} - \mathbf{u}^k\|^2 \\
&= -\left\langle \overline{\boldsymbol{\lambda}}_1^{k+1}, \mathbf{A}\mathbf{x} - \mathbf{A}\mathbf{x}^{k+1} \right\rangle - \left( L + \frac{\beta \|\mathbf{A}\|_2^2}{\theta_k} \right) \left\langle \mathbf{x}^{k+1} - \mathbf{u}^k, \mathbf{x} - \mathbf{x}^{k+1} \right\rangle \\
&\quad - \frac{L}{2} \|\mathbf{x}^{k+1} - \mathbf{u}^k\|^2
\end{aligned}
$$

and similarly

$$
\begin{aligned}
g(\mathbf{y}) &- g(\mathbf{y}^{k+1}) \\
&\geq -\left\langle \overline{\boldsymbol{\lambda}}_2^{k+1}, \mathbf{B}\mathbf{y} - \mathbf{B}\mathbf{y}^{k+1} \right\rangle - \left( L + \frac{\beta \|\mathbf{B}\|_2^2}{\theta_k} \right) \left\langle \mathbf{y}^{k+1} - \mathbf{v}^k, \mathbf{y} - \mathbf{y}^{k+1} \right\rangle \\
&\quad - \frac{L}{2} \|\mathbf{y}^{k+1} - \mathbf{v}^k\|^2.
\end{aligned}
$$

Adding them together, we have

$$
\begin{aligned}
f(\mathbf{x}^{k+1}) &+ g(\mathbf{y}^{k+1}) - f(\mathbf{x}) - g(\mathbf{y}) \\
&\leq \left\langle \overline{\boldsymbol{\lambda}}_1^{k+1}, \mathbf{A}\mathbf{x} - \mathbf{A}\mathbf{x}^{k+1} \right\rangle + \left( L + \frac{\beta \|\mathbf{A}\|_2^2}{\theta_k} \right) \left\langle \mathbf{x}^{k+1} - \mathbf{u}^k, \mathbf{x} - \mathbf{x}^{k+1} \right\rangle \\
&\quad + \frac{L}{2} \|\mathbf{x}^{k+1} - \mathbf{u}^k\|^2 + \left\langle \overline{\boldsymbol{\lambda}}_2^{k+1}, \mathbf{B}\mathbf{y} - \mathbf{B}\mathbf{y}^{k+1} \right\rangle \\
&\quad + \left( L + \frac{\beta \|\mathbf{B}\|_2^2}{\theta_k} \right) \left\langle \mathbf{y}^{k+1} - \mathbf{v}^k, \mathbf{y} - \mathbf{y}^{k+1} \right\rangle + \frac{L}{2} \|\mathbf{y}^{k+1} - \mathbf{v}^k\|^2.
\end{aligned}
$$

Letting $(\mathbf{x}, \mathbf{y}) = (\mathbf{x}^k, \mathbf{y}^k)$ and $(\mathbf{x}, \mathbf{y}) = (\mathbf{x}^*, \mathbf{y}^*)$, respectively, we have

$$
f(\mathbf{x}^{k+1}) + g(\mathbf{y}^{k+1}) - f(\mathbf{x}^*) - g(\mathbf{y}^*)
$$

$$
\leq \left\langle \bar{\boldsymbol{\lambda}}_1^{k+1}, \mathbf{A}\mathbf{x}^* - \mathbf{A}\mathbf{x}^{k+1} \right\rangle + \left( L + \frac{\beta\|\mathbf{A}\|_2^2}{\theta_k} \right) \left\langle \mathbf{x}^{k+1} - \mathbf{u}^k, \mathbf{x}^* - \mathbf{x}^{k+1} \right\rangle
$$

$$
+ \frac{L}{2}\|\mathbf{x}^{k+1} - \mathbf{u}^k\|^2 + \left\langle \bar{\boldsymbol{\lambda}}_2^{k+1}, \mathbf{B}\mathbf{y}^* - \mathbf{B}\mathbf{y}^{k+1} \right\rangle
$$

$$
+ \left( L + \frac{\beta\|\mathbf{B}\|_2^2}{\theta_k} \right) \left\langle \mathbf{y}^{k+1} - \mathbf{v}^k, \mathbf{y}^* - \mathbf{y}^{k+1} \right\rangle + \frac{L}{2}\|\mathbf{y}^{k+1} - \mathbf{v}^k\|^2
$$

and

$$
f(\mathbf{x}^{k+1}) + g(\mathbf{y}^{k+1}) - f(\mathbf{x}^k) - g(\mathbf{y}^k)
$$

$$
\leq \left\langle \bar{\boldsymbol{\lambda}}_1^{k+1}, \mathbf{A}\mathbf{x}^k - \mathbf{A}\mathbf{x}^{k+1} \right\rangle + \left( L + \frac{\beta\|\mathbf{A}\|_2^2}{\theta_k} \right) \left\langle \mathbf{x}^{k+1} - \mathbf{u}^k, \mathbf{x}^k - \mathbf{x}^{k+1} \right\rangle
$$

$$
+ \frac{L}{2}\|\mathbf{x}^{k+1} - \mathbf{u}^k\|^2 + \left\langle \bar{\boldsymbol{\lambda}}_2^{k+1}, \mathbf{B}\mathbf{y}^k - \mathbf{B}\mathbf{y}^{k+1} \right\rangle
$$

$$
+ \left( L + \frac{\beta\|\mathbf{B}\|_2^2}{\theta_k} \right) \left\langle \mathbf{y}^{k+1} - \mathbf{v}^k, \mathbf{y}^k - \mathbf{y}^{k+1} \right\rangle + \frac{L}{2}\|\mathbf{y}^{k+1} - \mathbf{v}^k\|^2.
$$

Multiplying the first inequality by $\theta_k$, multiplying the second by $1 - \theta_k$, and adding them together, we have

$$
f(\mathbf{x}^{k+1}) + g(\mathbf{y}^{k+1}) - f(\mathbf{x}^*) - g(\mathbf{y}^*)
$$

$$
- (1 - \theta_k)\left( f(\mathbf{x}^k) + g(\mathbf{y}^k) - f(\mathbf{x}^*) - g(\mathbf{y}^*) \right)
$$

$$
\leq \left\langle \bar{\boldsymbol{\lambda}}_1^{k+1}, \theta_k \mathbf{A}\mathbf{x}^* + (1 - \theta_k)\mathbf{A}\mathbf{x}^k - \mathbf{A}\mathbf{x}^{k+1} \right\rangle
$$

$$
+ \left\langle \bar{\boldsymbol{\lambda}}_2^{k+1}, \theta_k \mathbf{B}\mathbf{y}^* + (1 - \theta_k)\mathbf{B}\mathbf{y}^k - \mathbf{B}\mathbf{y}^{k+1} \right\rangle
$$

$$
+ \left( L + \frac{\beta\|\mathbf{A}\|_2^2}{\theta_k} \right) \left\langle \mathbf{x}^{k+1} - \mathbf{u}^k, \theta_k \mathbf{x}^* + (1 - \theta_k)\mathbf{x}^k - \mathbf{x}^{k+1} \right\rangle
$$

$$
+ \left( L + \frac{\beta\|\mathbf{B}\|_2^2}{\theta_k} \right) \left\langle \mathbf{y}^{k+1} - \mathbf{v}^k, \theta_k \mathbf{y}^* + (1 - \theta_k)\mathbf{y}^k - \mathbf{y}^{k+1} \right\rangle
$$

$$
+ \frac{L}{2}\|\mathbf{x}^{k+1} - \mathbf{u}^k\|^2 + \frac{L}{2}\|\mathbf{y}^{k+1} - \mathbf{v}^k\|^2.
$$

Adding

$$\left\langle \boldsymbol{\lambda}^*, \mathbf{A}\mathbf{x}^{k+1} + \mathbf{B}\mathbf{y}^{k+1} - (1 - \theta_k)(\mathbf{A}\mathbf{x}^k + \mathbf{B}\mathbf{y}^k) - \theta_k \mathbf{b} \right\rangle$$

to both sides, we have

$$
\begin{aligned}
&f(\mathbf{x}^{k+1}) + g(\mathbf{y}^{k+1}) - f(\mathbf{x}^*) - g(\mathbf{y}^*) + \left\langle \boldsymbol{\lambda}^*, \mathbf{A}\mathbf{x}^{k+1} + \mathbf{B}\mathbf{y}^{k+1} - \mathbf{b} \right\rangle \\
&\quad - (1 - \theta_k)\left( f(\mathbf{x}^k) + g(\mathbf{y}^k) - f(\mathbf{x}^*) - g(\mathbf{y}^*) + \left\langle \boldsymbol{\lambda}^*, \mathbf{A}\mathbf{x}^k + \mathbf{B}\mathbf{y}^k - \mathbf{b} \right\rangle \right) \\
&\leq \left\langle \bar{\boldsymbol{\lambda}}_1^{k+1} - \boldsymbol{\lambda}^*, \theta_k \mathbf{A}\mathbf{x}^* + (1 - \theta_k)\mathbf{A}\mathbf{x}^k - \mathbf{A}\mathbf{x}^{k+1} \right\rangle \\
&\quad + \left\langle \bar{\boldsymbol{\lambda}}_2^{k+1} - \boldsymbol{\lambda}^*, \theta_k \mathbf{B}\mathbf{y}^* + (1 - \theta_k)\mathbf{B}\mathbf{y}^k - \mathbf{B}\mathbf{y}^{k+1} \right\rangle \\
&\quad + \left( L + \frac{\beta \|\mathbf{A}\|_2^2}{\theta_k} \right) \left\langle \mathbf{x}^{k+1} - \mathbf{u}^k, \theta_k \mathbf{x}^* + (1 - \theta_k)\mathbf{x}^k - \mathbf{x}^{k+1} \right\rangle \\
&\quad + \left( L + \frac{\beta \|\mathbf{B}\|_2^2}{\theta_k} \right) \left\langle \mathbf{y}^{k+1} - \mathbf{v}^k, \theta_k \mathbf{y}^* + (1 - \theta_k)\mathbf{y}^k - \mathbf{y}^{k+1} \right\rangle \\
&\quad + \frac{L}{2}\|\mathbf{x}^{k+1} - \mathbf{u}^k\|^2 + \frac{L}{2}\|\mathbf{y}^{k+1} - \mathbf{v}^k\|^2 \\
&= \left\langle \bar{\boldsymbol{\lambda}}_1^{k+1} - \bar{\boldsymbol{\lambda}}_2^{k+1}, \theta_k \mathbf{A}\mathbf{x}^* + (1 - \theta_k)\mathbf{A}\mathbf{x}^k - \mathbf{A}\mathbf{x}^{k+1} \right\rangle \\
&\quad + \left\langle \bar{\boldsymbol{\lambda}}_2^{k+1} - \boldsymbol{\lambda}^*, \theta_k \mathbf{b} + (1 - \theta_k)(\mathbf{A}\mathbf{x}^k + \mathbf{B}\mathbf{y}^k) - (\mathbf{A}\mathbf{x}^{k+1} + \mathbf{B}\mathbf{y}^{k+1}) \right\rangle \\
&\quad + \left( L + \frac{\beta \|\mathbf{A}\|_2^2}{\theta_k} \right) \left\langle \mathbf{x}^{k+1} - \mathbf{u}^k, \theta_k \mathbf{x}^* + (1 - \theta_k)\mathbf{x}^k - \mathbf{x}^{k+1} \right\rangle \\
&\quad + \left( L + \frac{\beta \|\mathbf{B}\|_2^2}{\theta_k} \right) \left\langle \mathbf{y}^{k+1} - \mathbf{v}^k, \theta_k \mathbf{y}^* + (1 - \theta_k)\mathbf{y}^k - \mathbf{y}^{k+1} \right\rangle \\
&\quad + \frac{L}{2}\|\mathbf{x}^{k+1} - \mathbf{u}^k\|^2 + \frac{L}{2}\|\mathbf{y}^{k+1} - \mathbf{v}^k\|^2.
\end{aligned}
$$

From Lemma 3.13, for the first inner product, we have

$$\left\langle \bar{\boldsymbol{\lambda}}_1^{k+1} - \bar{\boldsymbol{\lambda}}_2^{k+1}, \theta_k \mathbf{A}\mathbf{x}^* + (1 - \theta_k)\mathbf{A}\mathbf{x}^k - \mathbf{A}\mathbf{x}^{k+1} \right\rangle$$

$$= \frac{\beta}{\theta_k} \left\langle \mathbf{A}\mathbf{u}^k - \mathbf{A}\mathbf{x}^{k+1}, \theta_k \mathbf{A}\mathbf{x}^* + (1 - \theta_k)\mathbf{A}\mathbf{x}^k - \mathbf{A}\mathbf{x}^{k+1} \right\rangle$$

$$\overset{a}{=} \frac{\beta}{2\theta_k} \left( \|\theta_k \mathbf{A}\mathbf{x}^* + (1 - \theta_k)\mathbf{A}\mathbf{x}^k - \mathbf{A}\mathbf{x}^{k+1}\|^2 \right.$$

$$\left. -\|\theta_k \mathbf{A}\mathbf{x}^* + (1 - \theta_k)\mathbf{A}\mathbf{x}^k - \mathbf{A}\mathbf{u}^k\|^2 \right) + \frac{\beta}{2\theta_k} \|\mathbf{A}\mathbf{u}^k - \mathbf{A}\mathbf{x}^{k+1}\|^2$$

$$= \frac{\beta\theta_k}{2} \left( \|\mathbf{A}\mathbf{x}^* - \mathbf{A}\mathbf{r}^{k+1}\|^2 - \|\mathbf{A}\mathbf{x}^* - \mathbf{A}\mathbf{r}^k\|^2 \right) + \frac{\beta}{2\theta_k} \|\mathbf{A}\mathbf{u}^k - \mathbf{A}\mathbf{x}^{k+1}\|^2$$

$$\leq \frac{\beta\theta_k}{2} \left( \|\mathbf{A}\mathbf{x}^* - \mathbf{A}\mathbf{r}^{k+1}\|^2 - \|\mathbf{A}\mathbf{x}^* - \mathbf{A}\mathbf{r}^k\|^2 \right) + \frac{\beta\|\mathbf{A}\|_2^2}{2\theta_k} \|\mathbf{u}^k - \mathbf{x}^{k+1}\|^2,$$

where $\overset{a}{=}$ uses (A.1); for the second inner product, we have

$$\left\langle \bar{\boldsymbol{\lambda}}_2^{k+1} - \boldsymbol{\lambda}^*, \theta_k \mathbf{b} + (1 - \theta_k)(\mathbf{A}\mathbf{x}^k + \mathbf{B}\mathbf{y}^k) - (\mathbf{A}\mathbf{x}^{k+1} + \mathbf{B}\mathbf{y}^{k+1}) \right\rangle$$

$$= \frac{\theta_k}{\beta} \left\langle \bar{\boldsymbol{\lambda}}_2^{k+1} - \boldsymbol{\lambda}^*, \hat{\boldsymbol{\lambda}}^k - \hat{\boldsymbol{\lambda}}^{k+1} \right\rangle$$

$$\overset{a}{=} \frac{\theta_k}{2\beta} \left( \|\hat{\boldsymbol{\lambda}}^k - \boldsymbol{\lambda}^*\|^2 - \|\hat{\boldsymbol{\lambda}}^{k+1} - \boldsymbol{\lambda}^*\|^2 - \left\|\bar{\boldsymbol{\lambda}}_2^{k+1} - \hat{\boldsymbol{\lambda}}^k\right\|^2 + \left\|\bar{\boldsymbol{\lambda}}_2^{k+1} - \hat{\boldsymbol{\lambda}}^{k+1}\right\|^2 \right)$$

$$\leq \frac{\theta_k}{2\beta} \left( \|\hat{\boldsymbol{\lambda}}^k - \boldsymbol{\lambda}^*\|^2 - \|\hat{\boldsymbol{\lambda}}^{k+1} - \boldsymbol{\lambda}^*\|^2 \right) + \frac{\beta\|\mathbf{B}\|_2^2}{2\theta_k} \|\mathbf{v}^k - \mathbf{y}^{k+1}\|^2,$$

where $\overset{a}{=}$ uses (A.3); and for the third and the fourth inner products, we have

$$\left\langle \mathbf{x}^{k+1} - \mathbf{u}^k, \theta_k \mathbf{x}^* + (1 - \theta_k)\mathbf{x}^k - \mathbf{x}^{k+1} \right\rangle$$

$$\overset{a}{=} \frac{1}{2} \|\theta_k \mathbf{x}^* + (1 - \theta_k)\mathbf{x}^k - \mathbf{u}^k\|^2 - \frac{1}{2} \|\theta_k \mathbf{x}^* + (1 - \theta_k)\mathbf{x}^k - \mathbf{x}^{k+1}\|^2$$

$$- \frac{1}{2} \|\mathbf{x}^{k+1} - \mathbf{u}^k\|^2$$

$$= \frac{1}{2}\theta_k^2 \|\mathbf{x}^* - \mathbf{r}^k\|^2 - \frac{1}{2}\theta_k^2 \|\mathbf{x}^* - \mathbf{r}^{k+1}\|^2 - \frac{1}{2} \|\mathbf{x}^{k+1} - \mathbf{u}^k\|^2$$

and

$$\left\langle \mathbf{y}^{k+1} - \mathbf{v}^k, \theta_k \mathbf{y}^* + (1 - \theta_k)\mathbf{y}^k - \mathbf{y}^{k+1} \right\rangle$$

$$\overset{b}{=} \frac{1}{2} \|\theta_k \mathbf{y}^* + (1 - \theta_k)\mathbf{y}^k - \mathbf{v}^k\|^2 - \frac{1}{2} \|\theta_k \mathbf{y}^* + (1 - \theta_k)\mathbf{y}^k - \mathbf{y}^{k+1}\|^2$$

$$- \frac{1}{2} \|\mathbf{y}^{k+1} - \mathbf{v}^k\|^2$$

$$= \frac{1}{2} \theta_k^2 \|\mathbf{y}^* - \mathbf{s}^k\|^2 - \frac{1}{2} \theta_k^2 \|\mathbf{y}^* - \mathbf{s}^{k+1}\|^2 - \frac{1}{2} \|\mathbf{y}^{k+1} - \mathbf{v}^k\|^2,$$

where both $\overset{a}{=}$ and $\overset{b}{=}$ use (A.2). So we have

$$f(\mathbf{x}^{k+1}) + g(\mathbf{y}^{k+1}) - f(\mathbf{x}^*) - g(\mathbf{y}^*) + \left\langle \boldsymbol{\lambda}^*, \mathbf{A}\mathbf{x}^{k+1} + \mathbf{B}\mathbf{y}^{k+1} - \mathbf{b} \right\rangle$$

$$- (1 - \theta_k) \left( f(\mathbf{x}^k) + g(\mathbf{y}^k) - f(\mathbf{x}^*) - g(\mathbf{y}^*) + \left\langle \boldsymbol{\lambda}^*, \mathbf{A}\mathbf{x}^k + \mathbf{B}\mathbf{y}^k - \mathbf{b} \right\rangle \right)$$

$$\leq \frac{\beta \theta_k}{2} \left( \|\mathbf{A}\mathbf{x}^* - \mathbf{A}\mathbf{r}^{k+1}\|^2 - \|\mathbf{A}\mathbf{x}^* - \mathbf{A}\mathbf{r}^k\|^2 \right)$$

$$+ \frac{\theta_k}{2\beta} \left( \|\hat{\boldsymbol{\lambda}}^k - \boldsymbol{\lambda}^*\|^2 - \|\hat{\boldsymbol{\lambda}}^{k+1} - \boldsymbol{\lambda}^*\|^2 \right)$$

$$+ \frac{1}{2} \left( L\theta_k^2 + \beta \theta_k \|\mathbf{A}\|_2^2 \right) \left( \|\mathbf{x}^* - \mathbf{r}^k\|^2 - \|\mathbf{x}^* - \mathbf{r}^{k+1}\|^2 \right)$$

$$+ \frac{1}{2} \left( L\theta_k^2 + \beta \theta_k \|\mathbf{B}\|_2^2 \right) \left( \|\mathbf{y}^* - \mathbf{s}^k\|^2 - \|\mathbf{y}^* - \mathbf{s}^{k+1}\|^2 \right).$$

Dividing both sides by $\theta_k$ and using $\frac{1 - \theta_k}{\theta_k} = \frac{1}{\theta_{k-1}} - \tau$, we have

$$\frac{1}{\theta_k} \left( f(\mathbf{x}^{k+1}) + g(\mathbf{y}^{k+1}) - f(\mathbf{x}^*) - g(\mathbf{y}^*) + \left\langle \boldsymbol{\lambda}^*, \mathbf{A}\mathbf{x}^{k+1} + \mathbf{B}\mathbf{y}^{k+1} - \mathbf{b} \right\rangle \right)$$

$$- \frac{1}{\theta_{k-1}} \left( f(\mathbf{x}^k) + g(\mathbf{y}^k) - f(\mathbf{x}^*) - g(\mathbf{y}^*) + \left\langle \boldsymbol{\lambda}^*, \mathbf{A}\mathbf{x}^k + \mathbf{B}\mathbf{y}^k - \mathbf{b} \right\rangle \right)$$

$$+ \tau \left( f(\mathbf{x}^k) + g(\mathbf{y}^k) - f(\mathbf{x}^*) - g(\mathbf{y}^*) + \left\langle \boldsymbol{\lambda}^*, \mathbf{A}\mathbf{x}^k + \mathbf{B}\mathbf{y}^k - \mathbf{b} \right\rangle \right)$$

$$\leq \frac{\beta}{2} \left( \|\mathbf{A}\mathbf{x}^* - \mathbf{A}\mathbf{r}^{k+1}\|^2 - \|\mathbf{A}\mathbf{x}^* - \mathbf{A}\mathbf{r}^k\|^2 \right)$$

$$+ \frac{1}{2\beta} \left( \|\hat{\boldsymbol{\lambda}}^k - \boldsymbol{\lambda}^*\|^2 - \|\hat{\boldsymbol{\lambda}}^{k+1} - \boldsymbol{\lambda}^*\|^2 \right)$$

$$+\frac{1}{2}\left(L\theta_k + \beta\|\mathbf{A}\|_2^2\right)\left(\|\mathbf{x}^* - \mathbf{r}^k\|^2 - \|\mathbf{x}^* - \mathbf{r}^{k+1}\|^2\right)$$

$$+\frac{1}{2}\left(L\theta_k + \beta\|\mathbf{B}\|_2^2\right)\left(\|\mathbf{y}^* - \mathbf{s}^k\|^2 - \|\mathbf{y}^* - \mathbf{s}^{k+1}\|^2\right).$$

From $\theta_{k+1} \le \theta_k$, we have the conclusion.                                    $\square$

**Lemma 3.15** *Suppose that $f_1$, $f_2$, $g_1$, and $g_2$ are convex, and $f_2$ and $g_2$ are L-smooth. With the definitions in (3.53) and (3.54), for Algorithm 3.5 we have*

$$f(\mathbf{x}^{K+1}) + g(\mathbf{y}^{K+1}) - f(\mathbf{x}^*) - g(\mathbf{y}^*) + \left\langle \boldsymbol{\lambda}^*, \mathbf{A}\mathbf{x}^{K+1} + \mathbf{B}\mathbf{y}^{K+1} - \mathbf{b}\right\rangle$$

$$\le \theta_K C \tag{3.58}$$

*and*

$$\left\|\frac{1}{\theta_K}\left(\mathbf{A}\mathbf{x}^{K+1} + \mathbf{B}\mathbf{y}^{K+1} - \mathbf{b}\right) + \tau \sum_{k=1}^{K}\left(\mathbf{A}\mathbf{x}^k + \mathbf{B}\mathbf{y}^k - \mathbf{b}\right)\right\|$$

$$\le \frac{1}{\beta}\left\|\hat{\boldsymbol{\lambda}}^0 - \boldsymbol{\lambda}^*\right\| + \sqrt{\frac{2C}{\beta}}, \tag{3.59}$$

*where*

$$C = \frac{1}{2\beta}\left\|\hat{\boldsymbol{\lambda}}^0 - \boldsymbol{\lambda}^*\right\|^2 - \frac{\beta}{2}\left\|\mathbf{A}\mathbf{x}^* - \mathbf{A}\mathbf{r}^0\right\|^2$$

$$+\frac{1}{2}\left(L + \beta\|\mathbf{A}\|_2^2\right)\left\|\mathbf{x}^* - \mathbf{r}^0\right\|^2 + \frac{1}{2}\left(L + \beta\|\mathbf{B}\|_2^2\right)\left\|\mathbf{y}^* - \mathbf{s}^0\right\|^2.$$

***Proof*** Summing (3.57) over $k = 0, 1, \cdots, K$, we have

$$\frac{1}{\theta_K}\left(f(\mathbf{x}^{K+1}) + g(\mathbf{y}^{K+1}) - f(\mathbf{x}^*) - g(\mathbf{y}^*) + \left\langle \boldsymbol{\lambda}^*, \mathbf{A}\mathbf{x}^{K+1} + \mathbf{B}\mathbf{y}^{K+1} - \mathbf{b}\right\rangle\right)$$

$$+ \tau \sum_{k=1}^{K}\left(f(\mathbf{x}^k) + g(\mathbf{y}^k) - f(\mathbf{x}^*) - g(\mathbf{y}^*) + \left\langle \boldsymbol{\lambda}^*, \mathbf{A}\mathbf{x}^k + \mathbf{B}\mathbf{y}^k - \mathbf{b}\right\rangle\right)$$

$$\le \frac{1}{2\beta}\left(\|\hat{\boldsymbol{\lambda}}^0 - \boldsymbol{\lambda}^*\|^2 - \|\hat{\boldsymbol{\lambda}}^{K+1} - \boldsymbol{\lambda}^*\|^2\right) - \frac{\beta}{2}\|\mathbf{A}\mathbf{x}^* - \mathbf{A}\mathbf{r}^0\|^2$$

$$+\frac{1}{2}\left(L + \beta\|\mathbf{A}\|_2^2\right)\|\mathbf{x}^* - \mathbf{r}^0\|^2 + \frac{1}{2}\left(L + \beta\|\mathbf{B}\|_2^2\right)\|\mathbf{y}^* - \mathbf{s}^0\|^2$$

$$= C - \frac{1}{2\beta}\|\hat{\boldsymbol{\lambda}}^{K+1} - \boldsymbol{\lambda}^*\|^2,$$

where we use

$$\frac{1}{\theta_{-1}} = \tau, \quad \theta_0 = 1, \quad \text{and}$$

$$\frac{1}{2}\left(L\theta_{K+1} + \beta\|\mathbf{A}\|_2^2\right)\|\mathbf{x}^* - \mathbf{r}^{K+1}\|^2 \geq \frac{\beta}{2}\|\mathbf{A}\mathbf{x}^* - \mathbf{A}\mathbf{r}^{K+1}\|^2.$$

From Lemma 3.1, we have (3.58) and

$$\left\|\hat{\boldsymbol{\lambda}}^{K+1} - \boldsymbol{\lambda}^*\right\| \leq \sqrt{2\beta C},$$

$$\left\|\hat{\boldsymbol{\lambda}}^{K+1} - \hat{\boldsymbol{\lambda}}^0\right\| \leq \left\|\hat{\boldsymbol{\lambda}}^0 - \boldsymbol{\lambda}^*\right\| + \sqrt{2\beta C}.$$

From Lemma 3.13, we can have (3.59).                                    □

We need to bound the violation of constraint in the form of $\|\mathbf{A}\mathbf{x} + \mathbf{B}\mathbf{y} - \mathbf{b}\|$, rather than (3.59). The following lemma provides a useful tool for our purpose.

**Lemma 3.16** *Consider a sequence* $\{\mathbf{a}^k\}_{k=1}^\infty$ *of vectors. If* $\{\mathbf{a}^k\}$ *satisfies*

$$\left\|[1/\tau + K(1/\tau - 1)]\mathbf{a}^{K+1} + \sum_{k=1}^K \mathbf{a}^k\right\| \leq c, \quad \forall K = 0, 1, 2, \cdots, \tag{3.60}$$

*where* $0 < \tau < 1$, *then*

$$\left\|\sum_{k=1}^K \mathbf{a}^k\right\| < c, \quad \forall K = 1, 2, \cdots.$$

**Proof** We define

$$\mathbf{b}^K = \eta_K \mathbf{a}^{K+1} + \sum_{k=1}^K \mathbf{a}^k \quad \text{and} \quad \mathbf{s}^K = \sum_{k=1}^K \mathbf{a}^k,$$

where $\eta_K = 1/\tau + K(1/\tau - 1)$. Then

$$\mathbf{b}^K = \eta_K(\mathbf{s}^{K+1} - \mathbf{s}^K) + \mathbf{s}^K = \eta_K \mathbf{s}^{K+1} + (1 - \eta_K)\mathbf{s}^K.$$

Thus

$$\mathbf{s}^{K+1} = \frac{1}{\eta_K}\mathbf{b}^K + \left(1 - \frac{1}{\eta_K}\right)\mathbf{s}^K.$$

Therefore, since $\|\mathbf{b}^K\| \leq c$ is assumed and $\frac{1}{\eta_K} \in (0,1)$, if $\|\mathbf{s}^K\| < c$ we have

$$\|\mathbf{s}^{K+1}\| \leq \frac{1}{\eta_K}\|\mathbf{b}^K\| + \left(1 - \frac{1}{\eta_K}\right)\|\mathbf{s}^K\| < c.$$

On the other hand, letting $K = 0$ in (3.60), we have $\|\mathbf{s}^1\| = \|\mathbf{a}^1\| \leq \tau c < c$. So by mathematical induction, the lemma is proven. □

Now, based on the previous results, we are ready to present the convergence rate.

**Theorem 3.11** *Suppose that $f_1$, $f_2$, $g_1$, and $g_2$ are convex, and $f_2$ and $g_2$ are L-smooth. With the definitions in (3.53) and (3.54), for Algorithm 3.5, we have*

$$-\frac{2C_1\|\boldsymbol{\lambda}^*\|}{1 + K(1 - \tau)} \leq f(\mathbf{x}^{K+1}) + g(\mathbf{y}^{K+1}) - f(\mathbf{x}^*) - g(\mathbf{y}^*)$$

$$\leq \frac{2C_1\|\boldsymbol{\lambda}^*\|}{1 + K(1 - \tau)} + \frac{C}{1 + K(1 - \tau)}$$

*and*

$$\left\|\mathbf{A}\mathbf{x}^{K+1} + \mathbf{B}\mathbf{y}^{K+1} - \mathbf{b}\right\| \leq \frac{2C_1}{1 + K(1 - \tau)},$$

*where*

$$C = \frac{1}{2\beta}\left\|\hat{\boldsymbol{\lambda}}^0 - \boldsymbol{\lambda}^*\right\|^2 - \frac{\beta}{2}\left\|\mathbf{A}\mathbf{x}^* - \mathbf{A}\mathbf{r}^0\right\|^2$$

$$+ \frac{1}{2}\left(L + \beta\|\mathbf{A}\|_2^2\right)\left\|\mathbf{x}^* - \mathbf{r}^0\right\|^2 + \frac{1}{2}\left(L + \beta\|\mathbf{B}\|_2^2\right)\left\|\mathbf{y}^* - \mathbf{s}^0\right\|^2$$

*and $C_1 = \frac{1}{\beta}\left\|\hat{\boldsymbol{\lambda}}^0 - \boldsymbol{\lambda}^*\right\| + \sqrt{\frac{2C}{\beta}}$.*

**Proof** Since

$$\frac{1}{\theta_k} = \frac{1}{\theta_{k-1}} + 1 - \tau = \frac{1}{\theta_0} + k(1 - \tau),$$

we have

$$\theta_k = \frac{1}{\frac{1}{\theta_0} + k(1 - \tau)} = \frac{1}{1 + k(1 - \tau)}.$$

For simplicity, let $\mathbf{a}^k = \mathbf{A}\mathbf{x}^k + \mathbf{B}\mathbf{y}^k - \mathbf{b}$. Then from (3.59) we have

$$
\left\| [1/\tau + K(1/\tau - 1)]\mathbf{a}^{K+1} + \sum_{k=1}^{K} \mathbf{a}^k \right\|
$$

$$
\leq \frac{1}{\tau\beta}\|\hat{\boldsymbol{\lambda}}^0 - \boldsymbol{\lambda}^*\| + \frac{1}{\tau}\sqrt{\frac{2C}{\beta}} \equiv \frac{1}{\tau}C_1, \quad \forall K = 0, 1, \cdots . \tag{3.61}
$$

From Lemma 3.16 we have

$$
\left\| \sum_{k=1}^{K} \mathbf{a}^k \right\| \leq \frac{1}{\tau}C_1, \quad \forall K = 1, 2, \cdots .
$$

So

$$
\|\mathbf{a}^{K+1}\| \leq \frac{2\frac{1}{\tau}C_1}{1/\tau + K(1/\tau - 1)}, \quad \forall K = 1, 2, \cdots .
$$

Moreover, letting $K = 0$ in (3.61), we have

$$
\|\mathbf{a}^1\| \leq C_1 \leq \frac{2\frac{1}{\tau}C_1}{1/\tau + 0(1/\tau - 1)}.
$$

So

$$
\left\| \mathbf{A}\mathbf{x}^{K+1} + \mathbf{B}\mathbf{y}^{K+1} - \mathbf{b} \right\| \leq \frac{2C_1}{1 + K(1 - \tau)}, \quad \forall K = 0, 1, \cdots .
$$

Then from (3.58) and Lemma 3.2, we can have the conclusion. □

### 3.3.2  Linear Convergence Rate

In this section, we further assume that $g$ is $\mu_g$-strongly convex. We want to accelerate the following linearized ADMM:

$$
\mathbf{x}^{k+1} = \underset{\mathbf{x}}{\operatorname{argmin}} \left( f(\mathbf{x}) + \left\langle \boldsymbol{\lambda}^k, \mathbf{A}\mathbf{x} + \mathbf{B}\mathbf{y}^k - \mathbf{b} \right\rangle + \frac{\beta}{2}\|\mathbf{A}\mathbf{x} + \mathbf{B}\mathbf{y}^k - \mathbf{b}\|^2 \right),
$$

$$
\mathbf{y}^{k+1} = \underset{\mathbf{y}}{\operatorname{argmin}} \left( g(\mathbf{y}^k) + \left\langle \nabla g(\mathbf{y}^k), \mathbf{y} - \mathbf{y}^k \right\rangle + \left\langle \boldsymbol{\lambda}^k, \mathbf{A}\mathbf{x}^{k+1} + \mathbf{B}\mathbf{y} - \mathbf{b} \right\rangle \right.
$$

$$+ \beta \left\langle \mathbf{B}^T (\mathbf{A}\mathbf{x}^{k+1} + \mathbf{B}\mathbf{y}^k - \mathbf{b}), \mathbf{y} - \mathbf{y}^k \right\rangle + \frac{L_g + \beta \|\mathbf{B}\|_2^2}{2} \|\mathbf{y} - \mathbf{y}^k\|^2 \right),$$

$$\boldsymbol{\lambda}^{k+1} = \boldsymbol{\lambda}^k + \beta (\mathbf{A}\mathbf{x}^{k+1} + \mathbf{B}\mathbf{y}^{k+1} - \mathbf{b}),$$

where $f$ and $g$ are both convex, and $g$ is $L_g$-smooth. Recall from Remark 3.2 (by setting $\phi = 0$) that its complexity is $O\left( \left( \frac{\|\mathbf{B}\|_2^2}{\sigma^2} + \frac{L_g}{\mu_g} \right) \log \frac{1}{\epsilon} \right)$.

We give the following accelerated linearized ADMM, which is adapted from the accelerated Lagrangian method proposed in [15]. The iterations are shown in (3.62a)–(3.62f) and we present the algorithm in Algorithm 3.6.

$$\mathbf{w}^k = \theta \mathbf{y}^k + (1 - \theta) \widetilde{\mathbf{y}}^k, \tag{3.62a}$$

$$\mathbf{x}^{k+1} = \underset{\mathbf{x}}{\operatorname{argmin}} \left( f(\mathbf{x}) + \left\langle \boldsymbol{\lambda}^k, \mathbf{A}\mathbf{x} + \mathbf{B}\mathbf{y}^k - \mathbf{b} \right\rangle + \frac{\beta \theta}{2} \|\mathbf{A}\mathbf{x} + \mathbf{B}\mathbf{y}^k - \mathbf{b}\|^2 \right), \tag{3.62b}$$

$$\mathbf{y}^{k+1} = \underset{\mathbf{y}}{\operatorname{argmin}} \left( \left\langle \nabla g(\mathbf{w}^k), \mathbf{y} \right\rangle + \left\langle \boldsymbol{\lambda}^k, \mathbf{B}\mathbf{y} \right\rangle + \beta \theta \left\langle \mathbf{B}^T (\mathbf{A}\mathbf{x}^{k+1} + \mathbf{B}\mathbf{y}^k - \mathbf{b}), \mathbf{y} \right\rangle \right.$$

$$\left. + \frac{1}{2} \left( \frac{\theta}{\alpha} + \mu_g \right) \left\| \mathbf{y} - \frac{1}{\frac{\theta}{\alpha} + \mu_g} \left( \frac{\theta}{\alpha} \mathbf{y}^k + \mu_g \mathbf{w}^k \right) \right\|^2 \right)$$

$$= \frac{1}{\frac{\theta}{\alpha} + \mu_g} \left\{ \mu_g \mathbf{w}^k + \frac{\theta}{\alpha} \mathbf{y}^k \right.$$

$$\left. - \left[ \nabla g(\mathbf{w}^k) + \mathbf{B}^T \boldsymbol{\lambda}^k + \beta \theta \mathbf{B}^T (\mathbf{A}\mathbf{x}^{k+1} + \mathbf{B}\mathbf{y}^k - \mathbf{b}) \right] \right\}, \tag{3.62c}$$

$$\widetilde{\mathbf{x}}^{k+1} = \theta \mathbf{x}^{k+1} + (1 - \theta) \widetilde{\mathbf{x}}^k, \tag{3.62d}$$

$$\widetilde{\mathbf{y}}^{k+1} = \theta \mathbf{y}^{k+1} + (1 - \theta) \widetilde{\mathbf{y}}^k, \tag{3.62e}$$

$$\boldsymbol{\lambda}^{k+1} = \boldsymbol{\lambda}^k + \beta \theta (\mathbf{A}\mathbf{x}^{k+1} + \mathbf{B}\mathbf{y}^{k+1} - \mathbf{b}). \tag{3.62f}$$

---

**Algorithm 3.6** Accelerated linearized ADMM for strongly convex problems

Initialize $\mathbf{x}^0 = \widetilde{\mathbf{x}}^0, \mathbf{y}^0 = \widetilde{\mathbf{y}}^0, \boldsymbol{\lambda}^0$.
**for** $k = 0, 1, 2, 3, \cdots$ **do**
    Update the variables by (3.62a)–(3.62f), respectively.
**end for**

---

**Table 3.2** Complexity comparisons between ADMM, linearized ADMM (LADMM), and its accelerated version

| Method | ADMM | LADMM | Accelerated LADMM |
|---|---|---|---|
| Rates | $O\left(\sqrt{\frac{L_g\|\mathbf{B}\|_2^2}{\mu_g\sigma^2}}\log\frac{1}{\epsilon}\right)$ | $O\left(\left(\frac{\|\mathbf{B}\|_2^2}{\sigma^2}+\frac{L_g}{\mu_g}\right)\log\frac{1}{\epsilon}\right)$ | $O\left(\sqrt{\frac{L_g\|\mathbf{B}\|_2^2}{\mu_g\sigma^2}}\log\frac{1}{\epsilon}\right)$ |

The complexity of Algorithm 3.6 is $O\left(\sqrt{\frac{L_g\|\mathbf{B}\|_2^2}{\mu_g\sigma^2}}\log\frac{1}{\epsilon}\right)$, which is lower than

the $O\left(\left(\frac{\|\mathbf{B}\|_2^2}{\sigma^2}+\frac{L_g}{\mu_g}\right)\log\frac{1}{\epsilon}\right)$ one of linearized ADMM. We list the comparisons in Table 3.2. Note that ADMM needs to solve a subproblem in the update of $\mathbf{y}$, while the linearized ADMM and accelerated linearized ADMM only performs a gradient descent. Thus, it is reasonable that ADMM has a faster rate than linearized ADMM but the accelerated linearized ADMM fills in this gap of rate.

**Lemma 3.17** *Suppose that $f(\mathbf{x})$ is convex and $g(\mathbf{y})$ is $\mu_g$-strongly convex and $L_g$-smooth. Let $\theta \leq 1$. Then for Algorithm 3.6, we have*

$$f(\widetilde{\mathbf{x}}^{k+1}) - f(\mathbf{x}^*) + g(\widetilde{\mathbf{y}}^{k+1}) - g(\mathbf{y}^*) + \left\langle \boldsymbol{\lambda}^*, \mathbf{A}\widetilde{\mathbf{x}}^{k+1} + \mathbf{B}\widetilde{\mathbf{y}}^{k+1} - \mathbf{b} \right\rangle$$

$$\leq (1-\theta)\left(f(\widetilde{\mathbf{x}}^k) - f(\mathbf{x}^*) + g(\widetilde{\mathbf{y}}^k) - g(\mathbf{y}^*) + \left\langle \boldsymbol{\lambda}^*, \mathbf{A}\widetilde{\mathbf{x}}^k + \mathbf{B}\widetilde{\mathbf{y}}^k - \mathbf{b} \right\rangle\right)$$

$$+ \frac{\theta^2}{2\alpha}\|\mathbf{y}^k - \mathbf{y}^*\|^2 - \left(\frac{\theta^2}{2\alpha} + \frac{\mu_g\theta}{2}\right)\|\mathbf{y}^{k+1} - \mathbf{y}^*\|^2 - \frac{\mu_g\theta}{2}\|\mathbf{y}^{k+1} - \mathbf{w}^k\|^2$$

$$+ \frac{1}{2\beta}\left(\|\boldsymbol{\lambda}^k - \boldsymbol{\lambda}^*\|^2 - \|\boldsymbol{\lambda}^{k+1} - \boldsymbol{\lambda}^*\|^2\right)$$

$$- \left(\frac{\theta^2}{2\alpha} - \frac{L_g\theta^2}{2} - \frac{\beta\theta^2\|\mathbf{B}\|_2^2}{2}\right)\|\mathbf{y}^{k+1} - \mathbf{y}^k\|^2. \tag{3.63}$$

***Proof*** We write $\mu_g$ and $L_g$ as $\mu$ and $L$, respectively, for notation simplicity. Similar to (3.49) and (3.50), we have

$$f(\widetilde{\mathbf{x}}^{k+1}) - f(\mathbf{x}^*)$$

$$\leq \theta\left(f(\mathbf{x}^{k+1}) - f(\mathbf{x}^*)\right) + (1-\theta)\left(f(\widetilde{\mathbf{x}}^k) - f(\mathbf{x}^*)\right)$$

$$\overset{a}{\leq} \theta\left\langle \boldsymbol{\lambda}^{k+1}, \mathbf{A}\mathbf{x}^* - \mathbf{A}\mathbf{x}^{k+1} \right\rangle + \beta\theta^2\left\langle \mathbf{B}\mathbf{y}^{k+1} - \mathbf{B}\mathbf{y}^k, \mathbf{A}\mathbf{x}^{k+1} - \mathbf{A}\mathbf{x}^* \right\rangle$$

$$+ (1-\theta)\left(f(\widetilde{\mathbf{x}}^k) - f(\mathbf{x}^*)\right) \tag{3.64}$$

and

$$g(\widetilde{\mathbf{y}}^{k+1})$$

$$\leq (1-\theta)g(\widetilde{\mathbf{y}}^k) + \theta\left(g(\mathbf{w}^k) + \left\langle\nabla g(\mathbf{w}^k), \mathbf{y}^* - \mathbf{w}^k\right\rangle + \left\langle\nabla g(\mathbf{w}^k), \mathbf{y}^{k+1} - \mathbf{y}^*\right\rangle\right)$$

$$+ \frac{L\theta^2}{2}\|\mathbf{y}^{k+1} - \mathbf{y}^k\|^2$$

$$\leq (1-\theta)g(\widetilde{\mathbf{y}}^k) + \theta g(\mathbf{y}^*) - \frac{\mu\theta}{2}\|\mathbf{w}^k - \mathbf{y}^*\|^2 + \theta\left\langle\nabla g(\mathbf{w}^k), \mathbf{y}^{k+1} - \mathbf{y}^*\right\rangle$$

$$+ \frac{L\theta^2}{2}\|\mathbf{y}^{k+1} - \mathbf{y}^k\|^2, \tag{3.65}$$

where $\overset{a}{\leq}$ has an additional $\theta$ compared to (3.49) due to the additional $\theta$ in (3.62b) and (3.62f). From (3.62c), we have

$$-\mu(\mathbf{y}^{k+1} - \mathbf{w}^k) - \frac{\theta}{\alpha}(\mathbf{y}^{k+1} - \mathbf{y}^k)$$

$$= \nabla g(\mathbf{w}^k) + \mathbf{B}^T\boldsymbol{\lambda}^{k+1} - \beta\theta\mathbf{B}^T(\mathbf{B}\mathbf{y}^{k+1} - \mathbf{B}\mathbf{y}^k). \tag{3.66}$$

So we have

$$\theta\left\langle\nabla g(\mathbf{w}^k), \mathbf{y}^{k+1} - \mathbf{y}^*\right\rangle$$

$$= -\theta\left\langle\mu(\mathbf{y}^{k+1} - \mathbf{w}^k) + \frac{\theta}{\alpha}(\mathbf{y}^{k+1} - \mathbf{y}^k) + \mathbf{B}^T\boldsymbol{\lambda}^{k+1} - \beta\theta\mathbf{B}^T(\mathbf{B}\mathbf{y}^{k+1} - \mathbf{B}\mathbf{y}^k),\right.$$

$$\left. \mathbf{y}^{k+1} - \mathbf{y}^*\right\rangle$$

$$= \frac{\mu\theta}{2}\left(\|\mathbf{w}^k - \mathbf{y}^*\|^2 - \|\mathbf{y}^{k+1} - \mathbf{y}^*\|^2 - \|\mathbf{y}^{k+1} - \mathbf{w}^k\|^2\right)$$

$$+ \frac{\theta^2}{2\alpha}\left(\|\mathbf{y}^k - \mathbf{y}^*\|^2 - \|\mathbf{y}^{k+1} - \mathbf{y}^*\|^2 - \|\mathbf{y}^{k+1} - \mathbf{y}^k\|^2\right)$$

$$- \theta\left\langle\boldsymbol{\lambda}^{k+1}, \mathbf{B}\mathbf{y}^{k+1} - \mathbf{B}\mathbf{y}^*\right\rangle + \beta\theta^2\left\langle\mathbf{B}\mathbf{y}^{k+1} - \mathbf{B}\mathbf{y}^k, \mathbf{B}\mathbf{y}^{k+1} - \mathbf{B}\mathbf{y}^*\right\rangle. \tag{3.67}$$

Combining (3.64), (3.65), and (3.67), we have

$$f(\widetilde{\mathbf{x}}^{k+1}) - f(\mathbf{x}^*) + g(\widetilde{\mathbf{y}}^{k+1}) - g(\mathbf{y}^*)$$

$$\leq (1-\theta)\left(f(\widetilde{\mathbf{x}}^k) - f(\mathbf{x}^*) + g(\widetilde{\mathbf{y}}^k) - g(\mathbf{y}^*)\right)$$

$$+ \frac{\theta^2}{2\alpha} \|\mathbf{y}^k - \mathbf{y}^*\|^2 - \left( \frac{\theta^2}{2\alpha} + \frac{\mu\theta}{2} \right) \|\mathbf{y}^{k+1} - \mathbf{y}^*\|^2$$

$$- \frac{\mu\theta}{2} \|\mathbf{y}^{k+1} - \mathbf{w}^k\|^2 - \left( \frac{\theta^2}{2\alpha} - \frac{L\theta^2}{2} \right) \|\mathbf{y}^{k+1} - \mathbf{y}^k\|^2$$

$$- \theta \left\langle \boldsymbol{\lambda}^{k+1}, \mathbf{A}\mathbf{x}^{k+1} + \mathbf{B}\mathbf{y}^{k+1} - \mathbf{b} \right\rangle$$

$$+ \beta\theta^2 \left\langle \mathbf{B}\mathbf{y}^{k+1} - \mathbf{B}\mathbf{y}^k, \mathbf{A}\mathbf{x}^{k+1} + \mathbf{B}\mathbf{y}^{k+1} - \mathbf{b} \right\rangle.$$

Adding $\left\langle \boldsymbol{\lambda}^*, \mathbf{A}\widetilde{\mathbf{x}}^{k+1} + \mathbf{B}\widetilde{\mathbf{y}}^{k+1} - \mathbf{b} \right\rangle$ to both sides, and using (3.62d) and (3.62e), we have

$$f(\widetilde{\mathbf{x}}^{k+1}) - f(\mathbf{x}^*) + g(\widetilde{\mathbf{y}}^{k+1}) - g(\mathbf{y}^*) + \left\langle \boldsymbol{\lambda}^*, \mathbf{A}\widetilde{\mathbf{x}}^{k+1} + \mathbf{B}\widetilde{\mathbf{y}}^{k+1} - \mathbf{b} \right\rangle$$

$$\leq (1 - \theta) \left( f(\widetilde{\mathbf{x}}^k) - f(\mathbf{x}^*) + g(\widetilde{\mathbf{y}}^k) - g(\mathbf{y}^*) + \left\langle \boldsymbol{\lambda}^*, \mathbf{A}\widetilde{\mathbf{x}}^k + \mathbf{B}\widetilde{\mathbf{y}}^k - \mathbf{b} \right\rangle \right)$$

$$+ \frac{\theta^2}{2\alpha} \|\mathbf{y}^k - \mathbf{y}^*\|^2 - \left( \frac{\theta^2}{2\alpha} + \frac{\mu\theta}{2} \right) \|\mathbf{y}^{k+1} - \mathbf{y}^*\|^2$$

$$- \frac{\mu\theta}{2} \|\mathbf{y}^{k+1} - \mathbf{w}^k\|^2 - \left( \frac{\theta^2}{2\alpha} - \frac{L\theta^2}{2} \right) \|\mathbf{y}^{k+1} - \mathbf{y}^k\|^2$$

$$- \theta \left\langle \boldsymbol{\lambda}^{k+1} - \boldsymbol{\lambda}^*, \mathbf{A}\mathbf{x}^{k+1} + \mathbf{B}\mathbf{y}^{k+1} - \mathbf{b} \right\rangle$$

$$+ \beta\theta^2 \left\langle \mathbf{B}\mathbf{y}^{k+1} - \mathbf{B}\mathbf{y}^k, \mathbf{A}\mathbf{x}^{k+1} + \mathbf{B}\mathbf{y}^{k+1} - \mathbf{b} \right\rangle$$

$$\overset{a}{=} (1 - \theta) \left( f(\widetilde{\mathbf{x}}^k) - f(\mathbf{x}^*) + g(\widetilde{\mathbf{y}}^k) - g(\mathbf{y}^*) + \left\langle \boldsymbol{\lambda}^*, \mathbf{A}\widetilde{\mathbf{x}}^k + \mathbf{B}\widetilde{\mathbf{y}}^k - \mathbf{b} \right\rangle \right)$$

$$+ \frac{\theta^2}{2\alpha} \|\mathbf{y}^k - \mathbf{y}^*\|^2 - \left( \frac{\theta^2}{2\alpha} + \frac{\mu\theta}{2} \right) \|\mathbf{y}^{k+1} - \mathbf{y}^*\|^2$$

$$- \frac{\mu\theta}{2} \|\mathbf{y}^{k+1} - \mathbf{w}^k\|^2 - \left( \frac{\theta^2}{2\alpha} - \frac{L\theta^2}{2} \right) \|\mathbf{y}^{k+1} - \mathbf{y}^k\|^2$$

$$- \frac{1}{\beta} \left\langle \boldsymbol{\lambda}^{k+1} - \boldsymbol{\lambda}^*, \boldsymbol{\lambda}^{k+1} - \boldsymbol{\lambda}^k \right\rangle + \beta\theta^2 \left\langle \mathbf{B}\mathbf{y}^{k+1} - \mathbf{B}\mathbf{y}^k, \mathbf{A}\mathbf{x}^{k+1} + \mathbf{B}\mathbf{y}^{k+1} - \mathbf{b} \right\rangle$$

$$= (1 - \theta) \left( f(\widetilde{\mathbf{x}}^k) - f(\mathbf{x}^*) + g(\widetilde{\mathbf{y}}^k) - g(\mathbf{y}^*) + \left\langle \boldsymbol{\lambda}^*, \mathbf{A}\widetilde{\mathbf{x}}^k + \mathbf{B}\widetilde{\mathbf{y}}^k - \mathbf{b} \right\rangle \right)$$

$$+ \frac{\theta^2}{2\alpha} \|\mathbf{y}^k - \mathbf{y}^*\|^2 - \left( \frac{\theta^2}{2\alpha} + \frac{\mu\theta}{2} \right) \|\mathbf{y}^{k+1} - \mathbf{y}^*\|^2$$

$$- \frac{\mu\theta}{2} \|\mathbf{y}^{k+1} - \mathbf{w}^k\|^2 - \left( \frac{\theta^2}{2\alpha} - \frac{L\theta^2}{2} \right) \|\mathbf{y}^{k+1} - \mathbf{y}^k\|^2$$

$$+ \frac{1}{2\beta} \left( \|\boldsymbol{\lambda}^k - \boldsymbol{\lambda}^*\|^2 - \|\boldsymbol{\lambda}^{k+1} - \boldsymbol{\lambda}^*\|^2 - \|\boldsymbol{\lambda}^{k+1} - \boldsymbol{\lambda}^k\|^2 \right)$$

$$+ \frac{\beta\theta^2}{2} \left( \|\mathbf{A}\mathbf{x}^{k+1} + \mathbf{B}\mathbf{y}^{k+1} - \mathbf{b}\|^2 + \|\mathbf{B}(\mathbf{y}^{k+1} - \mathbf{y}^k)\|^2 \right.$$

$$\left. - \|\mathbf{A}\mathbf{x}^{k+1} + \mathbf{B}\mathbf{y}^k - \mathbf{b}\|^2 \right)$$

$$\overset{b}{\leq} (1-\theta) \left( f(\widetilde{\mathbf{x}}^k) - f(\mathbf{x}^*) + g(\widetilde{\mathbf{y}}^k) - g(\mathbf{y}^*) + \left\langle \boldsymbol{\lambda}^*, \mathbf{A}\widetilde{\mathbf{x}}^k + \mathbf{B}\widetilde{\mathbf{y}}^k - \mathbf{b} \right\rangle \right)$$

$$+ \frac{\theta^2}{2\alpha} \|\mathbf{y}^k - \mathbf{y}^*\|^2 - \left( \frac{\theta^2}{2\alpha} + \frac{\mu\theta}{2} \right) \|\mathbf{y}^{k+1} - \mathbf{y}^*\|^2$$

$$- \frac{\mu\theta}{2} \|\mathbf{y}^{k+1} - \mathbf{w}^k\|^2 - \left( \frac{\theta^2}{2\alpha} - \frac{L\theta^2}{2} - \frac{\beta\theta^2\|\mathbf{B}\|_2^2}{2} \right) \|\mathbf{y}^{k+1} - \mathbf{y}^k\|^2$$

$$+ \frac{1}{2\beta} \left( \|\boldsymbol{\lambda}^k - \boldsymbol{\lambda}^*\|^2 - \|\boldsymbol{\lambda}^{k+1} - \boldsymbol{\lambda}^*\|^2 \right),$$

where we use (3.62f) in $\overset{a}{=}$ and $\overset{b}{\leq}$. So we have the conclusion. $\qquad\square$

Denote

$$\ell_k = (1-\theta) \left( f(\widetilde{\mathbf{x}}^k) - f(\mathbf{x}^*) + g(\widetilde{\mathbf{y}}^k) - g(\mathbf{y}^*) + \left\langle \boldsymbol{\lambda}^*, \mathbf{A}\widetilde{\mathbf{x}}^k + \mathbf{B}\widetilde{\mathbf{y}}^k - \mathbf{b} \right\rangle \right)$$

$$+ \frac{\theta^2}{2\alpha} \|\mathbf{y}^k - \mathbf{y}^*\|^2 + \frac{1}{2\beta} \|\boldsymbol{\lambda}^k - \boldsymbol{\lambda}^*\|^2.$$

**Theorem 3.12** *Suppose that $f(\mathbf{x})$ is convex and $g(\mathbf{y})$ is $\mu_g$-strongly convex and $L_g$-smooth. Assume that*

$$\frac{\|\mathbf{B}\|_2^2}{\sigma^2} \leq \frac{L_g}{\mu_g} \quad \text{and} \quad \|\mathbf{B}^T\boldsymbol{\lambda}\| \geq \sigma \|\boldsymbol{\lambda}\|, \forall \boldsymbol{\lambda}, \text{ where } \sigma > 0.$$

*Let*

$$\alpha = \frac{1}{4L_g}, \quad \beta = \frac{L_g}{\|\mathbf{B}\|_2^2}, \quad \text{and} \quad \theta = \sqrt{\frac{\mu_g\|\mathbf{B}\|_2^2}{L_g\sigma^2}} \leq 1.$$

*Then for Algorithm 3.6, we have*

$$\ell_{k+1} \leq \left( 1 - \sqrt{\frac{\mu_g\sigma^2}{L_g\|\mathbf{B}\|_2^2}} \right) \ell_k.$$

***Proof*** Again, we write $\mu_g$ and $L_g$ as $\mu$ and $L$, respectively, for notation simplicity. For the algorithm, we have

$$f(\widetilde{\mathbf{x}}^{k+1}) - f(\mathbf{x}^*) + g(\widetilde{\mathbf{y}}^{k+1}) - g(\mathbf{y}^*) + \left\langle \boldsymbol{\lambda}^*, \mathbf{A}\widetilde{\mathbf{x}}^{k+1} + \mathbf{B}\widetilde{\mathbf{y}}^{k+1} - \mathbf{b} \right\rangle$$

$$\stackrel{a}{=} f(\widetilde{\mathbf{x}}^{k+1}) - f(\mathbf{x}^*) - \left\langle -\mathbf{A}^T\boldsymbol{\lambda}^*, \widetilde{\mathbf{x}}^{k+1} - \mathbf{x}^* \right\rangle$$

$$+ g(\widetilde{\mathbf{y}}^{k+1}) - g(\mathbf{y}^*) - \left\langle -\mathbf{B}^T\boldsymbol{\lambda}^*, \widetilde{\mathbf{y}}^{k+1} - \mathbf{y}^* \right\rangle$$

$$\stackrel{b}{\geq} \frac{1}{2L} \| \nabla g(\widetilde{\mathbf{y}}^{k+1}) - \nabla g(\mathbf{y}^*) \|^2$$

$$\stackrel{c}{=} \frac{1}{2L} \| \nabla g(\widetilde{\mathbf{y}}^{k+1}) + \mathbf{B}^T\boldsymbol{\lambda}^* \|^2$$

$$\stackrel{d}{=} \frac{1}{2L} \left\| \mu(\mathbf{y}^{k+1} - \mathbf{w}^k) + \frac{\theta}{\alpha}(\mathbf{y}^{k+1} - \mathbf{y}^k) + \mathbf{B}^T(\boldsymbol{\lambda}^{k+1} - \boldsymbol{\lambda}^*) \right.$$

$$\left. - \beta\theta\mathbf{B}^T(\mathbf{B}\mathbf{y}^{k+1} - \mathbf{B}\mathbf{y}^k) + \nabla g(\mathbf{w}^k) - \nabla g(\widetilde{\mathbf{y}}^{k+1}) \right\|^2$$

$$\stackrel{e}{\geq} \frac{1-\nu}{2L} \| \mathbf{B}^T(\boldsymbol{\lambda}^{k+1} - \boldsymbol{\lambda}^*) \|^2 - \frac{1}{2L} \left( \frac{1}{\nu} - 1 \right) \left\| \mu(\mathbf{y}^{k+1} - \mathbf{w}^k) \right.$$

$$\left. + \frac{\theta}{\alpha}(\mathbf{y}^{k+1} - \mathbf{y}^k) - \beta\theta\mathbf{B}^T(\mathbf{B}\mathbf{y}^{k+1} - \mathbf{B}\mathbf{y}^k) + \nabla g(\mathbf{w}^k) - \nabla g(\widetilde{\mathbf{y}}^{k+1}) \right\|^2$$

$$\stackrel{f}{\geq} \frac{(1-\nu)\sigma^2}{2L} \| \boldsymbol{\lambda}^{k+1} - \boldsymbol{\lambda}^* \|^2 - \frac{1}{2L} \left( \frac{1}{\nu} - 1 \right) \left( 4\mu^2 \| \mathbf{y}^{k+1} - \mathbf{w}^k \|^2 \right.$$

$$+ \frac{4\theta^2}{\alpha^2} \| \mathbf{y}^{k+1} - \mathbf{y}^k \|^2 + 4\beta^2\theta^2 \| \mathbf{B} \|_2^4 \| \mathbf{y}^{k+1} - \mathbf{y}^k \|^2$$

$$\left. + 4L^2\theta^2 \| \mathbf{y}^{k+1} - \mathbf{y}^k \|^2 \right), \tag{3.68}$$

where $\stackrel{a}{=}$ uses (3.6), $\stackrel{b}{\geq}$ uses the convexity of $f$ and (3.4) and the convexity and $L$-smoothness of $g$, (3.5), and (A.5), $\stackrel{c}{=}$ uses (3.5), $\stackrel{d}{=}$ uses (3.66), $\stackrel{e}{\geq}$ uses $\|\mathbf{u} + \mathbf{v}\|^2 \geq (1 - \nu)\|\mathbf{u}\|^2 - (\frac{1}{\nu} - 1)\|\mathbf{v}\|^2$, and $\stackrel{f}{\geq}$ uses (3.62a) and (3.62e).

Multiplying both sides of (3.68) by $\frac{\theta}{2}$ and plugging it into (3.63), we have

$$\left(1 - \frac{\theta}{2}\right) \left(f(\widetilde{\mathbf{x}}^{k+1}) - f(\mathbf{x}^*) + g(\widetilde{\mathbf{y}}^{k+1}) - g(\mathbf{y}^*) \right.$$

$$\left. + \left\langle \boldsymbol{\lambda}^*, \mathbf{A}\widetilde{\mathbf{x}}^{k+1} + \mathbf{B}\widetilde{\mathbf{y}}^{k+1} - \mathbf{b} \right\rangle \right)$$

$$\leq (1 - \theta) \left( f(\widetilde{\mathbf{x}}^k) - f(\mathbf{x}^*) + g(\widetilde{\mathbf{y}}^k) - g(\mathbf{y}^*) + \left\langle \boldsymbol{\lambda}^*, \mathbf{A}\widetilde{\mathbf{x}}^k + \mathbf{B}\widetilde{\mathbf{y}}^k - \mathbf{b} \right\rangle \right)$$

$$+ \frac{\theta^2}{2\alpha} \|\mathbf{y}^k - \mathbf{y}^*\|^2 - \left(\frac{\theta^2}{2\alpha} + \frac{\mu\theta}{2}\right) \|\mathbf{y}^{k+1} - \mathbf{y}^*\|^2$$

$$+ \frac{1}{2\beta} \|\boldsymbol{\lambda}^k - \boldsymbol{\lambda}^*\|^2 - \left[\frac{1}{2\beta} + \frac{(1-\nu)\sigma^2\theta}{4L}\right] \|\boldsymbol{\lambda}^{k+1} - \boldsymbol{\lambda}^*\|^2$$

$$- \left[\frac{\mu\theta}{2} - \left(\frac{1}{\nu} - 1\right) \frac{\theta\mu^2}{L}\right] \|\mathbf{y}^{k+1} - \mathbf{w}^k\|^2$$

$$- \theta^2 \left[\frac{1}{2\alpha} - \frac{L}{2} - \frac{\beta\|\mathbf{B}\|_2^2}{2} - \frac{\theta}{L}\left(\frac{1}{\nu} - 1\right)\left(\frac{1}{\alpha^2} + \beta^2\|\mathbf{B}\|_2^4 + L^2\right)\right]$$

$$\times \|\mathbf{y}^{k+1} - \mathbf{y}^k\|^2$$

$$\leq (1-\theta)\left(f(\widetilde{\mathbf{x}}^k) - f(\mathbf{x}^*) + g(\widetilde{\mathbf{y}}^k) - g(\mathbf{y}^*) + \left\langle\boldsymbol{\lambda}^*, \mathbf{A}\widetilde{\mathbf{x}}^k + \mathbf{B}\widetilde{\mathbf{y}}^k - \mathbf{b}\right\rangle\right)$$

$$+ \frac{\theta^2}{2\alpha} \|\mathbf{y}^k - \mathbf{y}^*\|^2 - \left(\frac{\theta^2}{2\alpha} + \frac{\mu\theta}{2}\right) \|\mathbf{y}^{k+1} - \mathbf{y}^*\|^2$$

$$+ \frac{1}{2\beta} \|\boldsymbol{\lambda}^k - \boldsymbol{\lambda}^*\|^2 - \left[\frac{1}{2\beta} + \frac{(1-\nu)\sigma^2\theta}{4L}\right] \|\boldsymbol{\lambda}^{k+1} - \boldsymbol{\lambda}^*\|^2,$$

where we let

$$\nu = \frac{18}{19}, \quad \beta = \frac{L}{\|\mathbf{B}\|_2^2}, \quad \text{and} \quad \alpha = \frac{1}{4L},$$

such that

$$\frac{\mu\theta}{2} - \left(\frac{1}{\nu} - 1\right)\frac{\theta\mu^2}{L} \geq 0, \text{ and}$$

$$\frac{1}{2\alpha} - \frac{L}{2} - \frac{\beta\|\mathbf{B}\|_2^2}{2} - \frac{\theta}{L}\left(\frac{1}{\nu} - 1\right)\left(\frac{1}{\alpha^2} + \beta^2\|\mathbf{B}\|_2^4 + L^2\right) \geq 0.$$

Thus, we have

$$\min\left\{\frac{1-\theta/2}{1-\theta}, 1 + \frac{\alpha\mu}{\theta}, 1 + \frac{(1-\nu)\sigma^2\beta\theta}{2L}\right\}$$

$$\times \left[(1-\theta)\left(f(\widetilde{\mathbf{x}}^{k+1}) - f(\mathbf{x}^*) + g(\widetilde{\mathbf{y}}^{k+1}) - g(\mathbf{y}^*)\right.\right.$$

$$\left.\left. + \left\langle\boldsymbol{\lambda}^*, \mathbf{A}\widetilde{\mathbf{x}}^{k+1} + \mathbf{B}\widetilde{\mathbf{y}}^{k+1} - \mathbf{b}\right\rangle\right) + \frac{\theta^2}{2\alpha}\|\mathbf{y}^{k+1} - \mathbf{y}^*\|^2 + \frac{1}{2\beta}\|\boldsymbol{\lambda}^{k+1} - \boldsymbol{\lambda}^*\|^2\right]$$

$$\leq (1 - \theta) \left( f(\widetilde{\mathbf{x}}^k) - f(\mathbf{x}^*) + g(\widetilde{\mathbf{y}}^k) - g(\mathbf{y}^*) + \left\langle \boldsymbol{\lambda}^*, \mathbf{A}\widetilde{\mathbf{x}}^k + \mathbf{B}\widetilde{\mathbf{y}}^k - \mathbf{b} \right\rangle \right)$$

$$+ \frac{\theta^2}{2\alpha} \|\mathbf{y}^k - \mathbf{y}^*\|^2 + \frac{1}{2\beta} \|\boldsymbol{\lambda}^k - \boldsymbol{\lambda}^*\|^2.$$

Thus, we have

$$\ell_{k+1} \leq \max \left\{ \frac{1 - \theta}{1 - \theta/2}, \frac{1}{1 + \frac{\alpha\mu}{\theta}}, \frac{1}{1 + \frac{(1-v)\sigma^2\beta\theta}{2L}} \right\} \ell_k$$

$$\overset{a}{\leq} \max \left\{ 1 - \frac{\theta}{2}, 1 - \frac{\alpha\mu}{2\theta}, 1 - \frac{(1-v)\sigma^2\beta\theta}{4L} \right\} \ell_k$$

$$= O \left( \max \left\{ 1 - \theta, 1 - \frac{\mu}{L\theta}, 1 - \frac{\sigma^2\theta}{\|\mathbf{B}\|_2^2} \right\} \right) \ell_k$$

$$= O \left( \max \left\{ 1 - \frac{\mu}{L\theta}, 1 - \frac{\sigma^2\theta}{\|\mathbf{B}\|_2^2} \right\} \right) \ell_k,$$

where we let

$$\frac{\alpha\mu}{\theta} \leq 1 \quad \text{and} \quad \frac{(1-v)\sigma^2\beta\theta}{2L} \leq 1$$

such that $\overset{a}{\leq}$ holds. This can be fulfilled by the following choice of parameters. Letting $\theta = \sqrt{\frac{\mu\|\mathbf{B}\|_2^2}{L\sigma^2}}$, we have $\frac{\mu}{L\theta} = \sqrt{\frac{\mu\sigma^2}{L\|\mathbf{B}\|_2^2}}$. From the assumption of $\frac{\|\mathbf{B}\|_2^2}{\sigma^2} \leq \frac{L}{\mu}$, we have $\theta \leq 1$, then

$$\frac{\alpha\mu}{\theta} \leq 1/4 \quad \text{and} \quad \frac{(1-v)\sigma^2\beta\theta}{2L} \leq (1-v)/2 < 1.$$

The proof is complete.                                                                 □

## 3.4  Special Case: Linearized Augmented Lagrangian Method and Its Acceleration

In this section, we consider the following simpler problem, which will be used in Sect. 6 for distributed optimization,

$$\min_{\mathbf{y}} g(\mathbf{y}), \quad s.t. \quad \mathbf{B}\mathbf{y} = \mathbf{b},$$

which is a special case of Problem (2.13) with $f(\mathbf{x}) = 0$ and $\mathbf{A} = \mathbf{0}$. Consider the following Bregman augmented Lagrangian method (ALM), which is a special case of the Bregman ADMM (Algorithm 3.1) and is presented in Algorithm 3.7,

$$\mathbf{y}^{k+1} = \underset{\mathbf{y}}{\operatorname{argmin}} \left( g(\mathbf{y}) + \left\langle \boldsymbol{\lambda}^k, \mathbf{By} - \mathbf{b} \right\rangle + \frac{\beta}{2} \|\mathbf{By} - \mathbf{b}\|^2 + D_\psi(\mathbf{y}, \mathbf{y}^k) \right), \quad (3.69a)$$

$$\boldsymbol{\lambda}^{k+1} = \boldsymbol{\lambda}^k + \beta(\mathbf{By}^{k+1} - \mathbf{b}). \quad (3.69b)$$

---

**Algorithm 3.7** Bregman ALM

Initialize $\mathbf{y}^0$ and $\boldsymbol{\lambda}^0$.
**for** $k = 0, 1, 2, 3, \cdots$ **do**
    Update $\mathbf{y}^{k+1}$ and $\boldsymbol{\lambda}^{k+1}$ by (3.69a) and (3.69b), respectively.
**end for**

---

As a special case of Theorems 3.6 and 3.8, we have the following convergence rates for non-strongly convex and strongly convex problems, respectively.

**Theorem 3.13**  *Suppose that $g(\mathbf{y})$ is convex. Then for Algorithm 3.7, we have*

$$|g(\hat{\mathbf{y}}^{K+1}) - g(\mathbf{y}^*)| \leq \frac{D}{2(K+1)} + \frac{2\sqrt{D}\|\boldsymbol{\lambda}^*\|}{\sqrt{\beta}(K+1)},$$

$$\|\mathbf{B}\hat{\mathbf{y}}^{K+1} - \mathbf{b}\| \leq \frac{2\sqrt{D}}{\sqrt{\beta}(K+1)},$$

*where*

$$\hat{\mathbf{y}}^{K+1} = \frac{1}{K+1} \sum_{k=1}^{K+1} \mathbf{y}^k \text{ and}$$

$$D = \frac{1}{\beta}\|\boldsymbol{\lambda}^0 - \boldsymbol{\lambda}^*\|^2 + \beta\|\mathbf{By}^0 - \mathbf{By}^*\|^2 + 2D_\psi(\mathbf{y}^*, \mathbf{y}^0).$$

**Theorem 3.14** *Assume that $g(\mathbf{y})$ is $\mu_g$-strongly convex and $L_g$-smooth, $\psi(\mathbf{y})$ is convex and $L_\psi$-smooth. Assume that $\|\mathbf{B}^T\boldsymbol{\lambda}\| \geq \sigma\|\boldsymbol{\lambda}\|, \forall\boldsymbol{\lambda}$, where $\sigma > 0$. Then for Algorithm 3.7, we have*

$$\frac{1}{2\beta}\|\boldsymbol{\lambda}^{k+1} - \boldsymbol{\lambda}^*\|^2 + \frac{\beta}{2}\|\mathbf{B}\mathbf{y}^{k+1} - \mathbf{B}\mathbf{y}^*\|^2 + D_\psi(\mathbf{y}^*, \mathbf{y}^{k+1})$$

$$\leq \left(1 + \frac{1}{3}\min\left\{\frac{\beta\sigma^2}{L_g + L_\psi}, \frac{\mu_g}{\beta\|\mathbf{B}\|_2^2}, \frac{\mu_g}{L_\psi}\right\}\right)^{-1}$$

$$\times \left(\frac{1}{2\beta}\|\boldsymbol{\lambda}^k - \boldsymbol{\lambda}^*\|^2 + \frac{\beta}{2}\|\mathbf{B}\mathbf{y}^k - \mathbf{B}\mathbf{y}^*\|^2 + D_\psi(\mathbf{y}^*, \mathbf{y}^k)\right).$$

*Remark 3.4* Similar to Remark 3.2, when

$$\psi(\mathbf{y}) = \frac{L_g + \beta\|\mathbf{B}\|_2^2}{2}\|\mathbf{y}\|^2 - g(\mathbf{y}) - \frac{\beta}{2}\|\mathbf{B}\mathbf{y}\|^2,$$

letting $\beta = \frac{\sqrt{\mu_g L_g}}{\sigma\|\mathbf{B}\|_2}$, the complexity is

$$O\left(\left(\frac{\|\mathbf{B}\|_2^2}{\sigma^2} + \frac{L_g}{\mu_g}\right)\log\frac{1}{\epsilon}\right).$$

When

$$\psi(\mathbf{y}) = \frac{\beta\|\mathbf{B}\|_2^2}{2}\|\mathbf{y}\|^2 - \frac{\beta}{2}\|\mathbf{B}\mathbf{y}\|^2,$$

the complexity becomes

$$O\left(\left(\frac{\|\mathbf{B}\|_2}{\sigma}\sqrt{\frac{L_g}{\mu_g}} + \frac{\|\mathbf{B}\|_2^2}{\sigma^2}\right)\log\frac{1}{\epsilon}\right).$$

Consider the following accelerated linearized augmented Lagrangian method, which is presented in Algorithm 3.8,

$$\mathbf{w}^k = \theta\mathbf{y}^k + (1 - \theta)\tilde{\mathbf{y}}^k, \tag{3.70a}$$

$$\mathbf{y}^{k+1} = \underset{\mathbf{y}}{\operatorname{argmin}}\left(\left\langle\nabla g(\mathbf{w}^k), \mathbf{y}\right\rangle + \left\langle\boldsymbol{\lambda}^k, \mathbf{B}\mathbf{y}\right\rangle + \beta\theta\left\langle\mathbf{B}^T(\mathbf{B}\mathbf{y}^k - \mathbf{b}), \mathbf{y}\right\rangle\right.$$

$$\left. + \frac{1}{2}\left(\frac{\theta}{\alpha} + \mu_g\right)\left\|\mathbf{y} - \frac{1}{\frac{\theta}{\alpha} + \mu_g}\left(\frac{\theta}{\alpha}\mathbf{y}^k + \mu_g\mathbf{w}^k\right)\right\|^2\right)$$

$$= \frac{1}{\frac{\theta}{\alpha} + \mu_g} \left\{ \mu_g \mathbf{w}^k + \frac{\theta}{\alpha} \mathbf{y}^k \right.$$

$$\left. - \left[ \nabla g(\mathbf{w}^k) + \mathbf{B}^T \boldsymbol{\lambda}^k + \beta\theta \mathbf{B}^T (\mathbf{B}\mathbf{y}^k - \mathbf{b}) \right] \right\}, \tag{3.70b}$$

$$\widetilde{\mathbf{y}}^{k+1} = \theta \mathbf{y}^{k+1} + (1-\theta) \widetilde{\mathbf{y}}^k, \tag{3.70c}$$

$$\boldsymbol{\lambda}^{k+1} = \boldsymbol{\lambda}^k + \beta\theta (\mathbf{B}\mathbf{y}^{k+1} - \mathbf{b}). \tag{3.70d}$$

---

**Algorithm 3.8** Accelerated linearized ALM

---

Initialize $\mathbf{y}^0 = \widetilde{\mathbf{y}}^0$ and $\boldsymbol{\lambda}^0$.
**for** $k = 0, 1, 2, 3, \cdots$ **do**
   Update $\mathbf{w}^k, \mathbf{y}^{k+1}, \widetilde{\mathbf{y}}^{k+1}$, and $\boldsymbol{\lambda}^{k+1}$ by (3.70a)–(3.70d), respectively.
**end for**

---

As a special case of Theorem 3.12, we have the following convergence rate theorem.

**Theorem 3.15** *Suppose that $g(\mathbf{y})$ is $\mu_g$-strongly convex and $L_g$-smooth. Assume that*

$$\frac{\|\mathbf{B}\|_2^2}{\sigma^2} \leq \frac{L_g}{\mu_g} \quad \text{and} \quad \|\mathbf{B}^T \boldsymbol{\lambda}\| \geq \sigma \|\boldsymbol{\lambda}\|, \forall \boldsymbol{\lambda}, \text{ where } \sigma > 0.$$

*Let*

$$\alpha = \frac{1}{4L_g}, \quad \beta = \frac{L_g}{\|\mathbf{B}\|_2^2}, \quad \text{and} \quad \theta = \sqrt{\frac{\mu_g \|\mathbf{B}\|_2^2}{L_g \sigma^2}}.$$

*Then for Algorithm 3.8, we have*

$$\ell_{k+1} \leq \left( 1 - \sqrt{\frac{\mu_g \sigma^2}{L_g \|\mathbf{B}\|_2^2}} \right) \ell_k,$$

*where we denote*

$$\ell_k = (1-\theta) \left( g(\widetilde{\mathbf{y}}^k) - g(\mathbf{y}^*) + \left\langle \boldsymbol{\lambda}^*, \mathbf{B}\widetilde{\mathbf{y}}^k - \mathbf{b} \right\rangle \right)$$

$$+ \frac{\theta^2}{2\alpha} \|\mathbf{y}^k - \mathbf{y}^*\|^2 + \frac{1}{2\beta} \|\boldsymbol{\lambda}^k - \boldsymbol{\lambda}^*\|^2.$$

## 3.5   Multi-block ADMM

In this section, we extend ADMM to solve the following problem with multi-blocks:

$$\min_{\mathbf{x}_i} \sum_{i=1}^{m} f_i(\mathbf{x}_i), \quad s.t. \quad \sum_{i=1}^{m} \mathbf{A}_i \mathbf{x}_i = \mathbf{b}, \tag{3.71}$$

while the previous sections only introduce the case of two blocks. Denote

$$f(\mathbf{x}) = \sum_{i=1}^{m} f_i(\mathbf{x}_i), \quad \mathbf{A} = [\mathbf{A}_1, \cdots, \mathbf{A}_m], \quad \text{and} \quad \mathbf{x} = (\mathbf{x}_1^T, \cdots, \mathbf{x}_m^T)^T.$$

The augmented Lagrangian function of Problem (3.71) is as follows:

$$L(\mathbf{x}_1, \cdots, \mathbf{x}_m, \boldsymbol{\lambda}) = \sum_{i=1}^{m} f_i(\mathbf{x}_i) + \left\langle \boldsymbol{\lambda}, \sum_{i=1}^{m} \mathbf{A}_i \mathbf{x}_i - \mathbf{b} \right\rangle + \frac{\beta}{2} \left\| \sum_{i=1}^{m} \mathbf{A}_i \mathbf{x}_i - \mathbf{b} \right\|^2.$$

A straightforward extension of two-block ADMM to the multi-block case is to update the primal variables $\mathbf{x}_1, \cdots, \mathbf{x}_m$ sequentially and then update the dual variable, yielding the following scheme:

$$\widetilde{\mathbf{x}}_1^{k+1} = \underset{\mathbf{x}_1}{\operatorname{argmin}} L\left(\mathbf{x}_1, \mathbf{x}_2^k, \cdots, \mathbf{x}_m^k, \boldsymbol{\lambda}^k\right), \tag{3.72a}$$

$$\widetilde{\mathbf{x}}_2^{k+1} = \underset{\mathbf{x}_2}{\operatorname{argmin}} L\left(\widetilde{\mathbf{x}}_1^{k+1}, \mathbf{x}_2, \mathbf{x}_3^k, \cdots, \mathbf{x}_m^k, \boldsymbol{\lambda}^k\right),$$

$$\vdots$$

$$\widetilde{\mathbf{x}}_i^{k+1} = \underset{\mathbf{x}_i}{\operatorname{argmin}} L\left(\widetilde{\mathbf{x}}_1^{k+1}, \cdots, \widetilde{\mathbf{x}}_{i-1}^{k+1}, \mathbf{x}_i, \mathbf{x}_{i+1}^k \cdots, \mathbf{x}_m^k, \boldsymbol{\lambda}^k\right),$$

$$\vdots$$

$$\widetilde{\mathbf{x}}_m^{k+1} = \underset{\mathbf{x}_m}{\operatorname{argmin}} L\left(\widetilde{\mathbf{x}}_1^{k+1}, \cdots, \widetilde{\mathbf{x}}_{m-1}^{k+1}, \mathbf{x}_m, \boldsymbol{\lambda}^k\right), \tag{3.72b}$$

$$\boldsymbol{\lambda}^{k+1} = \boldsymbol{\lambda}^k + \beta \left(\sum_{i=1}^{m} \mathbf{A}_i \widetilde{\mathbf{x}}_i^{k+1} - \mathbf{b}\right), \tag{3.72c}$$

with $\mathbf{x}_i^{k+1} = \widetilde{\mathbf{x}}_i^{k+1}$. However, Chen et al. [1] gave a counter-example to show that the above direct extension of ADMM for multi-block convex minimization problems is not necessarily convergent. Thus, we should make several modifications on the original ADMM for convergence guarantees.

### 3.5.1  Gaussian Back Substitution

In this section, we introduce the Gaussian back substitution scheme [9], which first predicts $\widetilde{\mathbf{x}}_i^{k+1}$ for all $i \in [m]$ using (3.72a)-(3.72b), and then corrects $\mathbf{x}_i^{k+1}$ from $\widetilde{\mathbf{x}}_i^{k+1}$.

Denote

$$
\mathbf{M} = \begin{pmatrix} \mathbf{A}_1^T \mathbf{A}_1 & \mathbf{0} & \cdots & \mathbf{0} \\ \mathbf{A}_2^T \mathbf{A}_1 & \mathbf{A}_2^T \mathbf{A}_2 & \ddots & \mathbf{0} \\ \vdots & \vdots & \ddots & \mathbf{0} \\ \mathbf{A}_m^T \mathbf{A}_1 & \mathbf{A}_m^T \mathbf{A}_2 & \cdots & \mathbf{A}_m^T \mathbf{A}_m \end{pmatrix} \quad \text{and} \quad \mathbf{H} = \begin{pmatrix} \mathbf{A}_1^T \mathbf{A}_1 & \mathbf{0} & \cdots & \mathbf{0} \\ \mathbf{0} & \mathbf{A}_2^T \mathbf{A}_2 & \ddots & \mathbf{0} \\ \vdots & \vdots & \ddots & \mathbf{0} \\ \mathbf{0} & \mathbf{0} & \cdots & \mathbf{A}_m^T \mathbf{A}_m \end{pmatrix}.
$$

**Lemma 3.18** *Suppose that $f_i(\mathbf{x}_i)$ is convex for all $i \in [m]$. Then for the above iterations (3.72a)–(3.72c), we have*

$$
f(\widetilde{\mathbf{x}}^{k+1}) - f(\mathbf{x}^*) + \left\langle \boldsymbol{\lambda}^*, \sum_{i=1}^m \mathbf{A}_i \widetilde{\mathbf{x}}_i^{k+1} - \mathbf{b} \right\rangle
$$

$$
\leq \frac{1}{2\beta} \left( \left\| \boldsymbol{\lambda}^k - \boldsymbol{\lambda}^* \right\|^2 - \left\| \boldsymbol{\lambda}^{k+1} - \boldsymbol{\lambda}^* \right\|^2 \right) - \frac{\beta}{2} \left\| \widetilde{\mathbf{x}}^{k+1} - \mathbf{x}^k \right\|_{\mathbf{H}}^2
$$

$$
- \beta \left( \mathbf{x}^k - \mathbf{x}^* \right)^T \mathbf{M} \left( \widetilde{\mathbf{x}}^{k+1} - \mathbf{x}^k \right). \tag{3.73}
$$

***Proof*** From the optimality condition of the update of $\widetilde{\mathbf{x}}_i^{k+1}$, we have

$$
\mathbf{0} \in \partial f_i(\widetilde{\mathbf{x}}_i^{k+1}) + \mathbf{A}_i^T \boldsymbol{\lambda}^k + \beta \mathbf{A}_i^T \left( \sum_{j=1}^i \mathbf{A}_j \widetilde{\mathbf{x}}_j^{k+1} + \sum_{j=i+1}^m \mathbf{A}_j \mathbf{x}_j^k - \mathbf{b} \right)
$$

$$
= \partial f_i(\widetilde{\mathbf{x}}_i^{k+1}) + \mathbf{A}_i^T \widetilde{\boldsymbol{\lambda}}^{k+1} + \beta \mathbf{A}_i^T \left[ \sum_{j=1}^i \mathbf{A}_j \left( \widetilde{\mathbf{x}}_j^{k+1} - \mathbf{x}_j^k \right) \right],
$$

where we denote

$$
\widetilde{\boldsymbol{\lambda}}^{k+1} = \boldsymbol{\lambda}^k + \beta \left( \sum_{j=1}^m \mathbf{A}_j \mathbf{x}_j^k - \mathbf{b} \right).
$$

So we have

$$
f_i(\widetilde{\mathbf{x}}_i^{k+1}) \leq f_i(\mathbf{x}_i^*) - \left\langle \mathbf{A}_i^T \widetilde{\boldsymbol{\lambda}}^{k+1} + \beta \mathbf{A}_i^T \left[ \sum_{j=1}^i \mathbf{A}_j \left( \widetilde{\mathbf{x}}_j^{k+1} - \mathbf{x}_j^k \right) \right], \widetilde{\mathbf{x}}_i^{k+1} - \mathbf{x}_i^* \right\rangle
$$

and thus

$$
f(\widetilde{\mathbf{x}}^{k+1}) \le f(\mathbf{x}^*) - \sum_{i=1}^{m} \left\langle \mathbf{A}_i^T \widetilde{\boldsymbol{\lambda}}^{k+1} + \beta \mathbf{A}_i^T \left[ \sum_{j=1}^{i} \mathbf{A}_j \left( \widetilde{\mathbf{x}}_j^{k+1} - \mathbf{x}_j^k \right) \right], \widetilde{\mathbf{x}}_i^{k+1} - \mathbf{x}_i^* \right\rangle
$$

$$
= f(\mathbf{x}^*) - \left\langle \widetilde{\boldsymbol{\lambda}}^{k+1}, \sum_{i=1}^{m} \mathbf{A}_i \left( \widetilde{\mathbf{x}}_i^{k+1} - \mathbf{x}_i^* \right) \right\rangle
$$

$$
- \beta \sum_{i=1}^{m} \left\langle \sum_{j=1}^{i} \mathbf{A}_j \left( \widetilde{\mathbf{x}}_j^{k+1} - \mathbf{x}_j^k \right), \mathbf{A}_i \left( \widetilde{\mathbf{x}}_i^{k+1} - \mathbf{x}_i^* \right) \right\rangle,
$$

which further yields

$$
f(\widetilde{\mathbf{x}}^{k+1}) - f(\mathbf{x}^*) + \left\langle \boldsymbol{\lambda}^*, \sum_{i=1}^{m} \mathbf{A}_i \widetilde{\mathbf{x}}_i^{k+1} - \mathbf{b} \right\rangle
$$

$$
\le - \left\langle \widetilde{\boldsymbol{\lambda}}^{k+1} - \boldsymbol{\lambda}^*, \sum_{i=1}^{m} \mathbf{A}_i \widetilde{\mathbf{x}}_i^{k+1} - \mathbf{b} \right\rangle
$$

$$
- \beta \sum_{i=1}^{m} \left\langle \sum_{j=1}^{i} \mathbf{A}_j \left( \widetilde{\mathbf{x}}_j^{k+1} - \mathbf{x}_j^k \right), \mathbf{A}_i \left( \widetilde{\mathbf{x}}_i^{k+1} - \mathbf{x}_i^* \right) \right\rangle
$$

$$
= - \frac{1}{\beta} \left\langle \widetilde{\boldsymbol{\lambda}}^{k+1} - \boldsymbol{\lambda}^*, \boldsymbol{\lambda}^{k+1} - \boldsymbol{\lambda}^k \right\rangle - \beta \left( \widetilde{\mathbf{x}}^{k+1} - \mathbf{x}^* \right)^T \mathbf{M} \left( \widetilde{\mathbf{x}}^{k+1} - \mathbf{x}^k \right)
$$

$$
\overset{a}{=} \frac{1}{2\beta} \left( \left\| \boldsymbol{\lambda}^k - \boldsymbol{\lambda}^* \right\|^2 - \left\| \boldsymbol{\lambda}^{k+1} - \boldsymbol{\lambda}^* \right\|^2 - \left\| \widetilde{\boldsymbol{\lambda}}^{k+1} - \boldsymbol{\lambda}^k \right\|^2 + \left\| \widetilde{\boldsymbol{\lambda}}^{k+1} - \boldsymbol{\lambda}^{k+1} \right\|^2 \right)
$$

$$
- \beta \left( \widetilde{\mathbf{x}}^{k+1} - \mathbf{x}^* \right)^T \mathbf{M} \left( \widetilde{\mathbf{x}}^{k+1} - \mathbf{x}^k \right),
$$

where $\overset{a}{=}$ uses (A.3).

On the other hand,

$$
- \beta \left( \widetilde{\mathbf{x}}^{k+1} - \mathbf{x}^* \right)^T \mathbf{M} \left( \widetilde{\mathbf{x}}^{k+1} - \mathbf{x}^k \right) + \frac{1}{2\beta} \left\| \widetilde{\boldsymbol{\lambda}}^{k+1} - \boldsymbol{\lambda}^{k+1} \right\|^2
$$

$$
= - \beta \left( \widetilde{\mathbf{x}}^{k+1} - \mathbf{x}^k \right)^T \mathbf{M} \left( \widetilde{\mathbf{x}}^{k+1} - \mathbf{x}^k \right) - \beta \left( \mathbf{x}^k - \mathbf{x}^* \right)^T \mathbf{M} \left( \widetilde{\mathbf{x}}^{k+1} - \mathbf{x}^k \right)
$$

$$
+ \frac{1}{2\beta} \left\| \widetilde{\boldsymbol{\lambda}}^{k+1} - \boldsymbol{\lambda}^{k+1} \right\|^2
$$

$$
\overset{a}{=} - \frac{\beta}{2} \left\| \widetilde{\mathbf{x}}^{k+1} - \mathbf{x}^k \right\|_{\mathbf{H}}^2 - \beta \left( \mathbf{x}^k - \mathbf{x}^* \right)^T \mathbf{M} \left( \widetilde{\mathbf{x}}^{k+1} - \mathbf{x}^k \right),
$$

where in $\overset{a}{=}$ we use

$$
\left(\tilde{\mathbf{x}}^{k+1} - \mathbf{x}^k\right)^T \mathbf{M} \left(\tilde{\mathbf{x}}^{k+1} - \mathbf{x}^k\right)
$$

$$
= \left(\tilde{\mathbf{x}}^{k+1} - \mathbf{x}^k\right)^T \mathbf{M}^T \left(\tilde{\mathbf{x}}^{k+1} - \mathbf{x}^k\right)
$$

$$
= \frac{1}{2} \left(\tilde{\mathbf{x}}^{k+1} - \mathbf{x}^k\right)^T \left(\mathbf{M} + \mathbf{M}^T\right) \left(\tilde{\mathbf{x}}^{k+1} - \mathbf{x}^k\right)
$$

$$
= \frac{1}{2} \left\| \sum_{j=1}^m \mathbf{A}_j \left(\tilde{\mathbf{x}}_j^{k+1} - \mathbf{x}_j^k\right) \right\|^2 + \frac{1}{2} \left(\tilde{\mathbf{x}}^{k+1} - \mathbf{x}^k\right)^T \mathbf{H} \left(\tilde{\mathbf{x}}^{k+1} - \mathbf{x}^k\right)
$$

$$
= \frac{1}{2\beta^2} \left\| \tilde{\lambda}^{k+1} - \lambda^{k+1} \right\|^2 + \frac{1}{2} \left(\tilde{\mathbf{x}}^{k+1} - \mathbf{x}^k\right)^T \mathbf{H} \left(\tilde{\mathbf{x}}^{k+1} - \mathbf{x}^k\right).
$$

Thus we have the conclusion. □

We need to make the right hand side of (3.73) in a form of $\Phi^k - \Phi^{k+1}$ so that a recursion can be established. To this end, we need to define $\mathbf{x}^{k+1}$ and find $\mathbf{G} \succeq \mathbf{0}$ so that

$$
-\frac{\beta}{2} \left\| \tilde{\mathbf{x}}^{k+1} - \mathbf{x}^k \right\|_{\mathbf{H}}^2 - \beta \left(\mathbf{x}^k - \mathbf{x}^*\right)^T \mathbf{M} \left(\tilde{\mathbf{x}}^{k+1} - \mathbf{x}^k\right)
$$

$$
= \frac{\beta}{2} \left( \left\| \mathbf{x}^k - \mathbf{x}^* \right\|_{\mathbf{G}}^2 - \left\| \mathbf{x}^{k+1} - \mathbf{x}^* \right\|_{\mathbf{G}}^2 \right) \tag{3.74}
$$

holds. Note that

$$
\left\| \mathbf{x}^k - \mathbf{x}^* \right\|_{\mathbf{G}}^2 - \left\| \mathbf{x}^k - \mathbf{x}^* + \mathbf{D}(\tilde{\mathbf{x}}^{k+1} - \mathbf{x}^k) \right\|_{\mathbf{G}}^2
$$

$$
= - \left(\tilde{\mathbf{x}}^{k+1} - \mathbf{x}^k\right)^T \mathbf{D}^T \mathbf{G} \mathbf{D} \left(\tilde{\mathbf{x}}^{k+1} - \mathbf{x}^k\right) - 2 \left(\mathbf{x}^k - \mathbf{x}^*\right)^T \mathbf{G} \mathbf{D} \left(\tilde{\mathbf{x}}^{k+1} - \mathbf{x}^k\right).
$$

By comparing the above identity with (3.74) we know that we should define

$$
\mathbf{x}^{k+1} = \mathbf{x}^k + \mathbf{D} \left(\tilde{\mathbf{x}}^{k+1} - \mathbf{x}^k\right) \tag{3.75}
$$

and find $\mathbf{G} \succeq \mathbf{0}$ and $\mathbf{D}$ such that

$$
\mathbf{D}^T \mathbf{G} \mathbf{D} = \mathbf{H} \quad \text{and} \quad \mathbf{G} \mathbf{D} = \mathbf{M}
$$

hold. Assume that $\mathbf{A}_i$'s are all of full column rank. Then $\mathbf{M}$ and $\mathbf{H}$ are invertible. Let

$$
\mathbf{D} = \mathbf{M}^{-T} \mathbf{H} \quad \text{and} \quad \mathbf{G} = \mathbf{M} \mathbf{H}^{-1} \mathbf{M}^T.
$$

Then we can check that the requirements on $\mathbf{G}$ and $\mathbf{D}$ are all fulfilled.

Note that with the above choice of $\mathbf{D}$, (3.75) can be rewritten as

$$\mathbf{x}^{k+1} = \mathbf{x}^k + \mathbf{M}^{-T}\mathbf{H}\left(\widetilde{\mathbf{x}}^{k+1} - \mathbf{x}^k\right), \tag{3.76}$$

which can be computed by the famous Gaussian back substitution efficiently due to the lower-block-triangular structure of $\mathbf{M}$. So the above correction scheme is called Gaussian back substitution. Finally, we get the method presented in Algorithm 3.9.

---

**Algorithm 3.9** ADMM with Gaussian back substitution

---

Initialize $\mathbf{x}_1^0, \cdots, \mathbf{x}_m^0, \boldsymbol{\lambda}^0$.
 **for** $k = 0, 1, 2, 3, \cdots$ **do**
    Update $\widetilde{\mathbf{x}}_1^{k+1}, \cdots, \widetilde{\mathbf{x}}_m^{k+1}$, and $\boldsymbol{\lambda}^{k+1}$ by (3.72a)–(3.72c).
    Update $\mathbf{x}_1^{k+1}, \cdots, \mathbf{x}_m^{k+1}$ by (3.76).
 **end for**

---

With (3.74), (3.73) becomes

$$f(\widetilde{\mathbf{x}}^{k+1}) - f(\mathbf{x}^*) + \left\langle \boldsymbol{\lambda}^*, \sum_{i=1}^m \mathbf{A}_i\widetilde{\mathbf{x}}_i^{k+1} - \mathbf{b} \right\rangle$$

$$\leq \frac{1}{2\beta}\left(\left\|\boldsymbol{\lambda}^k - \boldsymbol{\lambda}^*\right\|^2 - \left\|\boldsymbol{\lambda}^{k+1} - \boldsymbol{\lambda}^*\right\|^2\right) + \frac{\beta}{2}\left(\left\|\mathbf{x}^k - \mathbf{x}^*\right\|_{\mathbf{G}}^2 - \left\|\mathbf{x}^{k+1} - \mathbf{x}^*\right\|_{\mathbf{G}}^2\right).$$

Then similar to the proof of Theorem 3.3, we have the $O\left(\frac{1}{K}\right)$ convergence rate in the ergodic sense.

**Theorem 3.16** *Suppose that $f_i(\mathbf{x}_i)$ is convex for all $i \in [m]$. Then for Algorithm 3.9, we have*

$$\left|f(\hat{\mathbf{x}}^{K+1}) - f(\mathbf{x}^*)\right| \leq \frac{C}{2(K+1)} + \frac{2\sqrt{C}\|\boldsymbol{\lambda}^*\|}{\sqrt{\beta}(K+1)},$$

$$\left\|\mathbf{A}\hat{\mathbf{x}}^{K+1} - \mathbf{b}\right\| \leq \frac{2\sqrt{C}}{\sqrt{\beta}(K+1)},$$

*where*

$$\hat{\mathbf{x}}^{K+1} = \frac{1}{K+1}\sum_{k=0}^K \widetilde{\mathbf{x}}^{k+1} \quad and \quad C = \frac{1}{\beta}\left\|\boldsymbol{\lambda}^0 - \boldsymbol{\lambda}^*\right\|^2 + \beta\left\|\mathbf{x}^0 - \mathbf{x}^*\right\|_{\mathbf{G}}^2.$$

### 3.5.2 Prediction-Correction

The Gaussian back substitution (3.76) is actually not simple enough as it still requires to solve $m$ small-scale linear systems when computing (3.76). In the following, we give an improved strategy [12]. It is based on the key observation that when solving (3.72a)–(3.72b), we actually only need $\mathbf{A}_i\mathbf{x}_i^k$ rather than $\mathbf{x}_i^k$ itself. So we do not have to compute $\mathbf{x}_i^k$ explicitly as (3.76) does.

Denote

$$\mathbf{P} = \begin{pmatrix} \mathbf{A}_1 & \mathbf{0} & \cdots & \mathbf{0} \\ \mathbf{0} & \mathbf{A}_2 & \cdots & \mathbf{0} \\ \vdots & \vdots & \ddots & \vdots \\ \mathbf{0} & \mathbf{0} & \cdots & \mathbf{A}_m \end{pmatrix} \quad \text{and} \quad \mathbf{L} = \begin{pmatrix} \mathbf{I} & \mathbf{0} & \cdots & \mathbf{0} \\ \mathbf{I} & \mathbf{I} & \cdots & \mathbf{0} \\ \vdots & \vdots & \ddots & \vdots \\ \mathbf{I} & \mathbf{I} & \cdots & \mathbf{I} \end{pmatrix}, \tag{3.77}$$

then we can check that

$$\mathbf{L}^{-T} = \begin{pmatrix} \mathbf{I} & -\mathbf{I} & \mathbf{0} & \cdots & \mathbf{0} & \mathbf{0} \\ \mathbf{0} & \mathbf{I} & -\mathbf{I} & \cdots & \mathbf{0} & \mathbf{0} \\ \vdots & \vdots & \vdots & \vdots & \vdots & \vdots \\ \mathbf{0} & \mathbf{0} & \mathbf{0} & \cdots & \mathbf{I} & -\mathbf{I} \\ \mathbf{0} & \mathbf{0} & \mathbf{0} & \cdots & \mathbf{0} & \mathbf{I} \end{pmatrix}, \quad \mathbf{M} = \mathbf{P}^T\mathbf{L}\mathbf{P}, \quad \text{and} \quad \mathbf{H} = \mathbf{P}^T\mathbf{P}.$$

Similar to the deductions in Gaussian back substitution, we need to define $\mathbf{x}^{k+1}$ and find $\mathbf{G}' \succeq \mathbf{0}$ so that the following relationship holds:

$$-\frac{\beta}{2}\left\|\widetilde{\mathbf{x}}^{k+1} - \mathbf{x}^k\right\|_{\mathbf{H}}^2 - \beta\left(\mathbf{x}^k - \mathbf{x}^*\right)^T \mathbf{M}\left(\widetilde{\mathbf{x}}^{k+1} - \mathbf{x}^k\right)$$

$$= \frac{\beta}{2}\left(\left\|\mathbf{P}\mathbf{x}^k - \mathbf{P}\mathbf{x}^*\right\|_{\mathbf{G}'}^2 - \left\|\mathbf{P}\mathbf{x}^{k+1} - \mathbf{P}\mathbf{x}^*\right\|_{\mathbf{G}'}^2\right). \tag{3.78}$$

Note that

$$-\frac{\beta}{2}\left\|\widetilde{\mathbf{x}}^{k+1} - \mathbf{x}^k\right\|_{\mathbf{H}}^2 - \beta\left(\mathbf{x}^k - \mathbf{x}^*\right)^T \mathbf{M}\left(\widetilde{\mathbf{x}}^{k+1} - \mathbf{x}^k\right)$$

$$= -\frac{\beta}{2}\left\|\mathbf{P}\widetilde{\mathbf{x}}^{k+1} - \mathbf{P}\mathbf{x}^k\right\|^2 - \beta\left(\mathbf{P}\mathbf{x}^k - \mathbf{P}\mathbf{x}^*\right)^T \mathbf{L}\left(\mathbf{P}\widetilde{\mathbf{x}}^{k+1} - \mathbf{P}\mathbf{x}^k\right)$$

and

$$
\left\| \mathbf{P}\mathbf{x}^k - \mathbf{P}\mathbf{x}^* \right\|_{\mathbf{G}'}^2 - \left\| \mathbf{P}\mathbf{x}^k - \mathbf{P}\mathbf{x}^* + \mathbf{D}' \left( \mathbf{P}\widetilde{\mathbf{x}}^{k+1} - \mathbf{P}\mathbf{x}^k \right) \right\|_{\mathbf{G}'}^2
$$

$$
= - \left( \mathbf{P}\widetilde{\mathbf{x}}^{k+1} - \mathbf{P}\mathbf{x}^k \right)^T (\mathbf{D}')^T \mathbf{G}' \mathbf{D}' \left( \mathbf{P}\widetilde{\mathbf{x}}^{k+1} - \mathbf{P}\mathbf{x}^k \right)
$$

$$
- 2 \left( \mathbf{P}\mathbf{x}^k - \mathbf{P}\mathbf{x}^* \right)^T \mathbf{G}' \mathbf{D}' \left( \mathbf{P}\widetilde{\mathbf{x}}^{k+1} - \mathbf{P}\mathbf{x}^k \right).
$$

So we should define

$$
\mathbf{P}\mathbf{x}^{k+1} = \mathbf{P}\mathbf{x}^k + \mathbf{D}' \left( \mathbf{P}\widetilde{\mathbf{x}}^{k+1} - \mathbf{P}\mathbf{x}^k \right) \tag{3.79}
$$

and choose

$$
\mathbf{D}' = \mathbf{L}^{-T} \quad \text{and} \quad \mathbf{G}' = \mathbf{L}\mathbf{L}^T \tag{3.80}
$$

so that $(\mathbf{D}')^T \mathbf{G}' \mathbf{D}' = \mathbf{I}$ and $\mathbf{G}' \mathbf{D}' = \mathbf{L}$ hold.

(3.79) can be explicitly rewritten as

$$
\begin{pmatrix} \mathbf{A}_1 \mathbf{x}_1^{k+1} \\ \mathbf{A}_2 \mathbf{x}_2^{k+1} \\ \vdots \\ \mathbf{A}_{m-1} \mathbf{x}_{m-1}^{k+1} \\ \mathbf{A}_m \mathbf{x}_m^{k+1} \end{pmatrix} = \begin{pmatrix} \mathbf{A}_1 \mathbf{x}_1^k \\ \mathbf{A}_2 \mathbf{x}_2^k \\ \vdots \\ \mathbf{A}_{m-1} \mathbf{x}_{m-1}^k \\ \mathbf{A}_m \mathbf{x}_m^k \end{pmatrix}
$$

$$
+ \begin{pmatrix} \mathbf{I} & -\mathbf{I} & \mathbf{0} & \cdots & \mathbf{0} & \mathbf{0} \\ \mathbf{0} & \mathbf{I} & -\mathbf{I} & \cdots & \mathbf{0} & \mathbf{0} \\ \vdots & \vdots & \vdots & \vdots & \vdots & \vdots \\ \mathbf{0} & \mathbf{0} & \mathbf{0} & \cdots & \mathbf{I} & -\mathbf{I} \\ \mathbf{0} & \mathbf{0} & \mathbf{0} & \cdots & \mathbf{0} & \mathbf{I} \end{pmatrix} \begin{pmatrix} \mathbf{A}_1 \widetilde{\mathbf{x}}_1^{k+1} - \mathbf{A}_1 \mathbf{x}_1^k \\ \mathbf{A}_2 \widetilde{\mathbf{x}}_2^{k+1} - \mathbf{A}_2 \mathbf{x}_2^k \\ \vdots \\ \mathbf{A}_{m-1} \widetilde{\mathbf{x}}_{m-1}^{k+1} - \mathbf{A}_{m-1} \mathbf{x}_{m-1}^k \\ \mathbf{A}_m \widetilde{\mathbf{x}}_m^{k+1} - \mathbf{A}_m \mathbf{x}_m^k \end{pmatrix}
$$

$$
= \begin{pmatrix} \mathbf{A}_1 \widetilde{\mathbf{x}}_1^{k+1} + \mathbf{A}_2 \mathbf{x}_2^k - \mathbf{A}_2 \widetilde{\mathbf{x}}_2^{k+1} \\ \mathbf{A}_2 \widetilde{\mathbf{x}}_2^{k+1} + \mathbf{A}_3 \mathbf{x}_3^k - \mathbf{A}_3 \widetilde{\mathbf{x}}_3^{k+1} \\ \vdots \\ \mathbf{A}_{m-1} \widetilde{\mathbf{x}}_{m-1}^{k+1} + \mathbf{A}_m \mathbf{x}_m^k - \mathbf{A}_m \widetilde{\mathbf{x}}_m^{k+1} \\ \mathbf{A}_m \widetilde{\mathbf{x}}_m^{k+1} \end{pmatrix}. \tag{3.81}
$$

So we can obtain $\mathbf{A}_i \mathbf{x}_i^{k+1}$ conveniently without solving linear systems. However, $\mathbf{x}_i^{k+1}$ may not exist given $\mathbf{A}_i \mathbf{x}_i^{k+1}$, making the next round iteration and Lemma 3.18 invalid. So we need to revise the above algorithm and Lemma 3.18 accordingly.

Introducing variables $\boldsymbol{\xi}_i^k$, which play the role of $\mathbf{A}_i \mathbf{x}_i^k$ in (3.72a)–(3.72b), iterations (3.72a)–(3.72b) can be rewritten as follows:

$$\widetilde{\mathbf{x}}_1^{k+1} = \underset{\mathbf{x}_1}{\text{argmin}}\, \tilde{L}_1 \left( \mathbf{x}_1, \boldsymbol{\xi}_2^k, \cdots, \boldsymbol{\xi}_m^k, \boldsymbol{\lambda}^k \right), \tag{3.82a}$$

$$\widetilde{\mathbf{x}}_2^{k+1} = \underset{\mathbf{x}_2}{\text{argmin}}\, \tilde{L}_2 \left( \widetilde{\mathbf{x}}_1^{k+1}, \mathbf{x}_2, \boldsymbol{\xi}_3^k, \cdots, \boldsymbol{\xi}_m^k, \boldsymbol{\lambda}^k \right),$$

$$\vdots$$

$$\widetilde{\mathbf{x}}_i^{k+1} = \underset{\mathbf{x}_i}{\text{argmin}}\, \tilde{L}_i \left( \widetilde{\mathbf{x}}_1^{k+1}, \cdots, \widetilde{\mathbf{x}}_{i-1}^{k+1}, \mathbf{x}_i, \boldsymbol{\xi}_{i+1}^k \cdots, \boldsymbol{\xi}_m^k, \boldsymbol{\lambda}^k \right),$$

$$\vdots$$

$$\widetilde{\mathbf{x}}_m^{k+1} = \underset{\mathbf{x}_m}{\text{argmin}}\, \tilde{L}_m \left( \widetilde{\mathbf{x}}_1^{k+1}, \cdots, \widetilde{\mathbf{x}}_{m-1}^{k+1}, \mathbf{x}_m, \boldsymbol{\lambda}^k \right), \tag{3.82b}$$

$$\boldsymbol{\lambda}^{k+1} = \boldsymbol{\lambda}^k + \beta \left( \sum_{i=1}^m \mathbf{A}_i \widetilde{\mathbf{x}}_i^{k+1} - \mathbf{b} \right), \tag{3.82c}$$

$$\boldsymbol{\xi}^{k+1} = \boldsymbol{\xi}^k + \mathbf{L}^{-T} \left( \mathbf{P}\widetilde{\mathbf{x}}^{k+1} - \boldsymbol{\xi}^k \right), \tag{3.82d}$$

where

$$\tilde{L}_i(\mathbf{x}_1, \cdots, \mathbf{x}_i, \boldsymbol{\xi}_{i+1}, \cdots, \boldsymbol{\xi}_m, \boldsymbol{\lambda})$$

$$= \sum_{j=1}^i f_j(\mathbf{x}_j) + \left\langle \boldsymbol{\lambda}, \sum_{j=1}^i \mathbf{A}_j \mathbf{x}_j + \sum_{j=i+1}^m \boldsymbol{\xi}_j - \mathbf{b} \right\rangle + \frac{\beta}{2} \left\| \sum_{j=1}^i \mathbf{A}_j \mathbf{x}_j + \sum_{j=i+1}^m \boldsymbol{\xi}_j - \mathbf{b} \right\|^2$$

and $\boldsymbol{\xi} = \left( \boldsymbol{\xi}_1^T, \boldsymbol{\xi}_2^T, \cdots, \boldsymbol{\xi}_m^T \right)^T$.

---

**Algorithm 3.10** ADMM with prediction-correction

---

Initialize $\boldsymbol{\xi}_1^0, \cdots, \boldsymbol{\xi}_m^0, \boldsymbol{\lambda}^0$.
**for** $k = 0, 1, 2, 3, \cdots$ **do**
  Update $\widetilde{\mathbf{x}}_1^{k+1}, \cdots, \widetilde{\mathbf{x}}_m^{k+1}$, and $\boldsymbol{\lambda}^{k+1}$ by (3.82a)–(3.82c).
  Update $\boldsymbol{\xi}_1^{k+1}, \cdots, \boldsymbol{\xi}_m^{k+1}$ by (3.82d).
**end for**

---

With (3.78) and (3.80), Lemma 3.18 changes to the following one.

**Lemma 3.19** *Suppose that $f_i(\mathbf{x}_i)$ is convex for all $i \in [m]$. Then for Algorithm 3.10 we have*

$$f(\tilde{\mathbf{x}}^{k+1}) - f(\mathbf{x}^*) + \left\langle \boldsymbol{\lambda}^*, \sum_{i=1}^{m} \mathbf{A}_i \tilde{\mathbf{x}}_i^{k+1} - \mathbf{b} \right\rangle$$

$$\leq \frac{1}{2\beta} \left( \left\| \boldsymbol{\lambda}^k - \boldsymbol{\lambda}^* \right\|^2 - \left\| \boldsymbol{\lambda}^{k+1} - \boldsymbol{\lambda}^* \right\|^2 \right)$$

$$+ \frac{\beta}{2} \left( \left\| \boldsymbol{\xi}^k - \mathbf{P}\mathbf{x}^* \right\|_{\mathbf{LL}^T}^2 - \left\| \boldsymbol{\xi}^{k+1} - \mathbf{P}\mathbf{x}^* \right\|_{\mathbf{LL}^T}^2 \right).$$

And Theorem 3.16 still holds for Algorithm 3.10 with $C$ changing to

$$\frac{1}{\beta} \left\| \boldsymbol{\lambda}^0 - \boldsymbol{\lambda}^* \right\|^2 + \beta \left\| \boldsymbol{\xi}^0 - \mathbf{P}\mathbf{x}^* \right\|_{\mathbf{LL}^T}^2 .$$

### 3.5.3  Linearized ADMM with Parallel Splitting

Although the prediction-correction technique introduced in Sect. 3.5.2 resolves the convergence issue of naive multi-block ADMM, it nonetheless has drawbacks. First, it requires to maintain two groups of variables, the predicted and the corrected, thus increasing the memory cost (at least doubled). Second, for sparse or low-rank problems, neither the $\mathbf{x}_i^{k+1}$ in the Gaussian back substitution nor the $\boldsymbol{\xi}_i^{k+1}$ in the improved version can be sparse or low-rank, thus the memory consumption can be even more. Third, it may not be easy to solve subproblems (3.72a)–(3.72b) or (3.82a)–(3.82b). So in this section we use the linearized augmented Lagrangian method (Algorithm 3.7) to solve Problem (3.71), which consists of the following iterations:

$$\mathbf{x}^{k+1} = \underset{\mathbf{x}}{\operatorname{argmin}} \left( f(\mathbf{x}) + \left\langle \boldsymbol{\lambda}^k, \mathbf{A}\mathbf{x} - \mathbf{b} \right\rangle + \beta \left\langle \mathbf{A}^T (\mathbf{A}\mathbf{x}^k - \mathbf{b}), \mathbf{x} - \mathbf{x}^k \right\rangle \right.$$

$$\left. + \frac{\beta}{2} \|\mathbf{x} - \mathbf{x}^k\|_{\mathbf{L}}^2 \right), \tag{3.83}$$

$$\boldsymbol{\lambda}^{k+1} = \boldsymbol{\lambda}^k + \beta \left( \sum_{i=1}^{m} \mathbf{A}_i \mathbf{x}_i^{k+1} - \mathbf{b} \right), \tag{3.84}$$

where $\mathbf{L} = \text{Diag}([L_1\mathbf{I}_{n_1}, \cdots, L_m\mathbf{I}_{n_m}])$, in which $n_i$ is the dimension of $\mathbf{x}_i$ and $L_i$ is to be determined, and we use

$$\psi(\mathbf{x}) = \frac{\beta}{2}\|\mathbf{x}\|_{\mathbf{L}}^2 - \frac{\beta}{2}\|\mathbf{Ax}\|^2$$

in (3.69a), which results in

$$\frac{\beta}{2}\|\mathbf{Ax} - \mathbf{b}\|^2 + D_\psi(\mathbf{x}, \mathbf{x}^k)$$

$$= \frac{\beta}{2}\|\mathbf{Ax}^k - \mathbf{b}\|^2 + \beta\left\langle \mathbf{A}^T(\mathbf{Ax}^k - \mathbf{b}), \mathbf{x} - \mathbf{x}^k\right\rangle + \frac{\beta}{2}\|\mathbf{x} - \mathbf{x}^k\|_{\mathbf{L}}^2.$$

We only need to check that $\psi(\mathbf{x})$ is convex, which is equivalent to

$$\|\mathbf{x}\|_{\mathbf{L}}^2 \geq \|\mathbf{Ax}\|^2, \quad \forall \mathbf{x}.$$

We can easily check that

$$\|\mathbf{Ax}\|^2 = \left\|\sum_{i=1}^{m}\mathbf{A}_i\mathbf{x}_i\right\|^2 \leq m\sum_{i=1}^{m}\|\mathbf{A}_i\mathbf{x}_i\|^2 \leq m\sum_{i=1}^{m}\|\mathbf{A}_i\|_2^2\|\mathbf{x}_i\|^2.$$

So we may choose $L_i \geq m\|\mathbf{A}_i\|_2^2$. Thus, (3.83) becomes separable and the subproblem for each $\mathbf{x}_i$ is

$$\mathbf{x}_i^{k+1} = \underset{\mathbf{x}_i}{\text{argmin}}\left(f_i(\mathbf{x}_i) + \left\langle \boldsymbol{\lambda}^k, \mathbf{A}_i\mathbf{x}_i\right\rangle + \beta\left\langle \mathbf{A}_i^T\left(\sum_{j=1}^{m}\mathbf{A}_j\mathbf{x}_j^k - \mathbf{b}\right), \mathbf{x}_i - \mathbf{x}_i^k\right\rangle\right.$$

$$\left. + \frac{\beta L_i}{2}\|\mathbf{x}_i - \mathbf{x}_i^k\|^2\right), \tag{3.85}$$

which can be solved in parallel for $i \in [m]$. Thus, we call the method linearized ADMM with parallel splitting [18, 20] and present it in Algorithm 3.11.

---

**Algorithm 3.11** Linearized ADMM with parallel splitting

---
Initialize $\mathbf{x}_1^0, \cdots, \mathbf{x}_m^0, \boldsymbol{\lambda}^0$.
  **for** $k = 0, 1, 2, 3, \cdots$ **do**
    Update $\mathbf{x}_1^{k+1}, \cdots, \mathbf{x}_m^{k+1}$, and $\boldsymbol{\lambda}^{k+1}$ by (3.85) and (3.84), respectively.
  **end for**

---

From Theorems 3.13 and 3.14, we have the $O(1/K)$ and the linear convergence rates of Algorithm 3.11 under different conditions.

### 3.5.4   Combining the Serial and the Parallel Update Orders

In the previous section, we have seen that by replacing the serial update order in linearized ADMM by the parallel update order in the linearized augmented Lagrangian method (also called linearized ADMM with parallel splitting), we can prove the convergence for the multi-block case. Since for the two-block case, serial update order is faster than parallel update order, we may combine them for faster convergence when dealing with the multi-block case [23]. That is, divide the $m$ blocks into two partitions

$$(1, \cdots, m') \quad \text{and} \quad (m' + 1, \cdots, m)$$

and then update the two partitions in serial, while updating the blocks in the same partition in parallel. The algorithm is explicitly given as follows (for example, in [11, 23, 26]):

$$\mathbf{x}_i^{k+1} = \underset{\mathbf{x}_i}{\text{argmin}} \left( f_i(\mathbf{x}_i) + \left\langle \boldsymbol{\lambda}^k, \mathbf{A}_i \mathbf{x}_i \right\rangle + \beta \left\langle \mathbf{A}_i^T \left( \sum_{t=1}^m \mathbf{A}_t \mathbf{x}_t^k - \mathbf{b} \right), \mathbf{x}_i - \mathbf{x}_i^k \right\rangle \right.$$
$$\left. + \frac{m' \beta \|\mathbf{A}_i\|_2^2}{2} \|\mathbf{x}_i - \mathbf{x}_i^k\|^2 \right), \quad \forall 1 \leq i \leq m',$$

$$\mathbf{x}_j^{k+1} = \underset{\mathbf{x}_j}{\text{argmin}} \left( f_j(\mathbf{x}_j) + \left\langle \boldsymbol{\lambda}^k, \mathbf{A}_j \mathbf{x}_j \right\rangle \right.$$
$$+ \beta \left\langle \mathbf{A}_j^T \left( \sum_{t=1}^{m'} \mathbf{A}_t \mathbf{x}_t^{k+1} + \sum_{t=m'+1}^m \mathbf{A}_t \mathbf{x}_t^k - \mathbf{b} \right), \mathbf{x}_j - \mathbf{x}_j^k \right\rangle$$
$$\left. + \frac{(m - m') \beta \|\mathbf{A}_j\|_2^2}{2} \|\mathbf{x}_j - \mathbf{x}_j^k\|^2 \right), \quad \forall m' + 1 \leq j \leq m,$$

$$\boldsymbol{\lambda}^{k+1} = \boldsymbol{\lambda}^k + \beta \left( \sum_{i=1}^m \mathbf{A}_i \mathbf{x}_i^{k+1} - \mathbf{b} \right).$$

Denote

$$F_1(\mathbf{x}) = \sum_{i=1}^{m'} f_i(\mathbf{x}_i), \quad F_2(\mathbf{x}) = \sum_{i=m'+1}^m f_i(\mathbf{x}_i),$$

$$\mathbf{B} = [\mathbf{A}_1, \cdots, \mathbf{A}_{m'}], \quad \mathbf{C} = [\mathbf{A}_{m'+1}, \cdots, \mathbf{A}_m],$$

$$\mathbf{y} = (\mathbf{x}_1^T, \cdots, \mathbf{x}_{m'}^T)^T, \quad \text{and} \quad \mathbf{z} = (\mathbf{x}_{m'+1}^T, \cdots, \mathbf{x}_m^T)^T.$$

Similar to the previous section, we can rewrite the first two steps as

$$\mathbf{y}^{k+1} = \underset{\mathbf{y}}{\operatorname{argmin}} \left( F_1(\mathbf{y}) + \left\langle \boldsymbol{\lambda}^k, \mathbf{By} \right\rangle + \beta \left\langle \mathbf{B}^T \left( \mathbf{By}^k + \mathbf{Cz}^k - \mathbf{b} \right), \mathbf{y} - \mathbf{y}^k \right\rangle \right.$$
$$\left. + \frac{\beta}{2} \|\mathbf{y} - \mathbf{y}^k\|_{\mathbf{L}_1}^2 \right)$$
$$= \underset{\mathbf{y}}{\operatorname{argmin}} \left( F_1(\mathbf{y}) + \left\langle \boldsymbol{\lambda}^k, \mathbf{By} \right\rangle + \frac{\beta}{2} \|\mathbf{By} + \mathbf{Cz}^k - \mathbf{b}\|^2 + D_\phi(\mathbf{y}, \mathbf{y}^k) \right)$$

and

$$\mathbf{z}^{k+1} = \underset{\mathbf{z}}{\operatorname{argmin}} \left( F_2(\mathbf{z}) + \left\langle \boldsymbol{\lambda}^k, \mathbf{Cz} \right\rangle + \beta \left\langle \mathbf{C}^T \left( \mathbf{By}^{k+1} + \mathbf{Cz}^k - \mathbf{b} \right), \mathbf{z} - \mathbf{z}^k \right\rangle \right.$$
$$\left. + \frac{\beta}{2} \|\mathbf{z} - \mathbf{z}^k\|_{\mathbf{L}_2}^2 \right)$$
$$= \underset{\mathbf{z}}{\operatorname{argmin}} \left( F_2(\mathbf{z}) + \left\langle \boldsymbol{\lambda}^k, \mathbf{Cz} \right\rangle + \frac{\beta}{2} \|\mathbf{By}^{k+1} + \mathbf{Cz} - \mathbf{b}\|^2 + D_\psi(\mathbf{z}, \mathbf{z}^k) \right),$$

respectively, where

$$\mathbf{L}_1 = \operatorname{Diag} \left( \left[ m' \|\mathbf{A}_1\|_2^2 \mathbf{I}_{n_1}, \cdots, m' \|\mathbf{A}_{m'}\|_2^2 \mathbf{I}_{n_{m'}} \right] \right),$$
$$\mathbf{L}_2 = \operatorname{Diag} \left( \left[ (m - m') \|\mathbf{A}_{m'+1}\|_2^2 \mathbf{I}_{n_{m'+1}}, \cdots, (m - m') \|\mathbf{A}_m\|_2^2 \mathbf{I}_{n_m} \right] \right),$$
$$\phi(\mathbf{y}) = \frac{\beta}{2} \|\mathbf{y}\|_{\mathbf{L}_1}^2 - \frac{\beta}{2} \|\mathbf{By}\|^2,$$
$$\psi(\mathbf{z}) = \frac{\beta}{2} \|\mathbf{z}\|_{\mathbf{L}_2}^2 - \frac{\beta}{2} \|\mathbf{Cz}\|^2,$$

in which $n_i$ is the dimension of $\mathbf{x}_i$. We see that it is exactly the linearized ADMM in (3.27a)-(3.27c). So we can also have the sublinear and the linear convergence rates under different conditions from Theorems 3.6 and 3.8, respectively, which is faster than making all the $m$ blocks parallel.

## 3.6 Variational Inequality Perspective

Now we introduce the variational inequality viewpoint of ADMM, which has been frequently used in He's works (for example, see [9, 10, 12]). Introduce the Lagrangian function of Problem (2.13):

$$L(\mathbf{x}, \mathbf{y}, \boldsymbol{\lambda}) = f(\mathbf{x}) + g(\mathbf{y}) + \langle \boldsymbol{\lambda}, \mathbf{Ax} + \mathbf{By} - \mathbf{b} \rangle.$$

$(\mathbf{x}^*, \mathbf{y}^*, \boldsymbol{\lambda}^*)$ is a saddle point if it satisfies

$$L(\mathbf{x}^*, \mathbf{y}^*, \boldsymbol{\lambda}) \le L(\mathbf{x}^*, \mathbf{y}^*, \boldsymbol{\lambda}^*) \le L(\mathbf{x}, \mathbf{y}, \boldsymbol{\lambda}^*), \forall \mathbf{x}, \mathbf{y}, \boldsymbol{\lambda}.$$

For the left inequality, we have

$$\langle \boldsymbol{\lambda} - \boldsymbol{\lambda}^*, \mathbf{A}\mathbf{x}^* + \mathbf{B}\mathbf{y}^* - \mathbf{b} \rangle \le 0.$$

For the right inequality, we have

$$f(\mathbf{x}) + g(\mathbf{y}) - f(\mathbf{x}^*) - g(\mathbf{y}^*) + \langle \mathbf{x} - \mathbf{x}^*, \mathbf{A}^T \boldsymbol{\lambda}^* \rangle + \langle \mathbf{y} - \mathbf{y}^*, \mathbf{B}^T \boldsymbol{\lambda}^* \rangle \ge 0.$$

Combining them together, we have

$$f(\mathbf{x}) + g(\mathbf{y}) - f(\mathbf{x}^*) - g(\mathbf{y}^*) + \left\langle \begin{pmatrix} \mathbf{x} - \mathbf{x}^* \\ \mathbf{y} - \mathbf{y}^* \\ \boldsymbol{\lambda} - \boldsymbol{\lambda}^* \end{pmatrix}, \begin{pmatrix} \mathbf{A}^T \boldsymbol{\lambda}^* \\ \mathbf{B}^T \boldsymbol{\lambda}^* \\ -(\mathbf{A}\mathbf{x}^* + \mathbf{B}\mathbf{y}^* - \mathbf{b}) \end{pmatrix} \right\rangle$$

$$\ge 0, \quad \forall \mathbf{x}, \mathbf{y}, \boldsymbol{\lambda}. \tag{3.86}$$

Denote

$$\mathbf{w} = \begin{pmatrix} \mathbf{x} \\ \mathbf{y} \\ \boldsymbol{\lambda} \end{pmatrix}, \quad \mathbf{u}(\mathbf{w}) = \begin{pmatrix} \mathbf{x} \\ \mathbf{y} \end{pmatrix}, \quad \text{and} \quad F(\mathbf{w}) = \begin{pmatrix} \mathbf{A}^T \boldsymbol{\lambda} \\ \mathbf{B}^T \boldsymbol{\lambda} \\ -(\mathbf{A}\mathbf{x} + \mathbf{B}\mathbf{y} - \mathbf{b}) \end{pmatrix}.$$

For simplicity, in the sequel we write $\mathbf{u}$ instead of $\mathbf{u}(\mathbf{w})$ and define

$$\theta(\mathbf{u}) = f(\mathbf{x}) + g(\mathbf{y}).$$

Then (3.86) reduces to

$$\theta(\mathbf{u}) - \theta(\mathbf{u}^*) + \langle \mathbf{w} - \mathbf{w}^*, F(\mathbf{w}^*) \rangle \ge 0, \quad \forall \mathbf{w}. \tag{3.87}$$

(3.87) is called the variational inequality of Problem (2.13). The following lemma can be easily checked.

**Lemma 3.20** $(\mathbf{x}^*, \mathbf{y}^*, \boldsymbol{\lambda}^*)$ *is a saddle point of* $L(\mathbf{x}, \mathbf{y}, \boldsymbol{\lambda})$ *if and only if it satisfies* (3.86).

We can also easily check that

$$\langle \mathbf{w} - \hat{\mathbf{w}}, F(\mathbf{w}) - F(\hat{\mathbf{w}}) \rangle = 0, \quad \forall \mathbf{w}, \hat{\mathbf{w}}. \tag{3.88}$$

Thus, (3.87) is equivalent to

$$\theta(\mathbf{u}) - \theta(\mathbf{u}^*) + \langle \mathbf{w} - \mathbf{w}^*, F(\mathbf{w}) \rangle \geq 0, \quad \forall \mathbf{w}. \tag{3.89}$$

We say that $\widetilde{\mathbf{w}}$ is an approximate solution of the variational inequality problem (3.89) with accuracy $\epsilon$ if it satisfies

$$\theta(\widetilde{\mathbf{u}}) - \theta(\mathbf{u}) + \langle \widetilde{\mathbf{w}} - \mathbf{w}, F(\mathbf{w}) \rangle \leq \epsilon, \quad \forall \mathbf{w}. \tag{3.90}$$

Especially, $\epsilon = 0$ gives (3.89).

### 3.6.1 Unified Framework in Variational Inequality

We first write the original ADMM (Algorithm 2.1) in the form of (3.87) and then give the unified framework.

From (3.1), (3.2), and the convexity of $f$ and $g$, we have

$$f(\mathbf{x}) - f(\mathbf{x}^{k+1}) + \left\langle \mathbf{A}^T \widetilde{\boldsymbol{\lambda}}^{k+1}, \mathbf{x} - \mathbf{x}^{k+1} \right\rangle \geq 0, \tag{3.91}$$

$$g(\mathbf{y}) - g(\mathbf{y}^{k+1}) + \left\langle \mathbf{B}^T \widetilde{\boldsymbol{\lambda}}^{k+1} + \beta \mathbf{B}^T (\mathbf{B}\mathbf{y}^{k+1} - \mathbf{B}\mathbf{y}^k), \mathbf{y} - \mathbf{y}^{k+1} \right\rangle \geq 0, \tag{3.92}$$

where we denote

$$\widetilde{\mathbf{w}}^{k+1} = \begin{pmatrix} \widetilde{\mathbf{x}}^{k+1} \\ \widetilde{\mathbf{y}}^{k+1} \\ \widetilde{\boldsymbol{\lambda}}^{k+1} \end{pmatrix} = \begin{pmatrix} \mathbf{x}^{k+1} \\ \mathbf{y}^{k+1} \\ \boldsymbol{\lambda}^k + \beta(\mathbf{A}\mathbf{x}^{k+1} + \mathbf{B}\mathbf{y}^k - \mathbf{b}) \end{pmatrix}.$$

Adding (3.91) and (3.92) together and writing it in the form of (3.86), we have

$$f(\mathbf{x}) + g(\mathbf{y}) - f(\widetilde{\mathbf{x}}^{k+1}) - g(\widetilde{\mathbf{y}}^{k+1}) + \left\langle \begin{pmatrix} \mathbf{x} - \widetilde{\mathbf{x}}^{k+1} \\ \mathbf{y} - \widetilde{\mathbf{y}}^{k+1} \\ \boldsymbol{\lambda} - \widetilde{\boldsymbol{\lambda}}^{k+1} \end{pmatrix}, \begin{pmatrix} \mathbf{A}^T \widetilde{\boldsymbol{\lambda}}^{k+1} \\ \mathbf{B}^T \widetilde{\boldsymbol{\lambda}}^{k+1} \\ -(\mathbf{A}\widetilde{\mathbf{x}}^{k+1} + \mathbf{B}\widetilde{\mathbf{y}}^{k+1} - \mathbf{b}) \end{pmatrix} \right\rangle$$

$$\geq \left\langle \widetilde{\boldsymbol{\lambda}}^{k+1} - \boldsymbol{\lambda}, \mathbf{A}\widetilde{\mathbf{x}}^{k+1} + \mathbf{B}\widetilde{\mathbf{y}}^{k+1} - \mathbf{b} \right\rangle + \beta \left\langle \mathbf{B}\widetilde{\mathbf{y}}^{k+1} - \mathbf{B}\mathbf{y}^k, \mathbf{B}\widetilde{\mathbf{y}}^{k+1} - \mathbf{B}\mathbf{y} \right\rangle$$

$$= \begin{pmatrix} \mathbf{A}\widetilde{\mathbf{x}}^{k+1} - \mathbf{A}\mathbf{x} \\ \mathbf{B}\widetilde{\mathbf{y}}^{k+1} - \mathbf{B}\mathbf{y} \\ \widetilde{\boldsymbol{\lambda}}^{k+1} - \boldsymbol{\lambda} \end{pmatrix}^T \begin{pmatrix} \mathbf{0} & \mathbf{0} & \mathbf{0} \\ \mathbf{0} & \beta\mathbf{I} & \mathbf{0} \\ \mathbf{0} & \mathbf{I} & \frac{1}{\beta}\mathbf{I} \end{pmatrix} \begin{pmatrix} \mathbf{A}\widetilde{\mathbf{x}}^{k+1} - \mathbf{A}\mathbf{x}^k \\ \mathbf{B}\widetilde{\mathbf{y}}^{k+1} - \mathbf{B}\mathbf{y}^k \\ \widetilde{\boldsymbol{\lambda}}^{k+1} - \boldsymbol{\lambda}^k \end{pmatrix}$$

$$= (\widetilde{\mathbf{w}}^{k+1} - \mathbf{w})^T \mathbf{P}^T \mathbf{H} \mathbf{M} \mathbf{P} (\widetilde{\mathbf{w}}^{k+1} - \mathbf{w}^k),$$

where we denote

$$
\mathbf{H} = \begin{pmatrix} \mathbf{0} & \mathbf{0} & \mathbf{0} \\ \mathbf{0} & \beta\mathbf{I} & \mathbf{0} \\ \mathbf{0} & \mathbf{0} & \frac{1}{\beta}\mathbf{I} \end{pmatrix}, \quad \mathbf{M} = \begin{pmatrix} \mathbf{I} & \mathbf{0} & \mathbf{0} \\ \mathbf{0} & \mathbf{I} & \mathbf{0} \\ \mathbf{0} & \beta\mathbf{I} & \mathbf{I} \end{pmatrix}, \quad \text{and} \quad \mathbf{P} = \begin{pmatrix} \mathbf{A} & \mathbf{0} & \mathbf{0} \\ \mathbf{0} & \mathbf{B} & \mathbf{0} \\ \mathbf{0} & \mathbf{0} & \mathbf{I} \end{pmatrix}.
$$

From (3.3), we can also check that

$$
\mathbf{P}\mathbf{w}^{k+1} = \mathbf{P}\mathbf{w}^k - \mathbf{M}(\mathbf{P}\mathbf{w}^k - \mathbf{P}\widetilde{\mathbf{w}}^{k+1}).
$$

So ADMM belongs to the unified ADMM framework shown in Algorithm 3.12 [12], where we specify $\boldsymbol{\xi}^k = \mathbf{P}\mathbf{w}^k$ for ADMM.

---

**Algorithm 3.12** Unified ADMM framework in variational inequality

---

Initialize $\boldsymbol{\xi}^0$.
**for** $k = 0, 1, 2, 3, \cdots$ **do**
   Predict $\widetilde{\mathbf{w}}^{k+1}$ satisfying

$$
\theta(\mathbf{u}) - \theta(\widetilde{\mathbf{u}}^{k+1}) + \left\langle \mathbf{w} - \widetilde{\mathbf{w}}^{k+1}, F(\widetilde{\mathbf{w}}^{k+1}) \right\rangle \geq (\widetilde{\mathbf{w}}^{k+1} - \mathbf{w})^T \mathbf{P}^T \mathbf{H} \mathbf{M}(\mathbf{P}\widetilde{\mathbf{w}}^{k+1} - \boldsymbol{\xi}^k), \forall \mathbf{w}.
$$

   Correct $\boldsymbol{\xi}^{k+1}$ by

$$
\boldsymbol{\xi}^{k+1} = \boldsymbol{\xi}^k - \mathbf{M}(\boldsymbol{\xi}^k - \mathbf{P}\widetilde{\mathbf{w}}^{k+1}).
$$

**end for**

---

Next, we show that the linearized ADMM and ADMM with prediction-correction both belong to the unified framework. More examples can be found in [12]. Since the linearized ADMM with parallel splitting is a special case of the linearized ADMM, we omit the details.

For the linearized ADMM, from the updates (3.30) and (3.31), we have

$$
f(\mathbf{x}) + g(\mathbf{y}) - f(\widetilde{\mathbf{x}}^{k+1}) - g(\widetilde{\mathbf{y}}^{k+1}) + \left\langle \begin{pmatrix} \mathbf{x} - \widetilde{\mathbf{x}}^{k+1} \\ \mathbf{y} - \widetilde{\mathbf{y}}^{k+1} \\ \boldsymbol{\lambda} - \widetilde{\boldsymbol{\lambda}}^{k+1} \end{pmatrix}, \begin{pmatrix} \mathbf{A}^T \widetilde{\boldsymbol{\lambda}}^{k+1} \\ \mathbf{B}^T \widetilde{\boldsymbol{\lambda}}^{k+1} \\ -(\mathbf{A}\widetilde{\mathbf{x}}^{k+1} + \mathbf{B}\widetilde{\mathbf{y}}^{k+1} - \mathbf{b}) \end{pmatrix} \right\rangle
$$

$$
\geq \left\langle \widetilde{\boldsymbol{\lambda}}^{k+1} - \boldsymbol{\lambda}, \mathbf{A}\widetilde{\mathbf{x}}^{k+1} + \mathbf{B}\widetilde{\mathbf{y}}^{k+1} - \mathbf{b} \right\rangle + \beta \|\mathbf{A}\|_2^2 \left\langle \widetilde{\mathbf{x}}^{k+1} - \mathbf{x}^k, \widetilde{\mathbf{x}}^{k+1} - \mathbf{x} \right\rangle
$$

$$
- \beta \left\langle \mathbf{A}\widetilde{\mathbf{x}}^{k+1} - \mathbf{A}\mathbf{x}^k, \mathbf{A}\widetilde{\mathbf{x}}^{k+1} - \mathbf{A}\mathbf{x} \right\rangle + \beta \|\mathbf{B}\|_2^2 \left\langle \widetilde{\mathbf{y}}^{k+1} - \mathbf{y}^k, \widetilde{\mathbf{y}}^{k+1} - \mathbf{y} \right\rangle
$$

$$
= \begin{pmatrix} \widetilde{\mathbf{x}}^{k+1} - \mathbf{x} \\ \widetilde{\mathbf{y}}^{k+1} - \mathbf{y} \\ \widetilde{\boldsymbol{\lambda}}^{k+1} - \boldsymbol{\lambda} \end{pmatrix}^T \begin{pmatrix} \beta\|\mathbf{A}\|_2^2\mathbf{I} - \beta\mathbf{A}^T\mathbf{A} & \mathbf{0} & \mathbf{0} \\ \mathbf{0} & \beta\|\mathbf{B}\|_2^2\mathbf{I} & \mathbf{0} \\ \mathbf{0} & \mathbf{B} & \frac{1}{\beta}\mathbf{I} \end{pmatrix} \begin{pmatrix} \widetilde{\mathbf{x}}^{k+1} - \mathbf{x}^k \\ \widetilde{\mathbf{y}}^{k+1} - \mathbf{y}^k \\ \widetilde{\boldsymbol{\lambda}}^{k+1} - \boldsymbol{\lambda}^k \end{pmatrix}
$$

$$
= (\widetilde{\mathbf{w}}^{k+1} - \mathbf{w})^T \mathbf{H} \mathbf{M}(\widetilde{\mathbf{w}}^{k+1} - \mathbf{w}^k),
$$

where we denote

$$\widetilde{\mathbf{w}}^{k+1} = \begin{pmatrix} \widetilde{\mathbf{x}}^{k+1} \\ \widetilde{\mathbf{y}}^{k+1} \\ \widetilde{\boldsymbol{\lambda}}^{k+1} \end{pmatrix} = \begin{pmatrix} \mathbf{x}^{k+1} \\ \mathbf{y}^{k+1} \\ \boldsymbol{\lambda}^k + \beta(\mathbf{A}\mathbf{x}^{k+1} + \mathbf{B}\mathbf{y}^k - \mathbf{b}) \end{pmatrix},$$

$$\mathbf{H} = \begin{pmatrix} \beta\|\mathbf{A}\|_2^2\mathbf{I} - \beta\mathbf{A}^T\mathbf{A} & \mathbf{0} & \mathbf{0} \\ \mathbf{0} & \beta\|\mathbf{B}\|_2^2\mathbf{I} & \mathbf{0} \\ \mathbf{0} & \mathbf{0} & \frac{1}{\beta}\mathbf{I} \end{pmatrix}, \quad \text{and} \quad \mathbf{M} = \begin{pmatrix} \mathbf{I} & \mathbf{0} & \mathbf{0} \\ \mathbf{0} & \mathbf{I} & \mathbf{0} \\ \mathbf{0} & \beta\mathbf{B} & \mathbf{I} \end{pmatrix}. \tag{3.93}$$

We can also check that

$$\mathbf{w}^{k+1} = \mathbf{w}^k - \mathbf{M}(\mathbf{w}^k - \widetilde{\mathbf{w}}^{k+1}).$$

So the linearized ADMM belongs to the unified framework in Algorithm 3.12 with $\mathbf{P} = \mathbf{I}$, $\boldsymbol{\xi}^k = \mathbf{w}^k$, and $\mathbf{H}$ and $\mathbf{M}$ defined in (3.93).

For the multi-block problems, denote

$$\mathbf{w} = \begin{pmatrix} \mathbf{x}_1 \\ \vdots \\ \mathbf{x}_m \\ \boldsymbol{\lambda} \end{pmatrix}, \quad \mathbf{u} = \begin{pmatrix} \mathbf{x}_1 \\ \vdots \\ \mathbf{x}_m \end{pmatrix}, \quad F(\mathbf{w}) = \begin{pmatrix} \mathbf{A}_1^T\boldsymbol{\lambda} \\ \vdots \\ \mathbf{A}_m^T\boldsymbol{\lambda} \\ -\left(\sum_{i=1}^m \mathbf{A}_i\mathbf{x}_i - \mathbf{b}\right) \end{pmatrix}, \quad \theta(\mathbf{u}) = \sum_{i=1}^m f_i(\mathbf{x}_i),$$

and

$$\widetilde{\boldsymbol{\lambda}}^{k+1} = \boldsymbol{\lambda}^k + \beta\left(\sum_{i=1}^m \boldsymbol{\xi}_i^k - \mathbf{b}\right).$$

Then for the prediction (3.82a)–(3.82d), we have

$$\sum_{i=1}^m f_i(\mathbf{x}_i) - \sum_{i=1}^m f_i(\widetilde{\mathbf{x}}_i^{k+1}) + \left\langle \begin{pmatrix} \mathbf{x}_1 - \widetilde{\mathbf{x}}_1^{k+1} \\ \vdots \\ \mathbf{x}_m - \widetilde{\mathbf{x}}_m^{k+1} \\ \boldsymbol{\lambda} - \widetilde{\boldsymbol{\lambda}}^{k+1} \end{pmatrix}, \begin{pmatrix} \mathbf{A}_1^T\widetilde{\boldsymbol{\lambda}}^{k+1} \\ \vdots \\ \mathbf{A}_m^T\widetilde{\boldsymbol{\lambda}}^{k+1} \\ -\left(\sum_{i=1}^m \mathbf{A}_i\widetilde{\mathbf{x}}_i^{k+1} - \mathbf{b}\right) \end{pmatrix} \right\rangle$$

$$\geq \left\langle \widetilde{\boldsymbol{\lambda}}^{k+1} - \boldsymbol{\lambda}, \sum_{i=1}^m \mathbf{A}_i\widetilde{\mathbf{x}}_i^{k+1} - \mathbf{b} \right\rangle + \beta\sum_{i=1}^m \left\langle \sum_{j=1}^i \left(\mathbf{A}_j\widetilde{\mathbf{x}}_j^{k+1} - \boldsymbol{\xi}_j^k\right), \mathbf{A}_i\widetilde{\mathbf{x}}_i^{k+1} - \mathbf{A}_i\mathbf{x}_i \right\rangle$$

$$
= \begin{pmatrix} \mathbf{A}_1\widetilde{\mathbf{x}}_1^{k+1} - \mathbf{A}_1\mathbf{x}_1 \\ \vdots \\ \mathbf{A}_m\widetilde{\mathbf{x}}_m^{k+1} - \mathbf{A}_m\mathbf{x}_m \\ \widetilde{\boldsymbol{\lambda}}^{k+1} - \boldsymbol{\lambda} \end{pmatrix}^T \begin{pmatrix} \beta\mathbf{I} & \mathbf{0} & \cdots & \mathbf{0} & \mathbf{0} \\ \vdots & \vdots & \ddots & \vdots & \vdots \\ \beta\mathbf{I} & \beta\mathbf{I} & \cdots & \beta\mathbf{I} & \mathbf{0} \\ \mathbf{I} & \mathbf{I} & \mathbf{I} & \mathbf{I} & \frac{1}{\beta}\mathbf{I} \end{pmatrix} \begin{pmatrix} \mathbf{A}_1\widetilde{\mathbf{x}}_1^{k+1} - \boldsymbol{\xi}_1^k \\ \vdots \\ \mathbf{A}_m\widetilde{\mathbf{x}}_m^{k+1} - \boldsymbol{\xi}_m^k \\ \widetilde{\boldsymbol{\lambda}}^{k+1} - \boldsymbol{\lambda}^k \end{pmatrix}
$$

$$
= (\widetilde{\mathbf{w}}^{k+1} - \mathbf{w})^T \mathbf{P}^T \mathbf{H} \mathbf{M} (\mathbf{P}\widetilde{\mathbf{w}}^{k+1} - \boldsymbol{\xi}^k),
$$

where we denote

$$
\mathbf{H} = \begin{pmatrix} \beta\mathbf{L}\mathbf{L}^T & \mathbf{0} \\ \mathbf{0} & \frac{1}{\beta}\mathbf{I} \end{pmatrix}, \quad \mathbf{M} = \begin{pmatrix} \mathbf{L}^{-T} & \mathbf{0} \\ \beta(\mathbf{I}\cdots\mathbf{I}) & \mathbf{I} \end{pmatrix}, \quad \mathbf{P} = \begin{pmatrix} \mathbf{A}_1 & \cdots & \mathbf{0} & \mathbf{0} \\ \vdots & \ddots & \vdots & \vdots \\ \mathbf{0} & \cdots & \mathbf{A}_m & \mathbf{0} \\ \mathbf{0} & \cdots & \mathbf{0} & \mathbf{I} \end{pmatrix},
$$

$$
\widetilde{\mathbf{w}} = (\widetilde{\mathbf{x}}_1^T, \cdots, \widetilde{\mathbf{x}}_m^T, \widetilde{\boldsymbol{\lambda}}^T)^T, \quad \boldsymbol{\xi} = (\boldsymbol{\xi}_1^T, \cdots, \boldsymbol{\xi}_m^T, \boldsymbol{\lambda}^T)^T,
$$

and $\mathbf{L}$ is defied in (3.77). From (3.82d) and (3.82c), we can also check that the following relationship

$$
\boldsymbol{\xi}^{k+1} = \boldsymbol{\xi}^k - \mathbf{M}(\boldsymbol{\xi}^k - \mathbf{P}\widetilde{\mathbf{w}}^{k+1})
$$

holds. Thus ADMM with prediction-correction also belongs to the unified framework.

### 3.6.2  Unified Convergence Rate Analysis

In this section, we give the convergence rate of the unified framework in the sense of (3.90) [12].

**Theorem 3.17** *Suppose that $\theta(\mathbf{u})$ is convex. Define*

$$
\hat{\mathbf{w}}^{K+1} = \frac{1}{K+1}\sum_{k=0}^{K}\widetilde{\mathbf{w}}^{k+1} \quad and \quad \hat{\mathbf{u}}^{K+1} = \frac{1}{K+1}\sum_{k=0}^{K}\widetilde{\mathbf{u}}^{k+1}.
$$

*Suppose that*

$$
\mathbf{M}^T\mathbf{H}^T + \mathbf{H}\mathbf{M} - \mathbf{M}^T\mathbf{H}\mathbf{M} \succeq 0, \quad \mathbf{P}^T\mathbf{H}\mathbf{P} \succeq 0, \quad and \quad \mathbf{P}^T\mathbf{H}\mathbf{P} \neq 0. \qquad (3.94)
$$

*Then for Algorithm 3.12, we have*

$$
\theta(\hat{\mathbf{u}}^{K+1}) - \theta(\mathbf{u}) - \left\langle \mathbf{w} - \hat{\mathbf{w}}^{K+1}, F(\mathbf{w}) \right\rangle \leq \frac{1}{2(K+1)}\|\boldsymbol{\xi}^0 - \mathbf{P}\mathbf{w}\|_{\mathbf{H}}^2, \quad \forall\mathbf{w}.
$$

***Proof*** From the two conditions in Algorithm 3.12, we have that for any $\mathbf{w}$,

$$\theta(\mathbf{u}) - \theta(\widetilde{\mathbf{u}}^{k+1}) + \left\langle \mathbf{w} - \widetilde{\mathbf{w}}^{k+1}, F(\widetilde{\mathbf{w}}^{k+1}) \right\rangle$$

$$\geq (\widetilde{\mathbf{w}}^{k+1} - \mathbf{w})^T \mathbf{P}^T \mathbf{HM}(\mathbf{P}\widetilde{\mathbf{w}}^{k+1} - \xi^k)$$

$$= (\widetilde{\mathbf{w}}^{k+1} - \mathbf{w})^T \mathbf{P}^T \mathbf{H}(\xi^{k+1} - \xi^k)$$

$$\overset{a}{=} \frac{1}{2}\left( \|\xi^{k+1} - \mathbf{Pw}\|_{\mathbf{H}}^2 - \|\xi^k - \mathbf{Pw}\|_{\mathbf{H}}^2 + \|\mathbf{P}\widetilde{\mathbf{w}}^{k+1} - \xi^k\|_{\mathbf{H}}^2 \right.$$

$$\left. - \|\mathbf{P}\widetilde{\mathbf{w}}^{k+1} - \xi^{k+1}\|_{\mathbf{H}}^2 \right),$$

where $\overset{a}{=}$ uses a generalized version of (A.3).

We can check that

$$\|\mathbf{P}\widetilde{\mathbf{w}}^{k+1} - \xi^k\|_{\mathbf{H}}^2 - \|\mathbf{P}\widetilde{\mathbf{w}}^{k+1} - \xi^{k+1}\|_{\mathbf{H}}^2$$

$$= \|\mathbf{P}\widetilde{\mathbf{w}}^{k+1} - \xi^k\|_{\mathbf{H}}^2 - \|\mathbf{P}\widetilde{\mathbf{w}}^{k+1} - \xi^k - \mathbf{M}(\mathbf{P}\widetilde{\mathbf{w}}^{k+1} - \xi^k)\|_{\mathbf{H}}^2$$

$$= 2(\mathbf{P}\widetilde{\mathbf{w}}^{k+1} - \xi^k)^T \mathbf{HM}(\mathbf{P}\widetilde{\mathbf{w}}^{k+1} - \xi^k)$$

$$\quad - (\mathbf{P}\widetilde{\mathbf{w}}^{k+1} - \xi^k)^T \mathbf{M}^T \mathbf{HM}(\mathbf{P}\widetilde{\mathbf{w}}^{k+1} - \xi^k)$$

$$= (\mathbf{P}\widetilde{\mathbf{w}}^{k+1} - \xi^k)^T \left( \mathbf{HM} + \mathbf{M}^T \mathbf{H}^T - \mathbf{M}^T \mathbf{HM} \right) (\mathbf{P}\widetilde{\mathbf{w}}^{k+1} - \xi^k)$$

$$\geq 0.$$

Using (3.88), we have that for any $\mathbf{w}$,

$$\theta(\mathbf{u}) - \theta(\widetilde{\mathbf{u}}^{k+1}) + \left\langle \mathbf{w} - \widetilde{\mathbf{w}}^{k+1}, F(\mathbf{w}) \right\rangle$$

$$\geq \frac{1}{2}\left( \|\xi^{k+1} - \mathbf{Pw}\|_{\mathbf{H}}^2 - \|\xi^k - \mathbf{Pw}\|_{\mathbf{H}}^2 \right).$$

Summing over $k = 0, 1, \cdots, K$ and dividing both sides by $K + 1$, using the convexity of $f$ and $g$, we have the conclusion. $\qquad\square$

We can check that ADMM, linearized ADMM, and ADMM with prediction-correction all satisfy (3.94). So their convergence rates can be given by Theorem 3.17 in a unified way.

## 3.7   The Case of Nonlinear Constraints

In this section, we introduce how to extend ADMM to solve the generally convex program with both equality and inequality constraints. Consider problem:

$$\min_{\mathbf{x},\mathbf{y}} \ (f(\mathbf{x}) + g(\mathbf{y})),$$

$$s.t. \ h_0(\mathbf{x}) \leq 0,$$

$$p_0(\mathbf{y}) \leq 0,$$

$$\mathbf{Ax} + \mathbf{By} = \mathbf{b},$$

where $f$, $g$, $h_0$, and $p_0$ are convex functions. Define

$$h(\mathbf{x}) = \max\{0, h_0(\mathbf{x})\} \quad \text{and} \quad p(\mathbf{y}) = \max\{0, p_0(\mathbf{y})\}.$$

Then we can turn the inequality constraints into equality constraints. Thus, we consider the following problem instead:

$$\min_{\mathbf{x},\mathbf{y}} \ (f(\mathbf{x}) + g(\mathbf{y})),$$

$$s.t. \ h(\mathbf{x}) = 0,$$

$$p(\mathbf{y}) = 0,$$

$$\mathbf{Ax} + \mathbf{By} = \mathbf{b}. \tag{3.95}$$

Define the augmented Lagrangian function as follows:

$$L_{\rho_1,\rho_2,\beta}(\mathbf{x}, \mathbf{y}, \gamma, \tau, \boldsymbol{\lambda})$$

$$= f(\mathbf{x}) + g(\mathbf{y}) + \gamma h(\mathbf{x}) + \frac{\rho_1}{2} h^2(\mathbf{x}) + \tau p(\mathbf{y}) + \frac{\rho_2}{2} p^2(\mathbf{y})$$

$$+ \langle \boldsymbol{\lambda}, \mathbf{Ax} + \mathbf{By} - \mathbf{b} \rangle + \frac{\beta}{2} \|\mathbf{Ax} + \mathbf{By} - \mathbf{b}\|^2.$$

We can use the following ADMM to solve Problem (3.95) with $O\left(\frac{1}{K}\right)$ convergence rate [6]:

$$\mathbf{x}^{k+1} = \underset{\mathbf{x}}{\operatorname{argmin}} \ L_{\rho_1,\rho_2,\beta}(\mathbf{x}, \mathbf{y}^k, \gamma^k, \tau^k, \boldsymbol{\lambda}^k), \tag{3.96a}$$

$$\mathbf{y}^{k+1} = \underset{\mathbf{y}}{\operatorname{argmin}} \ L_{\rho_1,\rho_2,\beta}(\mathbf{x}^{k+1}, \mathbf{y}, \gamma^k, \tau^k, \boldsymbol{\lambda}^k), \tag{3.96b}$$

$$\gamma^{k+1} = \gamma^k + \rho_1 h(\mathbf{x}^{k+1}), \tag{3.96c}$$

$$\tau^{k+1} = \tau^k + \rho_2 p(\mathbf{y}^{k+1}), \tag{3.96d}$$

$$\boldsymbol{\lambda}^{k+1} = \boldsymbol{\lambda}^k + \beta(\mathbf{A}\mathbf{x}^{k+1} + \mathbf{B}\mathbf{y}^{k+1} - \mathbf{b}). \tag{3.96e}$$

We present the above method in Algorithm 3.13

---

**Algorithm 3.13** ADMM with nonlinear constraints

Initialize $\mathbf{x}^0, \mathbf{y}^0, \gamma^0, \tau^0$, and $\boldsymbol{\lambda}^0$.
**for** $k = 0, 1, 2, 3, \cdots$ **do**
   Update $\mathbf{x}^{k+1}, \mathbf{y}^{k+1}, \gamma^{k+1}, \tau^{k+1}$, and $\boldsymbol{\lambda}^{k+1}$ by (3.96a)–(3.96e), respectively.
**end for**

---

**Theorem 3.18**  *Suppose that $f(\mathbf{x})$, $g(\mathbf{y})$, $h_0(\mathbf{x})$, and $p_0(\mathbf{y})$ are all convex. Then for the above ADMM, we have*

$$|f(\hat{\mathbf{x}}^{K+1}) + g(\hat{\mathbf{y}}^{K+1}) - f(\mathbf{x}^*) - g(\mathbf{y}^*)| \le \frac{C}{2(K+1)} + \frac{2\sqrt{C}\|\boldsymbol{\lambda}^*\|}{\sqrt{\beta}(K+1)}$$

$$+ \frac{2\sqrt{C}|\gamma^*|}{\sqrt{\rho_1}(K+1)} + \frac{2\sqrt{C}|\tau^*|}{\sqrt{\rho_2}(K+1)},$$

$$\|\mathbf{A}\hat{\mathbf{x}}^{K+1} + \mathbf{B}\hat{\mathbf{y}}^{K+1} - \mathbf{b}\| \le \frac{2\sqrt{C}}{\sqrt{\beta}(K+1)},$$

$$h(\hat{\mathbf{x}}^{K+1}) \le \frac{2\sqrt{C}}{\sqrt{\rho_1}(K+1)},$$

$$p(\hat{\mathbf{y}}^{K+1}) \le \frac{2\sqrt{C}}{\sqrt{\rho_2}(K+1)},$$

*where*

$$\hat{\mathbf{x}}^{K+1} = \frac{1}{K+1}\sum_{k=1}^{K+1}\mathbf{x}^k, \quad \hat{\mathbf{y}}^{K+1} = \frac{1}{K+1}\sum_{k=1}^{K+1}\mathbf{y}^k, \quad and$$

$$C = \frac{1}{\beta}\|\boldsymbol{\lambda}^0 - \boldsymbol{\lambda}^*\|^2 + \frac{1}{\rho_1}(\gamma^0 - \gamma^*)^2 + \frac{1}{\rho_2}(\tau^0 - \tau^*)^2 + \beta\|\mathbf{B}\mathbf{y}^0 - \mathbf{B}\mathbf{y}^*\|^2.$$

***Proof***  Similar to Lemma 3.3, we have the following properties:

$$\mathbf{0} \in \partial f(\mathbf{x}^{k+1}) + \mathbf{A}^T\boldsymbol{\lambda}^k + \beta\mathbf{A}^T(\mathbf{A}\mathbf{x}^{k+1} + \mathbf{B}\mathbf{y}^k - \mathbf{b})$$

$$+ \gamma^k\partial h(\mathbf{x}^{k+1}) + \rho_1 h(\mathbf{x}^{k+1})\partial h(\mathbf{x}^{k+1}),$$

$$\mathbf{0} \in \partial g(\mathbf{y}^{k+1}) + \mathbf{B}^T\boldsymbol{\lambda}^k + \beta\mathbf{B}^T(\mathbf{A}\mathbf{x}^{k+1} + \mathbf{B}\mathbf{y}^{k+1} - \mathbf{b})$$

$$+\tau^k \partial p(\mathbf{y}^{k+1}) + \rho_2 p(\mathbf{y}^{k+1}) \partial p(\mathbf{y}^{k+1}),$$

$$\mathbf{0} \in \partial f(\mathbf{x}^*) + \mathbf{A}^T \boldsymbol{\lambda}^* + \gamma^* \partial h(\mathbf{x}^*),$$

$$\mathbf{0} \in \partial g(\mathbf{y}^*) + \mathbf{B}^T \boldsymbol{\lambda}^* + \tau^* \partial p(\mathbf{y}^*),$$

$$\mathbf{A}\mathbf{x}^* + \mathbf{B}\mathbf{y}^* = \mathbf{b},$$

$$h(\mathbf{x}^*) = 0,$$

$$p(\mathbf{y}^*) = 0,$$

where $(\mathbf{x}^*, \mathbf{y}^*, \gamma^*, \tau^*, \boldsymbol{\lambda}^*)$ is a KKT point. Then there exists

$$\hat{\nabla} h(\mathbf{x}^{k+1}) \in \partial h(\mathbf{x}^{k+1}) \quad \text{and} \quad \hat{\nabla} p(\mathbf{y}^{k+1}) \in \partial p(\mathbf{y}^{k+1}),$$

such that

$$\hat{\nabla} f(\mathbf{x}^{k+1}) \in \partial f(\mathbf{x}^{k+1}) \quad \text{and} \quad \hat{\nabla} g(\mathbf{y}^{k+1}) \in \partial g(\mathbf{y}^{k+1}),$$

where

$$\hat{\nabla} f(\mathbf{x}^{k+1}) = -\mathbf{A}^T \boldsymbol{\lambda}^k - \beta \mathbf{A}^T (\mathbf{A}\mathbf{x}^{k+1} + \mathbf{B}\mathbf{y}^k - \mathbf{b})$$

$$-\gamma^k \hat{\nabla} h(\mathbf{x}^{k+1}) - \rho_1 h(\mathbf{x}^{k+1}) \hat{\nabla} h(\mathbf{x}^{k+1})$$

$$= -\mathbf{A}^T \boldsymbol{\lambda}^{k+1} + \beta \mathbf{A}^T (\mathbf{B}\mathbf{y}^{k+1} - \mathbf{B}\mathbf{y}^k) - \gamma^{k+1} \hat{\nabla} h(\mathbf{x}^{k+1}),$$

$$\hat{\nabla} g(\mathbf{y}^{k+1}) = -\mathbf{B}^T \boldsymbol{\lambda}^k - \beta \mathbf{B}^T (\mathbf{A}\mathbf{x}^{k+1} + \mathbf{B}\mathbf{y}^{k+1} - \mathbf{b})$$

$$-\tau^k \hat{\nabla} p(\mathbf{y}^{k+1}) - \rho_2 p(\mathbf{y}^{k+1}) \hat{\nabla} p(\mathbf{y}^{k+1})$$

$$= -\mathbf{B}^T \boldsymbol{\lambda}^{k+1} - \tau^{k+1} \hat{\nabla} p(\mathbf{y}^{k+1}).$$

Similar to Lemma 3.4, we have

$$\left\langle \hat{\nabla} g(\mathbf{y}^{k+1}), \mathbf{y}^{k+1} - \mathbf{y} \right\rangle$$

$$= -\left\langle \boldsymbol{\lambda}^{k+1}, \mathbf{B}\mathbf{y}^{k+1} - \mathbf{B}\mathbf{y} \right\rangle - \tau^{k+1} \left\langle \hat{\nabla} p(\mathbf{y}^{k+1}), \mathbf{y}^{k+1} - \mathbf{y} \right\rangle$$

and

$$\left\langle \hat{\nabla} f(\mathbf{x}^{k+1}), \mathbf{x}^{k+1} - \mathbf{x} \right\rangle + \left\langle \hat{\nabla} g(\mathbf{y}^{k+1}), \mathbf{y}^{k+1} - \mathbf{y} \right\rangle$$

$$= -\left\langle \boldsymbol{\lambda}^{k+1}, \mathbf{A}\mathbf{x}^{k+1} + \mathbf{B}\mathbf{y}^{k+1} - \mathbf{A}\mathbf{x} - \mathbf{B}\mathbf{y} \right\rangle + \beta \left\langle \mathbf{B}\mathbf{y}^{k+1} - \mathbf{B}\mathbf{y}^k, \mathbf{A}\mathbf{x}^{k+1} - \mathbf{A}\mathbf{x} \right\rangle$$

$$- \gamma^{k+1} \left\langle \hat{\nabla} h(\mathbf{x}^{k+1}), \mathbf{x}^{k+1} - \mathbf{x} \right\rangle - \tau^{k+1} \left\langle \hat{\nabla} p(\mathbf{y}^{k+1}), \mathbf{y}^{k+1} - \mathbf{y} \right\rangle$$

$$\overset{a}{\leq} -\left\langle \boldsymbol{\lambda}^{k+1}, \mathbf{A}\mathbf{x}^{k+1} + \mathbf{B}\mathbf{y}^{k+1} - \mathbf{A}\mathbf{x} - \mathbf{B}\mathbf{y} \right\rangle + \beta \left\langle \mathbf{B}\mathbf{y}^{k+1} - \mathbf{B}\mathbf{y}^k, \mathbf{A}\mathbf{x}^{k+1} - \mathbf{A}\mathbf{x} \right\rangle$$

$$- \gamma^{k+1} \left( h(\mathbf{x}^{k+1}) - h(\mathbf{x}) \right) - \tau^{k+1} \left( p(\mathbf{y}^{k+1}) - p(\mathbf{y}) \right), \tag{3.97}$$

where we use the convexity of $h$ and $p$ ($h$ and $p$ are convex since $h_0$ and $p_0$ are convex) in $\overset{a}{\leq}$.

Letting $\mathbf{x} = \mathbf{x}^*$ and $\mathbf{y} = \mathbf{y}^*$ in (3.97) and adding

$$\left\langle \boldsymbol{\lambda}^*, \mathbf{A}\mathbf{x}^{k+1} + \mathbf{B}\mathbf{y}^{k+1} - \mathbf{b} \right\rangle + \gamma^* h(\mathbf{x}^{k+1}) + \tau^* p(\mathbf{y}^{k+1})$$

to both sides of (3.97), similar to Lemma 3.5 we have

$$\left\langle \hat{\nabla} f(\mathbf{x}^{k+1}), \mathbf{x}^{k+1} - \mathbf{x}^* \right\rangle + \left\langle \hat{\nabla} g(\mathbf{y}^{k+1}), \mathbf{y}^{k+1} - \mathbf{y}^* \right\rangle + \left\langle \boldsymbol{\lambda}^*, \mathbf{A}\mathbf{x}^{k+1} + \mathbf{B}\mathbf{y}^{k+1} - \mathbf{b} \right\rangle$$

$$+ \gamma^* h(\mathbf{x}^{k+1}) + \tau^* p(\mathbf{y}^{k+1})$$

$$\overset{a}{\leq} -\left\langle \boldsymbol{\lambda}^{k+1} - \boldsymbol{\lambda}^*, \mathbf{A}\mathbf{x}^{k+1} + \mathbf{B}\mathbf{y}^{k+1} - \mathbf{b} \right\rangle + \beta \left\langle \mathbf{B}\mathbf{y}^{k+1} - \mathbf{B}\mathbf{y}^k, \mathbf{A}\mathbf{x}^{k+1} - \mathbf{A}\mathbf{x}^* \right\rangle$$

$$- (\gamma^{k+1} - \gamma^*) h(\mathbf{x}^{k+1}) - (\tau^{k+1} - \tau^*) p(\mathbf{y}^{k+1})$$

$$= \frac{1}{2\beta} \|\boldsymbol{\lambda}^k - \boldsymbol{\lambda}^*\|^2 - \frac{1}{2\beta} \|\boldsymbol{\lambda}^{k+1} - \boldsymbol{\lambda}^*\|^2 - \frac{1}{2\beta} \|\boldsymbol{\lambda}^{k+1} - \boldsymbol{\lambda}^k\|^2$$

$$+ \frac{\beta}{2} \|\mathbf{B}\mathbf{y}^k - \mathbf{B}\mathbf{y}^*\|^2 - \frac{\beta}{2} \|\mathbf{B}\mathbf{y}^{k+1} - \mathbf{B}\mathbf{y}^*\|^2 - \frac{\beta}{2} \|\mathbf{B}\mathbf{y}^{k+1} - \mathbf{B}\mathbf{y}^k\|^2$$

$$+ \left\langle \mathbf{B}\mathbf{y}^{k+1} - \mathbf{B}\mathbf{y}^k, \boldsymbol{\lambda}^{k+1} - \boldsymbol{\lambda}^k \right\rangle$$

$$- (\gamma^{k+1} - \gamma^*) h(\mathbf{x}^{k+1}) - (\tau^{k+1} - \tau^*) p(\mathbf{y}^{k+1})$$

$$\overset{b}{=} \frac{1}{2\beta} \|\boldsymbol{\lambda}^k - \boldsymbol{\lambda}^*\|^2 - \frac{1}{2\beta} \|\boldsymbol{\lambda}^{k+1} - \boldsymbol{\lambda}^*\|^2 - \frac{1}{2\beta} \|\boldsymbol{\lambda}^{k+1} - \boldsymbol{\lambda}^k\|^2$$

$$+ \frac{\beta}{2} \|\mathbf{B}\mathbf{y}^k - \mathbf{B}\mathbf{y}^*\|^2 - \frac{\beta}{2} \|\mathbf{B}\mathbf{y}^{k+1} - \mathbf{B}\mathbf{y}^*\|^2 - \frac{\beta}{2} \|\mathbf{B}\mathbf{y}^{k+1} - \mathbf{B}\mathbf{y}^k\|^2$$

$$+ \left\langle \mathbf{B}\mathbf{y}^{k+1} - \mathbf{B}\mathbf{y}^k, \boldsymbol{\lambda}^{k+1} - \boldsymbol{\lambda}^k \right\rangle$$

$$+ \frac{1}{2\rho_1} (\gamma^k - \gamma^*)^2 - \frac{1}{2\rho_1} (\gamma^{k+1} - \gamma^*)^2 - \frac{1}{2\rho_1} (\gamma^{k+1} - \gamma^k)^2$$

$$+ \frac{1}{2\rho_2} (\tau^k - \tau^*)^2 - \frac{1}{2\rho_2} (\tau^{k+1} - \tau^*)^2 - \frac{1}{2\rho_2} (\tau^{k+1} - \tau^k)^2,$$

where we use $h(\mathbf{x}^*) = 0$ and $p(\mathbf{y}^*) = 0$ in $\overset{a}{\leq}$, $\gamma^{k+1} = \gamma^k + \rho_1 h(\mathbf{x}^{k+1})$, $\tau^{k+1} = \tau^k + \rho_2 p(\mathbf{y}^{k+1})$, and Lemma A.1 in $\overset{b}{=}$. Thus, using the convexity of $f$ and $g$, we have

$$
\begin{aligned}
& f(\mathbf{x}^{k+1}) + g(\mathbf{y}^{k+1}) - f(\mathbf{x}^*) - g(\mathbf{y}^*) + \left\langle \boldsymbol{\lambda}^*, \mathbf{A}^{k+1} + \mathbf{By}^{k+1} - \mathbf{b} \right\rangle \\
& \quad + \gamma^* h(\mathbf{x}^{k+1}) + \tau^* p(\mathbf{y}^{k+1}) \\
& \leq \frac{1}{2\beta} \|\boldsymbol{\lambda}^k - \boldsymbol{\lambda}^*\|^2 - \frac{1}{2\beta} \|\boldsymbol{\lambda}^{k+1} - \boldsymbol{\lambda}^*\|^2 \\
& \quad + \frac{\beta}{2} \|\mathbf{By}^k - \mathbf{By}^*\|^2 - \frac{\beta}{2} \|\mathbf{By}^{k+1} - \mathbf{By}^*\|^2 \\
& \quad + \frac{1}{2\rho_1}(\gamma^k - \gamma^*)^2 - \frac{1}{2\rho_1}(\gamma^{k+1} - \gamma^*)^2 - \frac{1}{2\rho_1}(\gamma^{k+1} - \gamma^k)^2 \\
& \quad + \frac{1}{2\rho_2}(\tau^k - \tau^*)^2 - \frac{1}{2\rho_2}(\tau^{k+1} - \tau^*)^2 - \frac{1}{2\rho_2}(\tau^{k+1} - \tau^k)^2,
\end{aligned}
$$

where we drop

$$
\left\langle \mathbf{By}^{k+1} - \mathbf{By}^k, \boldsymbol{\lambda}^{k+1} - \boldsymbol{\lambda}^k \right\rangle - \frac{1}{2\beta} \|\boldsymbol{\lambda}^{k+1} - \boldsymbol{\lambda}^k\|^2 - \frac{\beta}{2} \|\mathbf{By}^{k+1} - \mathbf{By}^k\|^2 \leq 0
$$

by $2 \langle \mathbf{a}, \mathbf{b} \rangle \leq \frac{1}{\beta} \|\mathbf{a}\|^2 + \beta \|\mathbf{b}\|^2$.

Note that Lemmas 3.1 and 3.2 still hold for Problem (3.95) with minor modification. Similar to Theorem 3.3, using the convexity of $f$, $g$, $h$, and $p$, we have

$$
\begin{aligned}
& f(\hat{\mathbf{x}}^{K+1}) + g(\hat{\mathbf{y}}^{K+1}) - f(\mathbf{x}^*) - g(\mathbf{y}^*) + \left\langle \boldsymbol{\lambda}^*, \mathbf{A}\hat{\mathbf{x}}^{K+1} + \mathbf{B}\hat{\mathbf{y}}^{K+1} - \mathbf{b} \right\rangle \\
& \quad + \gamma^* h(\hat{\mathbf{x}}^{K+1}) + \tau^* p(\hat{\mathbf{y}}^{K+1}) \leq \frac{C}{2(K+1)},
\end{aligned}
$$

$$
\|\boldsymbol{\lambda}^{K+1} - \boldsymbol{\lambda}^*\| \leq \sqrt{\beta C}, \quad |\gamma^{K+1} - \gamma^*| \leq \sqrt{\rho_1 C}, \quad |\tau^{K+1} - \tau^*| \leq \sqrt{\rho_2 C},
$$

$$
\|\mathbf{A}\hat{\mathbf{x}}^{K+1} + \mathbf{B}\hat{\mathbf{y}}^{K+1} - \mathbf{b}\| \leq \frac{2\sqrt{C}}{\sqrt{\beta}(K+1)},
$$

and

$$
h(\hat{\mathbf{x}}^{K+1}) \leq \frac{1}{K+1} \sum_{k=0}^{K} h(\mathbf{x}^{k+1})
$$

$$
= \frac{1}{\rho_1(K+1)} \sum_{k=0}^{K} (\gamma^{k+1} - \gamma^k)
$$

$$= \frac{1}{\rho_1(K+1)}(\gamma^{K+1} - \gamma^0)$$

$$\leq \frac{1}{\rho_1(K+1)}\left(|\gamma^0 - \gamma^*| + |\gamma^{K+1} - \gamma^*|\right)$$

$$\leq \frac{2\sqrt{\rho_1 C}}{\rho_1(K+1)}.$$

Similarly, we also have $p(\hat{\mathbf{y}}^{K+1}) \leq \frac{2\sqrt{\rho_2 C}}{\rho_2(K+1)}$. The proof completes. $\qquad\square$

At last, we extend to solving the following generally constrained convex problem

$$\min_{\mathbf{x}} \sum_{i=1}^{m} f_i(\mathbf{x}),$$

$$s.t. \ h_i(\mathbf{x}) \leq 0, \quad i \in [m],$$

$$\mathbf{A}\mathbf{x} = \mathbf{b}.$$

Reformulate it as the following problem

$$\min_{\{\mathbf{x}_i\}, \mathbf{z}} \sum_{i=1}^{m} f_i(\mathbf{x}_i),$$

$$s.t. \ h_i(\mathbf{x}_i) \leq 0, \quad i \in [m],$$

$$\begin{pmatrix} \mathbf{I} \\ \mathbf{I} \\ \vdots \\ \mathbf{I} \\ \mathbf{A} \end{pmatrix} \mathbf{z} - \begin{pmatrix} \mathbf{I} & 0 & \cdots & 0 \\ 0 & \mathbf{I} & \cdots & 0 \\ \vdots & \vdots & \ddots & \vdots \\ 0 & 0 & 0 & \mathbf{I} \\ 0 & 0 & 0 & 0 \end{pmatrix} \begin{pmatrix} \mathbf{x}_1 \\ \mathbf{x}_2 \\ \cdots \\ \mathbf{x}_m \end{pmatrix} = \begin{pmatrix} 0 \\ 0 \\ \vdots \\ 0 \\ \mathbf{b} \end{pmatrix}.$$

We can use ADMM to solve it, where we solve the first subproblem with $(\mathbf{x}_1^T, \cdots, \mathbf{x}_m^T)^T$, and then solve the second subproblem with $\mathbf{z}$. Note that the first subproblem can be decomposed into $m$ subproblems in parallel.

# References

1. C. Chen, B. He, Y. Ye, X. Yuan, The direct extension of ADMM for multi-block convex minimization problems is not necessarily convergent. Math. Program. 155(1–2), 57–79 (2016)
2. D. Davis, W. Yin, Convergence rate analysis of several splitting schemes, in *Splitting Methods in Communication, Imaging, Science, and Engineering* (Springer, Berlin, 2016), pp. 115–163
3. W. Deng, W. Yin, On the global and linear convergence of the generalized alternating direction method of multipliers. J. Sci. Comput. 66(3), 889–916 (2016)

4. J. Eckstein, D.P. Bertsekas, On the Douglas-Rachford splitting method and the proximal point algorithm for maximal monotone operators. Math. Program. **55**(1), 293–318 (1992)
5. D. Gabay, Applications of the method of multipliers to variational inequalities. Math. Appl. **15**, 299–331 (1983)
6. J. Giesen, S. Laue, Distributed convex optimization with many convex constraints (2018). ArXiv:1610.02967
7. B. He, X. Yuan, On non-ergodic convergence rate of Douglas-Rachford alternating directions method of multipliers. Numer. Math. **130**(3), 567–577 (2015)
8. B. He, L.-Z. Liao, D. Han, H. Yang, A new inexact alternating directions method for monotone variational inequalities. Math. Program. **92**(1), 103–118 (2002)
9. B. He, M. Tao, X. Yuan, Alternating direction method with Gaussian back substitution for separable convex programming. SIAM J. Optim. **22**(2), 313–340 (2012)
10. B. He, X. Yuan, On the $O(1/t)$ convergence rate of the Douglas-Rachford alternating direction method. SIAM J. Numer. Anal. **50**(2), 700–709 (2012)
11. B. He, M. Tao, X. Yuan, A splitting method for separable convex programming. IMA J. Numer. Anal. **35**(1), 394–426 (2015)
12. B. He, S. Xu, X. Yuan, Extensions of ADMM for separable convex optimization problems with linear equation or inequality constraints (2021). Arxiv:2107.01897
13. M. Hong, Z.-Q. Luo, On the linear convergence of the alternating direction method of multipliers. Math. Program. **162**(1–2), 165–199 (2017)
14. H. Li, Z. Lin, Accelerated alternating direction method of multipliers: an optimal $O(1/K)$ nonergodic analysis. J. Sci. Comput. **79**(2), 671–699 (2019)
15. H. Li, Z. Lin, Y. Fang, Variance reduced EXTRA and DIGing and their optimal acceleration for strongly convex decentralized optimization (2020). Arxiv:2009.04373
16. Z. Lin, M. Chen, Y. Ma, The augmented Lagrange multiplier method for exact recovery of corrupted low-rank matrices (2010). ArXiv:1009.5055
17. Z. Lin, R. Liu, Z. Su, Linearized alternating direction method with adaptive penalty for low-rank representation, in *Advances in Neural Information Processing Systems* (2011), pp. 612–620
18. Z. Lin, R. Liu, H. Li, Linearized alternating direction method with parallel splitting and adaptive penalty for separable convex programs in machine learning. Mach. Learn. **99**(2), 287–325 (2015)
19. G. Liu, Z. Lin, Y. Yu, Robust subspace segmentation by low-rank representation, in *International Conference on Machine Learning* (2010), pp. 663–670
20. R. Liu, Z. Lin, Z. Su, Linearized alternating direction method with parallel splitting and adaptive penalty for separable convex programs in machine learning, in *Asian Conference on Machine Learning* (2013), pp. 116–132
21. Y. Liu, X. Yuan, S. Zeng, J. Zhang, Partial error bound conditions and the linear convergence rate of the alternating direction method of multipliers. SIAM J. Numer. Anal. **56**(4), 2095–2123 (2018)
22. C. Lu, H. Li, Z. Lin, S. Yan, Fast proximal linearized alternating direction method of multiplier with parallel splitting, in *AAAI Conference on Artificial Intelligence 68* (2016), pp. 739–745
23. C. Lu, J. Feng, S. Yan, Z. Lin, A unified alternating direction method of multipliers by majorization minimization. IEEE Trans. Pattern Anal. Mach. Intell. **40**(3), 527–541 (2018)
24. Y. Ouyang, Y. Chen, G. Lan, E. Pasiliao Jr., An accelerated linearized alternating direction method of multipliers. SIAM J. Imaging Sci. **8**(1), 644–681 (2015)
25. R. Shefi, M. Teboulle, Rate of convergence analysis of decomposition methods based on the proximal method of multipliers for convex minimization. SIAM J. Optim. **24**(1), 269–297 (2014)
26. M. Tao, X. Yuan, Recovering low-rank and sparse components of matrices from incomplete and noisy observations. SIAM J. Optim. **21**(5), 57–81 (2011)
27. H. Wang, A. Banerjee, Bregman alternating direction method of multipliers, in *Advances in Neural Information Processing Systems* (2014), pp. 2816–2824

28. X. Wang, X. Yuan, The linearized alternating direction method for Dantzig selector. SIAM J. Sci. Comput. **34**(5), 2792–2811 (2012)
29. W. Yang, D. Han, Linear convergence of the alternating direction method of multipliers for a class of convex optimization problems. SIAM J. Imaging Sci. **54**(2), 625–640 (2016)
30. X. Yuan, S. Zeng, J. Zhang, Discerning the linear convergence of ADMM for structured convex optimization through the lens of variational analysis. J. Mach. Learn. Res. **21**, 1–75 (2020)

# Chapter 4
# ADMM for Nonconvex Optimization

In this chapter, we introduce ADMM for nonconvex optimization. We first introduce the convergence of Bregman ADMM for the general multi-block linearly constrained problems under mild assumptions, then we introduce a proximal ADMM with exponential averaging that imposes no assumptions on the linear constraint. At last, we introduce how to use ADMM to solve multilinearly constrained problems, especially the RPCA problem.

## 4.1 Multi-block Bregman ADMM

Consider the following multi-block linearly constrained problem

$$\min_{\mathbf{x}_1,\cdots,\mathbf{x}_m,\mathbf{y}} \sum_{i=1}^{m} f_i(\mathbf{x}_i) + g(\mathbf{y}),$$

$$s.t. \quad \sum_{i=1}^{m} \mathbf{A}_i\mathbf{x}_i + \mathbf{B}\mathbf{y} = \mathbf{b}, \tag{4.1}$$

under the following assumption:

**Assumption 1** Assume that $f_i$, $i \in [m]$, is a proper lower semicontinuous (Definition A.26) function and $g$ is $L$-smooth. Both $f_i$ and $g$ can be nonconvex.

---

**Algorithm 4.1** Multi-block Bregman ADMM

---

Initialize $\mathbf{x}_1^0, \cdots, \mathbf{x}_m^0, \mathbf{y}^0$, and $\boldsymbol{\lambda}^0$.
**for** $k = 0, 1, 2, 3, \cdots$ **do**
    Update $\mathbf{x}_1^{k+1}, \cdots, \mathbf{x}_m^{k+1}, \mathbf{y}^{k+1}$, and $\boldsymbol{\lambda}^{k+1}$ by (4.2a)–(4.2c), respectively.
**end for**

---

There has been a wide range of works to study the convergence properties of ADMM for solving Problem (4.1) (for example, see [1, 6–8, 11, 12]). We introduce the following multi-block Bregman ADMM which is presented in Algorithm 4.1:

$$\mathbf{x}_i^{k+1} = \operatorname*{argmin}_{\mathbf{x}_i} \left( f_i(\mathbf{x}_i) + \left\langle \boldsymbol{\lambda}^k, \mathbf{A}_i \mathbf{x}_i \right\rangle \right.$$

$$+ \frac{\beta}{2} \left\| \sum_{j<i} \mathbf{A}_j \mathbf{x}_j^{k+1} + \mathbf{A}_i \mathbf{x}_i + \sum_{j>i} \mathbf{A}_j \mathbf{x}_j^k + \mathbf{B}\mathbf{y}^k - \mathbf{b} \right\|^2$$

$$\left. + D_{\phi_i}(\mathbf{x}_i, \mathbf{x}_i^k) \right), \quad \text{for } i \in [m] \text{ sequentially,} \tag{4.2a}$$

$$\mathbf{y}^{k+1} = \operatorname*{argmin}_{\mathbf{y}} \left( g(\mathbf{y}) + \left\langle \boldsymbol{\lambda}^k, \mathbf{B}\mathbf{y} \right\rangle + \frac{\beta}{2} \left\| \sum_{i=1}^m \mathbf{A}_i \mathbf{x}_i^{k+1} + \mathbf{B}\mathbf{y} - \mathbf{b} \right\|^2 \right.$$

$$\left. + D_{\phi_0}(\mathbf{y}, \mathbf{y}^k) \right), \tag{4.2b}$$

$$\boldsymbol{\lambda}^{k+1} = \boldsymbol{\lambda}^k + \beta \left( \sum_{i=1}^m \mathbf{A}_i \mathbf{x}_i^{k+1} + \mathbf{B}\mathbf{y}^{k+1} - \mathbf{b} \right). \tag{4.2c}$$

Similar to the linearized ADMM in the convex case (see Sect. 3.2), we can choose suitable $\phi_i$ such that each subproblem can be solved easily. For example, let

$$\phi_0(\mathbf{y}) = \frac{L + \beta \|\mathbf{B}\|_2^2}{2} \|\mathbf{y}\|^2 - g(\mathbf{y}) - \frac{\beta}{2} \|\mathbf{B}\mathbf{y}\|^2,$$

then the $\mathbf{y}$ update step reduces to (see (3.36))

$$\mathbf{y}^{k+1} = \mathbf{y}^k - \frac{1}{L + \beta \|\mathbf{B}\|_2^2} \nabla_{\mathbf{y}} L_\beta(\mathbf{x}^{k+1}, \mathbf{y}^k, \boldsymbol{\lambda}^k),$$

where $L_\beta$ is the augmented Lagrangian function:

$$L_\beta(\mathbf{x}, \mathbf{y}, \boldsymbol{\lambda}) = \sum_{i=1}^{m} f_i(\mathbf{x}_i) + g(\mathbf{y}) + \left\langle \boldsymbol{\lambda}, \sum_{i=1}^{m} \mathbf{A}_i \mathbf{x}_i + \mathbf{B}\mathbf{y} - \mathbf{b} \right\rangle$$

$$+ \frac{\beta}{2} \left\| \sum_{i=1}^{m} \mathbf{A}_i \mathbf{x}_i + \mathbf{B}\mathbf{y} - \mathbf{b} \right\|^2,$$

in which we denote $\mathbf{x} = (\mathbf{x}_1^T, \cdots, \mathbf{x}_m^T)^T$. We further denote $\mathbf{A} = [\mathbf{A}_1, \cdots, \mathbf{A}_m]$.

Under Assumption 1 and the surjectiveness of $\mathbf{B}$, i.e., $\|\mathbf{B}^T\boldsymbol{\lambda}\| \geq \sigma\|\boldsymbol{\lambda}\|$ ($\sigma > 0$), the above Bregman ADMM needs $O(\frac{1}{\epsilon^2})$ iterations to find an $\epsilon$-approximate KKT point, as shown in the following theorem.

**Theorem 4.1** *Assume that Assumption 1 holds, there exists $\sigma > 0$ such that $\|\mathbf{B}^T\boldsymbol{\lambda}\| \geq \sigma\|\boldsymbol{\lambda}\|$ for all $\boldsymbol{\lambda}$, and $\phi_i$ is $\rho$-strongly convex and $L_i$-smooth with $\rho > \frac{12(L^2+2L_0^2)}{\sigma^2\beta}$, $i = 0, \cdots, m$. Suppose that the sequence $\{(\mathbf{x}^k, \mathbf{y}^k, \boldsymbol{\lambda}^k)\}_k$ is bounded and $\sum_{i=1}^{m} f_i(\mathbf{x}_i) + g(\mathbf{y})$ is bounded below with bounded $(\mathbf{x}, \mathbf{y})$. Then Algorithm 4.1 needs $O(\frac{1}{\epsilon^2})$ iterations to find an $\epsilon$-approximate KKT point $(\mathbf{x}^{k+1}, \mathbf{y}^{k+1}, \boldsymbol{\lambda}^{k+1})$. Namely,*

$$\left\| \sum_{i=1}^{m} \mathbf{A}_i \mathbf{x}_i^{k+1} + \mathbf{B}\mathbf{y}^{k+1} - \mathbf{b} \right\| \leq O(\epsilon),$$

$$\|\nabla g(\mathbf{y}^{k+1}) + \mathbf{B}^T\boldsymbol{\lambda}^{k+1}\| \leq O(\epsilon), \text{ and}$$

$$\text{dist}\left( -\mathbf{A}_i^T\boldsymbol{\lambda}^{k+1}, \partial f_i(\mathbf{x}_i^{k+1}) \right) \leq O(\epsilon), \quad \forall i \in [m].$$

**Proof** From the first step of the algorithm, we have

$$L_\beta\left( \mathbf{x}_{j\leq i}^{k+1}, \mathbf{x}_{j>i}^k, \mathbf{y}^k, \boldsymbol{\lambda}^k \right) + D_{\phi_i}\left( \mathbf{x}_i^{k+1}, \mathbf{x}_i^k \right) \leq L_\beta\left( \mathbf{x}_{j<i}^{k+1}, \mathbf{x}_{j\geq i}^k, \mathbf{y}^k, \boldsymbol{\lambda}^k \right).$$

Since $D_{\phi_i}(\mathbf{x}_i^{k+1}, \mathbf{x}_i^k) \geq \frac{\rho}{2}\|\mathbf{x}_i^{k+1} - \mathbf{x}_i^k\|^2$ (Lemma A.2), we have

$$\frac{\rho}{2}\|\mathbf{x}_i^{k+1} - \mathbf{x}_i^k\|^2 \leq L_\beta\left( \mathbf{x}_{j<i}^{k+1}, \mathbf{x}_{j\geq i}^k, \mathbf{y}^k, \boldsymbol{\lambda}^k \right) - L_\beta\left( \mathbf{x}_{j\leq i}^{k+1}, \mathbf{x}_{j>i}^k, \mathbf{y}^k, \boldsymbol{\lambda}^k \right).$$

Summing over $i = 1, \cdots, m$, we have

$$\frac{\rho}{2}\|\mathbf{x}^{k+1} - \mathbf{x}^k\|^2 \leq L_\beta(\mathbf{x}^k, \mathbf{y}^k, \boldsymbol{\lambda}^k) - L_\beta(\mathbf{x}^{k+1}, \mathbf{y}^k, \boldsymbol{\lambda}^k).$$

Similarly, for the second step, we have

$$\frac{\rho}{2}\|\mathbf{y}^{k+1} - \mathbf{y}^k\|^2 \leq L_\beta(\mathbf{x}^{k+1}, \mathbf{y}^k, \boldsymbol{\lambda}^k) - L_\beta(\mathbf{x}^{k+1}, \mathbf{y}^{k+1}, \boldsymbol{\lambda}^k).$$

From the update of $\boldsymbol{\lambda}$, we have

$$-\frac{1}{\beta}\|\boldsymbol{\lambda}^{k+1} - \boldsymbol{\lambda}^k\|^2 = L_\beta(\mathbf{x}^{k+1}, \mathbf{y}^{k+1}, \boldsymbol{\lambda}^k) - L_\beta(\mathbf{x}^{k+1}, \mathbf{y}^{k+1}, \boldsymbol{\lambda}^{k+1}).$$

Summing up, we have

$$\frac{\rho}{2}\|\mathbf{x}^{k+1} - \mathbf{x}^k\|^2 + \frac{\rho}{2}\|\mathbf{y}^{k+1} - \mathbf{y}^k\|^2 - \frac{1}{\beta}\|\boldsymbol{\lambda}^{k+1} - \boldsymbol{\lambda}^k\|^2$$

$$\leq L_\beta(\mathbf{x}^k, \mathbf{y}^k, \boldsymbol{\lambda}^k) - L_\beta(\mathbf{x}^{k+1}, \mathbf{y}^{k+1}, \boldsymbol{\lambda}^{k+1}). \qquad (4.3)$$

On the other hand, from the optimality condition of the $\mathbf{y}$ update, we have

$$\mathbf{0} = \nabla g(\mathbf{y}^{k+1}) + \mathbf{B}^T\boldsymbol{\lambda}^k + \beta\mathbf{B}^T(\mathbf{A}\mathbf{x}^{k+1} + \mathbf{B}\mathbf{y}^{k+1} - \mathbf{b}) + \nabla\phi_0(\mathbf{y}^{k+1}) - \nabla\phi_0(\mathbf{y}^k)$$

$$= \nabla g(\mathbf{y}^{k+1}) + \mathbf{B}^T\boldsymbol{\lambda}^{k+1} + \nabla\phi_0(\mathbf{y}^{k+1}) - \nabla\phi_0(\mathbf{y}^k). \qquad (4.4)$$

Thus we have

$$\sigma^2\|\boldsymbol{\lambda}^{k+1} - \boldsymbol{\lambda}^k\|^2$$

$$\leq \|\mathbf{B}^T(\boldsymbol{\lambda}^{k+1} - \boldsymbol{\lambda}^k)\|^2$$

$$= \left\| \nabla g(\mathbf{y}^{k+1}) - \nabla g(\mathbf{y}^k) + \nabla\phi_0(\mathbf{y}^{k+1}) - \nabla\phi_0(\mathbf{y}^k) \right.$$

$$\left. -(\nabla\phi_0(\mathbf{y}^k) - \nabla\phi_0(\mathbf{y}^{k-1})) \right\|^2$$

$$\leq 3\|\nabla g(\mathbf{y}^{k+1}) - \nabla g(\mathbf{y}^k)\|^2 + 3\|\nabla\phi_0(\mathbf{y}^{k+1}) - \nabla\phi_0(\mathbf{y}^k)\|^2$$

$$+ 3\|\nabla\phi_0(\mathbf{y}^k) - \nabla\phi_0(\mathbf{y}^{k-1})\|^2$$

$$\leq 3(L^2 + L_0^2)\|\mathbf{y}^{k+1} - \mathbf{y}^k\|^2 + 3L_0^2\|\mathbf{y}^k - \mathbf{y}^{k-1}\|^2. \qquad (4.5)$$

Adding (4.3) and (4.5), we have

$$\frac{\rho}{2}\|\mathbf{x}^{k+1} - \mathbf{x}^k\|^2 + \frac{\rho}{2}\|\mathbf{y}^{k+1} - \mathbf{y}^k\|^2 + \frac{1}{\beta}\|\boldsymbol{\lambda}^{k+1} - \boldsymbol{\lambda}^k\|^2$$

$$\leq L_\beta(\mathbf{x}^k, \mathbf{y}^k, \boldsymbol{\lambda}^k) - L_\beta(\mathbf{x}^{k+1}, \mathbf{y}^{k+1}, \boldsymbol{\lambda}^{k+1})$$

$$+ \frac{6(L^2 + L_0^2)}{\sigma^2\beta}\|\mathbf{y}^{k+1} - \mathbf{y}^k\|^2 + \frac{6L_0^2}{\sigma^2\beta}\|\mathbf{y}^k - \mathbf{y}^{k-1}\|^2.$$

Defining

$$\Phi^k = L_\beta(\mathbf{x}^k, \mathbf{y}^k, \boldsymbol{\lambda}^k) + \frac{6L_0^2}{\sigma^2\beta}\|\mathbf{y}^k - \mathbf{y}^{k-1}\|^2,$$

we have

$$\frac{\rho}{2}\|\mathbf{x}^{k+1} - \mathbf{x}^k\|^2 + \left(\frac{\rho}{2} - \frac{6L^2 + 12L_0^2}{\sigma^2 \beta}\right) \|\mathbf{y}^{k+1} - \mathbf{y}^k\|^2 + \frac{1}{\beta}\|\boldsymbol{\lambda}^{k+1} - \boldsymbol{\lambda}^k\|^2$$

$$\leq \Phi^k - \Phi^{k+1}.$$

From the assumptions that $\{(\mathbf{x}^k, \mathbf{y}^k, \boldsymbol{\lambda}^k)\}_k$ are bounded and $\sum_{i=1}^m f_i(\mathbf{x}_i) + g(\mathbf{y})$ is bounded below, we know that $\Phi^k$ is lower bounded by some $\Phi^*$. Summing the above inequality over $k = 0, \cdots, K$, we have

$$\min_{k \leq K} \left\{ \frac{\rho}{2}\|\mathbf{x}^{k+1} - \mathbf{x}^k\|^2 + \left(\frac{\rho}{2} - \frac{6L^2 + 12L_0^2}{\sigma^2 \beta}\right) \|\mathbf{y}^{k+1} - \mathbf{y}^k\|^2 + \frac{1}{\beta}\|\boldsymbol{\lambda}^{k+1} - \boldsymbol{\lambda}^k\|^2 \right\}$$

$$\leq \frac{\Phi^0 - \Phi^*}{K + 1}.$$

Thus, the algorithm needs $O(\frac{1}{\epsilon^2})$ iterations to find $(\mathbf{x}^{k+1}, \mathbf{y}^{k+1}, \boldsymbol{\lambda}^{k+1})$ such that

$$\|\mathbf{x}^{k+1} - \mathbf{x}^k\| \leq \epsilon, \quad \|\mathbf{y}^{k+1} - \mathbf{y}^k\| \leq \epsilon, \quad \text{and} \quad \|\boldsymbol{\lambda}^{k+1} - \boldsymbol{\lambda}^k\| \leq \epsilon.$$

From (4.4), we have

$$\|\nabla g(\mathbf{y}^{k+1}) + \mathbf{B}^T \boldsymbol{\lambda}^{k+1}\| = \|\nabla \phi_0(\mathbf{y}^{k+1}) - \nabla \phi_0(\mathbf{y}^k)\|$$

$$\leq L_0 \|\mathbf{y}^{k+1} - \mathbf{y}^k\| \leq O(\epsilon). \tag{4.6}$$

From the update of $\boldsymbol{\lambda}$, we have

$$\left\| \sum_{i=1}^m \mathbf{A}_i \mathbf{x}_i^{k+1} + \mathbf{B}\mathbf{y}^{k+1} - \mathbf{b} \right\| = \frac{1}{\beta}\|\boldsymbol{\lambda}^{k+1} - \boldsymbol{\lambda}^k\| \leq O(\epsilon).$$

From the optimality conditions of $\mathbf{x}_i$, we have that there exists $\widetilde{\nabla} f_i(\mathbf{x}_i^{k+1}) \in \partial f_i(\mathbf{x}_i^{k+1})$ such that

$$\widetilde{\nabla} f_i(\mathbf{x}_i^{k+1}) + \mathbf{A}_i^T \left[ \boldsymbol{\lambda}^k + \beta \left( \sum_{j \leq i} \mathbf{A}_j \mathbf{x}_j^{k+1} + \sum_{j > i} \mathbf{A}_j \mathbf{x}_j^k + \mathbf{B}\mathbf{y}^k - \mathbf{b} \right) \right]$$

$$+ \nabla \phi_i(\mathbf{x}_i^{k+1}) - \nabla \phi_i(\mathbf{x}_i^k) = \mathbf{0}.$$

Thus, we have

$$\left\| \widetilde{\nabla} f_i(\mathbf{x}_i^{k+1}) + \mathbf{A}_i^T \boldsymbol{\lambda}^{k+1} \right\|$$

$$\leq \| \nabla \phi_i(\mathbf{x}_i^{k+1}) - \nabla \phi_i(\mathbf{x}_i^k) \| + \beta \left\| \mathbf{A}_i^T \left( \sum_{j>i} \mathbf{A}_j(\mathbf{x}_j^{k+1} - \mathbf{x}_j^k) + \mathbf{B}(\mathbf{y}^{k+1} - \mathbf{y}^k) \right) \right\|$$

$$\leq O(\epsilon).$$

$\square$

*Remark 4.1* A crucial step in the above proof is to bound the dual variables by the primal ones in (4.5), which is established via the surjective assumption of $\|\mathbf{B}^T \boldsymbol{\lambda}\| \geq \sigma \|\boldsymbol{\lambda}\|$ such that we can get $\|\mathbf{B}^T(\boldsymbol{\lambda}^{k+1} - \boldsymbol{\lambda}^k)\| \geq \sigma \|\boldsymbol{\lambda}^{k+1} - \boldsymbol{\lambda}^k\|$. [12] replaced $\|\mathbf{B}^T \boldsymbol{\lambda}\| \geq \sigma \|\boldsymbol{\lambda}\|$ by the assumption of $\mathrm{Im}(\mathbf{A}_i) \subseteq \mathrm{Im}(\mathbf{B})$. In this case, we have

$$\boldsymbol{\lambda}^{k+1} - \boldsymbol{\lambda}^k = \beta \left( \sum_i \mathbf{A}_i \mathbf{x}_i^{k+1} + \mathbf{B} \mathbf{y}^{k+1} - \mathbf{b} \right) \in \mathrm{Im}(\mathbf{B}).$$

Suppose that the economical SVD of $\mathbf{B}$ is $\mathbf{B} = \mathbf{U} \boldsymbol{\Sigma} \mathbf{V}^T$. Then we may write $\boldsymbol{\lambda}^{k+1} - \boldsymbol{\lambda}^k = \mathbf{U} \boldsymbol{\alpha}$. So

$$\begin{aligned}
\|\mathbf{B}^T(\boldsymbol{\lambda}^{k+1} - \boldsymbol{\lambda}^k)\|^2 &= \|\mathbf{V} \boldsymbol{\Sigma} \mathbf{U}^T \mathbf{U} \boldsymbol{\alpha}\|^2 \\
&= \|\boldsymbol{\Sigma} \boldsymbol{\alpha}\|^2 \\
&\geq \lambda_+(\mathbf{B}\mathbf{B}^T) \|\boldsymbol{\alpha}\|^2 \\
&= \lambda_+(\mathbf{B}\mathbf{B}^T) \|\boldsymbol{\lambda}^{k+1} - \boldsymbol{\lambda}^k\|^2,
\end{aligned} \tag{4.7}$$

where $\lambda_+(\mathbf{B}\mathbf{B}^T)$ is the smallest strictly positive eigenvalue of $\mathbf{B}\mathbf{B}^T$. As a result, we do not need $\mathbf{B}$ to be of full row rank (i.e., $\|\mathbf{B}^T \boldsymbol{\lambda}\| \geq \sigma \|\boldsymbol{\lambda}\|$ for all $\boldsymbol{\lambda}$).

On the other hand, when we only consider the problem with one block of variables, that is, $f_i = 0, \forall i$, the assumptions of $\mathrm{Im}(\mathbf{A}_i) \subseteq \mathrm{Im}(\mathbf{B})$ can be further removed since $\boldsymbol{\lambda}^{k+1} - \boldsymbol{\lambda}^k = \beta(\mathbf{B}\mathbf{y}^{k+1} - \mathbf{b}) \in \mathrm{Im}(\mathbf{B})$ always holds.

In Theorem 4.1, we assume that the objectives are proper and lower semicontinuous such that the subdifferential (Definition A.28) is well defined. We assume that the objective is coercive (Definition A.27) over the entire space. [12] used a weaker assumption that the objective is coercive only over the constraint set $\{(\mathbf{x}, \mathbf{y}) | \mathbf{A}\mathbf{x} + \mathbf{B}\mathbf{y} = \mathbf{b}\}$.

When we remove the Bregman distance in the update of $\mathbf{x}_i$, we should assume that the smallest eigenvalue of $\mathbf{A}_i^T \mathbf{A}_i$ is positive and $\beta$ should be chosen large enough such that $L_\beta(\mathbf{x}, \mathbf{y}, \boldsymbol{\lambda})$ is $\mu$-strongly convex with respect to $\mathbf{x}_i$ (assume that $f_i$ is smooth if necessary). Then for the first step, we have

$$L_\beta\left(\mathbf{x}_{j<i}^{k+1}, \mathbf{x}_{j\geq i}^k, \mathbf{y}^k, \boldsymbol{\lambda}^k\right) - L_\beta\left(\mathbf{x}_{j\leq i}^{k+1}, \mathbf{x}_{j>i}^k, \mathbf{y}^k, \boldsymbol{\lambda}^k\right) \geq \frac{\mu}{2}\|\mathbf{x}_i^{k+1} - \mathbf{x}_i^k\|^2$$

due to $\mathbf{0} \in \partial_{\mathbf{x}_i} L_\beta(\mathbf{x}_{j\leq i}^{k+1}, \mathbf{x}_{j>i}^k, \mathbf{y}^k, \boldsymbol{\lambda}^k)$.

### 4.1.1 With More Assumptions on the Objectives

As discussed in Remark 4.1, the assumption on the linear constraint, either $\|\mathbf{B}^T\boldsymbol{\lambda}\| \geq \sigma\|\boldsymbol{\lambda}\|$ or $\text{Im}(\mathbf{A}_i) \subseteq \text{Im}(\mathbf{B})$, plays a critical role in Theorem 4.1. Sometimes such conditions may not be met. In this section, we introduce the convergence proof with more assumptions on the objectives instead. We further make the following assumption.

**Assumption 2** $f_i$'s and $g$ are all $L$-smooth.

Note that in Theorem 4.1, we only assume that $g$ is smooth, which allows $f_i$ to be nonsmooth and can be applied to problems such as sparse and low-rank optimization.

Then we have the following convergence theorem.

**Theorem 4.2** *Assume that Assumption 2 holds and $\phi_i$ is $\rho$-strongly convex and $L_i$-smooth with $\rho > \frac{4\max\{c_1+c_2,c_3+c_4\}}{\beta\lambda_+}$, $i = 0, 1, \cdots, m$, where $\lambda_+$ is the smallest strictly positive eigenvalue of $[\mathbf{A}, \mathbf{B}][\mathbf{A}, \mathbf{B}]^T$,*

$$c_1 = 5L^2 + 5L_{\max}^2 + 10\beta^2\|\mathbf{A}\|_2^2 \sum_{i=1}^m \|\mathbf{A}_i\|_2^2,$$

$$c_2 = 5L_{\max}^2 + 10\beta^2\|\mathbf{A}\|_2^2 \sum_{i=1}^m \|\mathbf{A}_i\|_2^2,$$

$$c_3 = 3L^2 + 3L_0^2 + 10\beta^2\|\mathbf{B}\|_2^2 \sum_{i=1}^m \|\mathbf{A}_i\|_2^2,$$

$$c_4 = 3L_0^2 + 10\beta^2\|\mathbf{B}\|_2^2 \sum_{i=1}^m \|\mathbf{A}_i\|_2^2, \text{ and}$$

$$L_{\max} = \max\{L_i, i \in [m]\}.$$

*Suppose that the sequence $\{(\mathbf{x}^k, \mathbf{y}^k, \boldsymbol{\lambda}^k)\}_k$ are bounded and $\sum_{i=1}^m f_i(\mathbf{x}_i) + g(\mathbf{y})$ is bounded below with bounded $(\mathbf{x}, \mathbf{y})$. Let $\boldsymbol{\lambda}^0 = \mathbf{0}$. Then Algorithm 4.1 needs $O(\frac{1}{\epsilon^2})$ iterations to find an $\epsilon$-approximate KKT point $(\mathbf{x}^{k+1}, \mathbf{y}^{k+1}, \boldsymbol{\lambda}^{k+1})$. Namely,*

$$\left\| \sum_{i=1}^m \mathbf{A}_i \mathbf{x}_i^{k+1} + \mathbf{B} \mathbf{y}^{k+1} - \mathbf{b} \right\| \leq O(\epsilon),$$

$$\|\nabla g(\mathbf{y}^{k+1}) + \mathbf{B}^T \boldsymbol{\lambda}^{k+1}\| \leq O(\epsilon), \ and$$

$$\left\| \nabla f_i(\mathbf{x}_i^{k+1}) + \mathbf{A}^T \boldsymbol{\lambda}^{k+1} \right\| \leq O(\epsilon).$$

**Proof** Recalling (4.3), we want to bound $\|\boldsymbol{\lambda}^{k+1} - \boldsymbol{\lambda}^k\|^2$ by $\|\mathbf{x}^{k+1} - \mathbf{x}^k\|^2$ and $\|\mathbf{y}^{k+1} - \mathbf{y}^k\|^2$. From (4.2c) and $\boldsymbol{\lambda}^0 = \mathbf{0}$, we know that $\boldsymbol{\lambda}^k$ belongs to the linear span of $[\mathbf{A}, \mathbf{B}]$. Thus, similar to (4.7) we have

$$\lambda_+ \|\boldsymbol{\lambda}^{k+1} - \boldsymbol{\lambda}^k\|^2$$

$$\leq \|[\mathbf{A}, \mathbf{B}]^T (\boldsymbol{\lambda}^{k+1} - \boldsymbol{\lambda}^k)\|^2$$

$$= \sum_{i=1}^m \|\mathbf{A}_i^T (\boldsymbol{\lambda}^{k+1} - \boldsymbol{\lambda}^k)\|^2 + \|\mathbf{B}^T (\boldsymbol{\lambda}^{k+1} - \boldsymbol{\lambda}^k)\|^2. \tag{4.8}$$

From the optimality condition of (4.2a), we have

$$\mathbf{0} = \nabla f_i(\mathbf{x}_i^{k+1}) + \mathbf{A}_i^T \left[ \boldsymbol{\lambda}^k + \beta \left( \sum_{j \leq i} \mathbf{A}_j \mathbf{x}_j^{k+1} + \sum_{j > i} \mathbf{A}_j \mathbf{x}_j^k + \mathbf{B} \mathbf{y}^k - \mathbf{b} \right) \right]$$

$$+ \nabla \phi_i(\mathbf{x}_i^{k+1}) - \nabla \phi_i(\mathbf{x}_i^k)$$

$$= \nabla f_i(\mathbf{x}_i^{k+1}) + \mathbf{A}_i^T \boldsymbol{\lambda}^{k+1} + \beta \mathbf{A}_i^T \left[ \sum_{j > i} \mathbf{A}_j (\mathbf{x}_j^k - \mathbf{x}_j^{k+1}) + \mathbf{B}(\mathbf{y}^k - \mathbf{y}^{k+1}) \right]$$

$$+ \nabla \phi_i(\mathbf{x}_i^{k+1}) - \nabla \phi_i(\mathbf{x}_i^k).$$

Thus, we have

$$\|\mathbf{A}_i^T (\boldsymbol{\lambda}^{k+1} - \boldsymbol{\lambda}^k)\|^2$$

$$= \left\| \left[ \nabla f_i(\mathbf{x}_i^{k+1}) - \nabla f_i(\mathbf{x}_i^k) \right] \right.$$

$$+ \beta \mathbf{A}_i^T \left[ \sum_{j > i} \mathbf{A}_j (\mathbf{x}_j^k - \mathbf{x}_j^{k+1}) + \mathbf{B}(\mathbf{y}^k - \mathbf{y}^{k+1}) \right]$$

$$-\beta\mathbf{A}_i^T\left[\sum_{j>i}\mathbf{A}_j(\mathbf{x}_j^{k-1}-\mathbf{x}_j^k)+\mathbf{B}(\mathbf{y}^{k-1}-\mathbf{y}^k)\right]$$

$$+\left[\nabla\phi_i(\mathbf{x}_i^{k+1})-\nabla\phi_i(\mathbf{x}_i^k)\right]-\left[\nabla\phi_i(\mathbf{x}_i^k)-\nabla\phi_i(\mathbf{x}_i^{k-1})\right]\Big\|^2$$

$$\leq5\|\nabla f_i(\mathbf{x}_i^{k+1})-\nabla f_i(\mathbf{x}_i^k)\|^2$$

$$+5\beta^2\|\mathbf{A}_i\|_2^2\left\|\sum_{j>i}\mathbf{A}_j(\mathbf{x}_j^k-\mathbf{x}_j^{k+1})+\mathbf{B}(\mathbf{y}^k-\mathbf{y}^{k+1})\right\|^2$$

$$+5\beta^2\|\mathbf{A}_i\|_2^2\left\|\sum_{j>i}\mathbf{A}_j(\mathbf{x}_j^{k-1}-\mathbf{x}_j^k)+\mathbf{B}(\mathbf{y}^{k-1}-\mathbf{y}^k)\right\|^2$$

$$+5\|\nabla\phi_i(\mathbf{x}_i^{k+1})-\nabla\phi_i(\mathbf{x}_i^k)\|^2+5\|\nabla\phi_i(\mathbf{x}_i^k)-\nabla\phi_i(\mathbf{x}_i^{k-1})\|^2$$

$$\overset{a}{\leq}5\left(L^2+L_i^2\right)\|\mathbf{x}_i^{k+1}-\mathbf{x}_i^k\|^2+5L_i^2\|\mathbf{x}_i^k-\mathbf{x}_i^{k-1}\|^2$$

$$+10\beta^2\|\mathbf{A}_i\|_2^2\left(\|\mathbf{A}\|_2^2\|\mathbf{x}^{k+1}-\mathbf{x}^k\|^2+\|\mathbf{B}\|_2^2\|\mathbf{y}^{k+1}-\mathbf{y}^k\|^2\right)$$

$$+10\beta^2\|\mathbf{A}_i\|_2^2\left(\|\mathbf{A}\|_2^2\|\mathbf{x}^k-\mathbf{x}^{k-1}\|^2+\|\mathbf{B}\|_2^2\|\mathbf{y}^k-\mathbf{y}^{k-1}\|^2\right),$$

where in $\overset{a}{\leq}$ we use

$$\left\|\sum_{j>i}\mathbf{A}_j(\mathbf{x}_j^k-\mathbf{x}_j^{k+1})\right\|^2\leq\|[\mathbf{A}_{i+1},\cdots,\mathbf{A}_m]\|_2^2\sum_{j>i}\|\mathbf{x}_j^k-\mathbf{x}_j^{k+1}\|^2$$

$$\leq\|\mathbf{A}\|_2^2\|\mathbf{x}^{k+1}-\mathbf{x}^k\|^2.$$

Similarly, from (4.4) we can deduce

$$\|\mathbf{B}^T(\boldsymbol{\lambda}^{k+1}-\boldsymbol{\lambda}^k)\|^2$$

$$\leq3\left(L^2+L_0^2\right)\|\mathbf{y}^{k+1}-\mathbf{y}^k\|^2+3L_0^2\|\mathbf{y}^k-\mathbf{y}^{k-1}\|^2. \tag{4.9}$$

Combining (4.8)–(4.9), we have

$$\lambda_+\|\boldsymbol{\lambda}^{k+1}-\boldsymbol{\lambda}^k\|^2$$

$$\leq\left(5L^2+5L_{\max}^2+10\beta^2\|\mathbf{A}\|_2^2\sum_{i=1}^m\|\mathbf{A}_i\|_2^2\right)\|\mathbf{x}^{k+1}-\mathbf{x}^k\|^2$$

$$+ \left( 5L_{\max}^2 + 10\beta^2 \|\mathbf{A}\|_2^2 \sum_{i=1}^{m} \|\mathbf{A}_i\|_2^2 \right) \|\mathbf{x}^k - \mathbf{x}^{k-1}\|^2$$

$$+ \left( 3L^2 + 3L_0^2 + 10\beta^2 \|\mathbf{B}\|_2^2 \sum_{i=1}^{m} \|\mathbf{A}_i\|_2^2 \right) \|\mathbf{y}^{k+1} - \mathbf{y}^k\|^2$$

$$+ \left( 3L_0^2 + 10\beta^2 \|\mathbf{B}\|_2^2 \sum_{i=1}^{m} \|\mathbf{A}_i\|_2^2 \right) \|\mathbf{y}^k - \mathbf{y}^{k-1}\|^2$$

$$= c_1 \|\mathbf{x}^{k+1} - \mathbf{x}^k\|^2 + c_2 \|\mathbf{x}^k - \mathbf{x}^{k-1}\|^2$$

$$+ c_3 \|\mathbf{y}^{k+1} - \mathbf{y}^k\|^2 + c_4 \|\mathbf{y}^k - \mathbf{y}^{k-1}\|^2.$$

Together with (4.3), we have

$$\frac{\rho}{2} \|\mathbf{x}^{k+1} - \mathbf{x}^k\|^2 + \frac{\rho}{2} \|\mathbf{y}^{k+1} - \mathbf{y}^k\|^2 + \frac{1}{\beta} \|\boldsymbol{\lambda}^{k+1} - \boldsymbol{\lambda}^k\|^2$$

$$\leq L_\beta(\mathbf{x}^k, \mathbf{y}^k, \boldsymbol{\lambda}^k) - L_\beta(\mathbf{x}^{k+1}, \mathbf{y}^{k+1}, \boldsymbol{\lambda}^{k+1})$$

$$+ \frac{2c_1}{\beta\lambda_+} \|\mathbf{x}^{k+1} - \mathbf{x}^k\|^2 + \frac{2c_3}{\beta\lambda_+} \|\mathbf{y}^{k+1} - \mathbf{y}^k\|^2$$

$$+ \frac{2c_2}{\beta\lambda_+} \|\mathbf{x}^k - \mathbf{x}^{k-1}\|^2 + \frac{2c_4}{\beta\lambda_+} \|\mathbf{y}^k - \mathbf{y}^{k-1}\|^2.$$

Defining

$$\Phi^k = L_\beta(\mathbf{x}^k, \mathbf{y}^k, \boldsymbol{\lambda}^k) + \frac{2}{\beta\lambda_+} \left( c_2 \|\mathbf{x}^k - \mathbf{x}^{k-1}\|^2 + c_4 \|\mathbf{y}^k - \mathbf{y}^{k-1}\|^2 \right),$$

we have

$$\left[ \frac{\rho}{2} - \frac{2(c_1 + c_2)}{\beta\lambda_+} \right] \|\mathbf{x}^{k+1} - \mathbf{x}^k\|^2 + \left[ \frac{\rho}{2} - \frac{2(c_3 + c_4)}{\beta\lambda_+} \right] \|\mathbf{y}^{k+1} - \mathbf{y}^k\|^2$$

$$+ \frac{1}{\beta} \|\boldsymbol{\lambda}^{k+1} - \boldsymbol{\lambda}^k\|^2$$

$$\leq \Phi^k - \Phi^{k+1}.$$

Then similar to the proof of Theorem 4.1, we have the conclusion.                    □

## 4.2 Proximal ADMM with Exponential Averaging

In the previous sections, the Bregman ADMM (4.2a)–(4.2c) uses the Bregman distance $D_{\phi_i}(\mathbf{x}_i, \mathbf{x}_i^k)$, which results in the proximal term $\frac{\beta'}{2}\|\mathbf{x} - \mathbf{x}^k\|^2$. In this section, we introduce another proximal ADMM proposed in [13], where the proximal term $\frac{\beta'}{2}\|\mathbf{x}_i - \mathbf{z}_i^k\|^2$ is used instead, in which $\mathbf{z}_i^k$ is an exponential averaging of $\mathbf{x}_i^0, \cdots, \mathbf{x}_i^k$.

Consider the following general problem

$$\min_{\mathbf{x}_1,\cdots,\mathbf{x}_m} \ f(\mathbf{x}_1, \cdots, \mathbf{x}_m),$$

$$s.t. \ \sum_{i=1}^m \mathbf{A}_i \mathbf{x}_i = \mathbf{b},$$

with a non-separable objective. Denote

$$\mathbf{x} = \left(\mathbf{x}_1^T, \cdots, \mathbf{x}_m^T\right)^T \quad \text{and} \quad \mathbf{A} = [\mathbf{A}_1, \cdots, \mathbf{A}_m].$$

Define the following proximal augmented Lagrangian function

$$P(\mathbf{x}, \mathbf{z}, \boldsymbol{\lambda}) = f(\mathbf{x}) + \langle \boldsymbol{\lambda}, \mathbf{A}\mathbf{x} - \mathbf{b}\rangle + \frac{\beta}{2}\|\mathbf{A}\mathbf{x} - \mathbf{b}\|^2 + \frac{\rho}{2}\|\mathbf{x} - \mathbf{z}\|^2.$$

The method proposed in [13] consists of the following steps:

$$\mathbf{x}_j^{k+1} = \mathbf{x}_j^k - \alpha_1 \nabla_j P\left(\mathbf{x}_1^{k+1}, \cdots, \mathbf{x}_{j-1}^{k+1}, \mathbf{x}_j^k, \cdots, \mathbf{x}_m^k, \mathbf{z}^k, \boldsymbol{\lambda}^k\right),$$

$$\text{for } j \in [m] \text{ sequentially,} \quad (4.10a)$$

$$\boldsymbol{\lambda}^{k+1} = \boldsymbol{\lambda}^k + \alpha_2(\mathbf{A}\mathbf{x}^{k+1} - \mathbf{b}), \quad (4.10b)$$

$$\mathbf{z}^{k+1} = \mathbf{z}^k + \alpha_3(\mathbf{x}^{k+1} - \mathbf{z}^k). \quad (4.10c)$$

From the last step, we have

$$\mathbf{z}^{k+1} = (1 - \alpha_3)\mathbf{z}^k + \alpha_3 \mathbf{x}^{k+1},$$

which gives

$$\mathbf{z}^{k+1} = \sum_{t=0}^k \alpha_3(1 - \alpha_3)^{k-t}\mathbf{x}^{t+1} + (1 - \alpha_3)^{k+1}\mathbf{z}^0.$$

The proximal term $\frac{\rho}{2}\|\mathbf{x} - \mathbf{z}\|^2$ in $P(\mathbf{x}, \mathbf{z}, \boldsymbol{\lambda})$ makes $\mathbf{x}^{k+1}$ not deviate too much from $\mathbf{z}^k$. The method is presented in Algorithm 4.2.

---

**Algorithm 4.2** Proximal ADMM with exponential averaging

---

Initialize $\mathbf{x}_1^0, \cdots, \mathbf{x}_m^0, \mathbf{z}^0$, and $\boldsymbol{\lambda}^0$.
**for** $k = 0, 1, 2, 3, \cdots$ **do**
  Update $\mathbf{x}_1^{k+1}, \cdots, \mathbf{x}_m^{k+1}, \boldsymbol{\lambda}^{k+1}$, and $\mathbf{z}^{k+1}$ by (4.10a)–(4.10c), respectively.
**end for**

---

For the convergence proof, define

$$M(\mathbf{z}) = \min_{\mathbf{Ax}=\mathbf{b}} \left( f(\mathbf{x}) + \frac{\rho}{2}\|\mathbf{x} - \mathbf{z}\|^2 \right), \tag{4.11}$$

$$\mathbf{x}^*(\mathbf{z}) = \underset{\mathbf{Ax}=\mathbf{b}}{\text{argmin}} \left( f(\mathbf{x}) + \frac{\rho}{2}\|\mathbf{x} - \mathbf{z}\|^2 \right),$$

$$d(\mathbf{z}, \boldsymbol{\lambda}) = \min_{\mathbf{x}} P(\mathbf{x}, \mathbf{z}, \boldsymbol{\lambda}),$$

$$\mathbf{x}(\mathbf{z}, \boldsymbol{\lambda}) = \underset{\mathbf{x}}{\text{argmin}} \, P(\mathbf{x}, \mathbf{z}, \boldsymbol{\lambda}),$$

and the potential function

$$\Phi^k = P(\mathbf{x}^k, \mathbf{z}^k, \boldsymbol{\lambda}^{k-1}) - 2d(\mathbf{z}^k, \boldsymbol{\lambda}^{k-1}) + 2M(\mathbf{z}^k).$$

We want to prove that $\Phi^k$ decreases sufficiently in order to establish the convergence.

We first give several error bounds in the following three lemmas.

**Lemma 4.1** *Assume that $f$ is L-smooth with respect to $\mathbf{x}$. Suppose $\rho > L$, then for Algorithm 4.2 we have*

$$\|\mathbf{x}^*(\mathbf{z}^k) - \mathbf{x}^*(\mathbf{z}^{k+1})\| \leq \frac{\rho}{\rho - L}\|\mathbf{z}^k - \mathbf{z}^{k+1}\|, \tag{4.12}$$

$$\|\mathbf{x}(\mathbf{z}^k, \boldsymbol{\lambda}^k) - \mathbf{x}(\mathbf{z}^{k+1}, \boldsymbol{\lambda}^k)\| \leq \frac{\rho}{\rho - L}\|\mathbf{z}^k - \mathbf{z}^{k+1}\|, \tag{4.13}$$

$$\|\mathbf{x}(\mathbf{z}^k, \boldsymbol{\lambda}^k) - \mathbf{x}^*(\mathbf{z}^k)\| \leq \frac{\widetilde{\sigma}}{\rho - L}\|\boldsymbol{\lambda}^k - \boldsymbol{\lambda}^*(\mathbf{z}^k)\|, \tag{4.14}$$

*where $\widetilde{\sigma} = \|\mathbf{A}\|_2$ and $\boldsymbol{\lambda}^*(\mathbf{z}^k)$ is the optimal dual variable of Problem (4.11).*

**Proof** Define

$$g(\mathbf{x}, \mathbf{z}) = f(\mathbf{x}) + I_{\mathbf{Ax}=\mathbf{b}}(\mathbf{x}) + \frac{\rho}{2}\|\mathbf{x} - \mathbf{z}\|^2$$

with

$$I_{\mathbf{Ax}=\mathbf{b}}(\mathbf{x}) = \begin{cases} 0, & \text{if } \mathbf{Ax} = \mathbf{b}, \\ \infty, & \text{otherwise.} \end{cases}$$

Then $g$ is $(\rho - L)$-strongly convex with respect to $\mathbf{x}$ and $\mathbf{x}^*(\mathbf{z})$ minimizes $g(\mathbf{x}, \mathbf{z})$. So by (A.7) we have

$$g(\mathbf{x}^*(\mathbf{z}^k), \mathbf{z}^{k+1}) - g(\mathbf{x}^*(\mathbf{z}^{k+1}), \mathbf{z}^{k+1}) \geq \frac{\rho - L}{2} \|\mathbf{x}^*(\mathbf{z}^k) - \mathbf{x}^*(\mathbf{z}^{k+1})\|^2.$$

On the other hand, we have

$$\begin{aligned}
&g(\mathbf{x}^*(\mathbf{z}^k), \mathbf{z}^{k+1}) - g(\mathbf{x}^*(\mathbf{z}^{k+1}), \mathbf{z}^{k+1}) \\
&= g(\mathbf{x}^*(\mathbf{z}^k), \mathbf{z}^k) - g(\mathbf{x}^*(\mathbf{z}^{k+1}), \mathbf{z}^k) \\
&\quad - \left( g(\mathbf{x}^*(\mathbf{z}^{k+1}), \mathbf{z}^{k+1}) - g(\mathbf{x}^*(\mathbf{z}^{k+1}), \mathbf{z}^k) \right) \\
&\quad + \left( g(\mathbf{x}^*(\mathbf{z}^k), \mathbf{z}^{k+1}) - g(\mathbf{x}^*(\mathbf{z}^k), \mathbf{z}^k) \right) \\
&= g(\mathbf{x}^*(\mathbf{z}^k), \mathbf{z}^k) - g(\mathbf{x}^*(\mathbf{z}^{k+1}), \mathbf{z}^k) \\
&\quad - \frac{\rho}{2} \left( -2 \left\langle \mathbf{z}^{k+1} - \mathbf{z}^k, \mathbf{x}^*(\mathbf{z}^{k+1}) \right\rangle + \|\mathbf{z}^{k+1}\|^2 - \|\mathbf{z}^k\|^2 \right) \\
&\quad + \frac{\rho}{2} \left( -2 \left\langle \mathbf{z}^{k+1} - \mathbf{z}^k, \mathbf{x}^*(\mathbf{z}^k) \right\rangle + \|\mathbf{z}^{k+1}\|^2 - \|\mathbf{z}^k\|^2 \right) \\
&= g(\mathbf{x}^*(\mathbf{z}^k), \mathbf{z}^k) - g(\mathbf{x}^*(\mathbf{z}^{k+1}), \mathbf{z}^k) + \rho \left\langle \mathbf{z}^{k+1} - \mathbf{z}^k, \mathbf{x}^*(\mathbf{z}^{k+1}) - \mathbf{x}^*(\mathbf{z}^k) \right\rangle \\
&\leq -\frac{\rho - L}{2} \|\mathbf{x}^*(\mathbf{z}^{k+1}) - \mathbf{x}^*(\mathbf{z}^k)\|^2 + \rho \left\langle \mathbf{z}^{k+1} - \mathbf{z}^k, \mathbf{x}^*(\mathbf{z}^{k+1}) - \mathbf{x}^*(\mathbf{z}^k) \right\rangle.
\end{aligned}$$

So we have

$$(\rho - L) \|\mathbf{x}^*(\mathbf{z}^{k+1}) - \mathbf{x}^*(\mathbf{z}^k)\|^2 \leq \rho \left\langle \mathbf{z}^{k+1} - \mathbf{z}^k, \mathbf{x}^*(\mathbf{z}^{k+1}) - \mathbf{x}^*(\mathbf{z}^k) \right\rangle.$$

By the Cauchy–Schwartz inequality (Proposition A.1), we have (4.12). Similarly, we also have (4.13) by replacing $g(\mathbf{x}, \mathbf{z})$ with $P(\mathbf{x}, \mathbf{z}, \boldsymbol{\lambda})$.

At last, we consider (4.14). By checking the KKT conditions, we know that $\mathbf{x}^*(\mathbf{z}^k)$ is a minimizer of $P(\mathbf{x}, \mathbf{z}^k, \boldsymbol{\lambda}^*(\mathbf{z}^k))$. Since $P(\mathbf{x}, \mathbf{z}^k, \boldsymbol{\lambda}^*(\mathbf{z}^k))$ is $(\rho - L)$-strongly convex with respect to $\mathbf{x}$, the minimizer is unique. So $\mathbf{x}^*(\mathbf{z}^k) = \mathbf{x}(\mathbf{z}^k, \boldsymbol{\lambda}^*(\mathbf{z}^k))$. Therefore, we only need to prove

$$\|\mathbf{x}(\mathbf{z}^k, \boldsymbol{\lambda}^k) - \mathbf{x}(\mathbf{z}^k, \boldsymbol{\lambda}^*(\mathbf{z}^k))\| \leq \frac{\tilde{\sigma}}{\rho - L} \|\boldsymbol{\lambda}^k - \boldsymbol{\lambda}^*(\mathbf{z}^k)\|.$$

We write $\boldsymbol{\lambda}^*$ for $\boldsymbol{\lambda}^*(\mathbf{z}^k)$ in the remaining of this proof for notation simplicity. Since $P(\mathbf{x}, \mathbf{z}, \boldsymbol{\lambda})$ is $(\rho - L)$-strongly convex with respect to $\mathbf{x}$ and $\mathbf{x}(\mathbf{z}, \boldsymbol{\lambda})$ minimizes $P(\mathbf{x}, \mathbf{z}, \boldsymbol{\lambda})$, by (A.7) we have

$$P(\mathbf{x}(\mathbf{z}^k, \boldsymbol{\lambda}^*), \mathbf{z}^k, \boldsymbol{\lambda}^k) - P(\mathbf{x}(\mathbf{z}^k, \boldsymbol{\lambda}^k), \mathbf{z}^k, \boldsymbol{\lambda}^k) \geq \frac{\rho - L}{2} \|\mathbf{x}(\mathbf{z}^k, \boldsymbol{\lambda}^*) - \mathbf{x}(\mathbf{z}^k, \boldsymbol{\lambda}^k)\|^2.$$

On the other hand, we have

$$P(\mathbf{x}(\mathbf{z}^k, \boldsymbol{\lambda}^*), \mathbf{z}^k, \boldsymbol{\lambda}^k) - P(\mathbf{x}(\mathbf{z}^k, \boldsymbol{\lambda}^k), \mathbf{z}^k, \boldsymbol{\lambda}^k)$$

$$= P(\mathbf{x}(\mathbf{z}^k, \boldsymbol{\lambda}^*), \mathbf{z}^k, \boldsymbol{\lambda}^*) - P(\mathbf{x}(\mathbf{z}^k, \boldsymbol{\lambda}^k), \mathbf{z}^k, \boldsymbol{\lambda}^*)$$

$$- \left( P(\mathbf{x}(\mathbf{z}^k, \boldsymbol{\lambda}^k), \mathbf{z}^k, \boldsymbol{\lambda}^k) - P(\mathbf{x}(\mathbf{z}^k, \boldsymbol{\lambda}^k), \mathbf{z}^k, \boldsymbol{\lambda}^*) \right)$$

$$+ \left( P(\mathbf{x}(\mathbf{z}^k, \boldsymbol{\lambda}^*), \mathbf{z}^k, \boldsymbol{\lambda}^k) - P(\mathbf{x}(\mathbf{z}^k, \boldsymbol{\lambda}^*), \mathbf{z}^k, \boldsymbol{\lambda}^*) \right)$$

$$= P(\mathbf{x}(\mathbf{z}^k, \boldsymbol{\lambda}^*), \mathbf{z}^k, \boldsymbol{\lambda}^*) - P(\mathbf{x}(\mathbf{z}^k, \boldsymbol{\lambda}^k), \mathbf{z}^k, \boldsymbol{\lambda}^*)$$

$$- \left\langle \mathbf{A}\mathbf{x}(\mathbf{z}^k, \boldsymbol{\lambda}^k) - \mathbf{A}\mathbf{x}(\mathbf{z}^k, \boldsymbol{\lambda}^*), \boldsymbol{\lambda}^k - \boldsymbol{\lambda}^* \right\rangle$$

$$\leq -\frac{\rho - L}{2} \|\mathbf{x}(\mathbf{z}^k, \boldsymbol{\lambda}^*) - \mathbf{x}(\mathbf{z}^k, \boldsymbol{\lambda}^k)\|^2$$

$$- \left\langle \mathbf{A}\mathbf{x}(\mathbf{z}^k, \boldsymbol{\lambda}^k) - \mathbf{A}\mathbf{x}(\mathbf{z}^k, \boldsymbol{\lambda}^*), \boldsymbol{\lambda}^k - \boldsymbol{\lambda}^* \right\rangle.$$

So we have

$$(\rho - L)\|\mathbf{x}(\mathbf{z}^k, \boldsymbol{\lambda}^k) - \mathbf{x}(\mathbf{z}^k, \boldsymbol{\lambda}^*)\|^2 \leq \tilde{\sigma}\|\mathbf{x}(\mathbf{z}^k, \boldsymbol{\lambda}^k) - \mathbf{x}(\mathbf{z}^k, \boldsymbol{\lambda}^*)\|\|\boldsymbol{\lambda}^k - \boldsymbol{\lambda}^*\|,$$

which gives (4.14).                                                                    □

**Lemma 4.2** *Assume that $f$ is $L$-smooth with respect to $\mathbf{x}$. Suppose $\rho > L$, then for Algorithm 4.2 we have*

$$\|\mathbf{x}^k - \mathbf{x}(\mathbf{z}^k, \boldsymbol{\lambda}^k)\| \leq \frac{1 + \alpha_1\sqrt{2mL^2 + (m-1)m\beta^2\sigma^4}}{\alpha_1(\rho - L)}\|\mathbf{x}^{k+1} - \mathbf{x}^k\|, \qquad (4.15)$$

*where $\sigma = \max\{\|\mathbf{A}_i\|_2, i \in [m]\}$.*

**Proof** The update of $\mathbf{x}_j$ can be rewritten as

$$\mathbf{x}_j^{k+1} = \mathbf{x}_j^k - \alpha_1 \nabla_j P(\mathbf{x}^k, \mathbf{z}^k, \boldsymbol{\lambda}^k)$$

$$+ \alpha_1 \nabla_j P(\mathbf{x}^k, \mathbf{z}^k, \boldsymbol{\lambda}^k) - \alpha_1 \nabla_j P(\mathbf{x}_1^{k+1}, \cdots, \mathbf{x}_{j-1}^{k+1}, \mathbf{x}_j^k, \cdots, \mathbf{x}_m^k, \mathbf{z}^k, \boldsymbol{\lambda}^k).$$

So

$$\alpha_1 \sqrt{\sum_{j=1}^m \left\| \nabla_j P(\mathbf{x}^k, \mathbf{z}^k, \boldsymbol{\lambda}^k) - \nabla_j P(\mathbf{x}_1^{k+1}, \cdots, \mathbf{x}_{j-1}^{k+1}, \mathbf{x}_j^k, \cdots, \mathbf{x}_m^k, \mathbf{z}^k, \boldsymbol{\lambda}^k) \right\|^2}$$

$$= \left\| \mathbf{x}^{k+1} - \mathbf{x}^k + \alpha_1 \nabla_{\mathbf{x}} P(\mathbf{x}^k, \mathbf{z}^k, \boldsymbol{\lambda}^k) \right\|.$$

On the other hand, we have

$$\sum_{j=1}^{m}\left\|\nabla_j P(\mathbf{x}^k, \mathbf{z}^k, \boldsymbol{\lambda}^k) - \nabla_j P(\mathbf{x}_1^{k+1}, \cdots, \mathbf{x}_{j-1}^{k+1}, \mathbf{x}_j^k, \cdots, \mathbf{x}_m^k, \mathbf{z}^k, \boldsymbol{\lambda}^k)\right\|^2$$

$$= \sum_{j=1}^{m}\left\|\nabla_j f(\mathbf{x}^k) - \nabla_j f(\mathbf{x}_1^{k+1}, \cdots, \mathbf{x}_{j-1}^{k+1}, \mathbf{x}_j^k, \cdots, \mathbf{x}_m^k)\right.$$

$$\left. - \beta \mathbf{A}_j^T \sum_{i=1}^{j-1} \mathbf{A}_i \left(\mathbf{x}_i^{k+1} - \mathbf{x}_i^k\right)\right\|^2$$

$$\leq 2L^2 \sum_{j=1}^{m}\sum_{i=1}^{j-1} \|\mathbf{x}_i^{k+1} - \mathbf{x}_i^k\|^2 + 2\beta^2\sigma^4 \sum_{j=1}^{m}(j-1)\sum_{i=1}^{j-1}\|\mathbf{x}_i^{k+1} - \mathbf{x}_i^k\|^2$$

$$\leq [2mL^2 + (m-1)m\beta^2\sigma^4]\|\mathbf{x}^{k+1} - \mathbf{x}^k\|^2.$$

Thus

$$\alpha_1\sqrt{2mL^2 + (m-1)m\beta^2\sigma^4}\|\mathbf{x}^{k+1} - \mathbf{x}^k\|$$

$$\geq \|\mathbf{x}^{k+1} - \mathbf{x}^k + \alpha_1\nabla_\mathbf{x} P(\mathbf{x}^k, \mathbf{z}^k, \boldsymbol{\lambda}^k)\|$$

$$\geq \alpha_1\|\nabla_\mathbf{x} P(\mathbf{x}^k, \mathbf{z}^k, \boldsymbol{\lambda}^k)\| - \|\mathbf{x}^{k+1} - \mathbf{x}^k\|. \tag{4.16}$$

Since $P(\mathbf{x}, \mathbf{z}, \boldsymbol{\lambda})$ is $(\rho - L)$-strongly convex with respect to $\mathbf{x}$, we have

$$\|\nabla_\mathbf{x} P(\mathbf{x}^k, \mathbf{z}^k, \boldsymbol{\lambda}^k)\| = \|\nabla_\mathbf{x} P(\mathbf{x}^k, \mathbf{z}^k, \boldsymbol{\lambda}^k) - \nabla_\mathbf{x} P(\mathbf{x}(\mathbf{z}^k, \boldsymbol{\lambda}^k), \mathbf{z}^k, \boldsymbol{\lambda}^k)\|$$

$$\overset{a}{\geq} (\rho - L)\|\mathbf{x}^k - \mathbf{x}(\mathbf{z}^k, \boldsymbol{\lambda}^k)\|,$$

where $\overset{a}{\geq}$ uses (A.9). So we have (4.15). $\qquad\square$

**Lemma 4.3** *Assume that $f$ is $L$-smooth with respect to $\mathbf{x}$. We have the following dual error bound for Algorithm 4.2:*

$$\mathrm{dist}(\boldsymbol{\lambda}, \Lambda^*(\mathbf{z})) \leq \frac{\theta^2(L + \rho + \beta\widetilde{\sigma}^2)^2}{\rho - L}\|\mathbf{A}\mathbf{x}(\mathbf{z}, \boldsymbol{\lambda}) - \mathbf{b}\|, \tag{4.17}$$

*where $\Lambda^*(\mathbf{z})$ is the solution set of the dual problem of Problem (4.11), and $\theta$ is the Hoffman's constant defined in Lemma A.3, which only depends on $\mathbf{A}$.*

***Proof*** The KKT condition of Problem (4.11) is

$$\nabla f(\mathbf{x}^*(\mathbf{z})) + \rho(\mathbf{x}^*(\mathbf{z}) - \mathbf{z}) + \mathbf{A}^T \boldsymbol{\lambda}^*(\mathbf{z}) = \mathbf{0},$$

$$\mathbf{A}\mathbf{x}^*(\mathbf{z}) = \mathbf{b},$$

where $\boldsymbol{\lambda}^*(\mathbf{z})$ is any dual variable in $\Lambda^*(\mathbf{z})$. From the definition of $\mathbf{x}(\mathbf{z}, \boldsymbol{\lambda})$, we have

$$\nabla f(\mathbf{x}(\mathbf{z}, \boldsymbol{\lambda})) + \rho(\mathbf{x}(\mathbf{z}, \boldsymbol{\lambda}) - \mathbf{z}) + \mathbf{A}^T \boldsymbol{\lambda} + \beta \mathbf{A}^T (\mathbf{A}\mathbf{x}(\mathbf{z}, \boldsymbol{\lambda}) - \mathbf{b}) = \mathbf{0}.$$

Consider the linear system:

$$\mathbf{A}^T \boldsymbol{\lambda}(\mathbf{z}) = - \left[ \nabla f(\mathbf{x}^*(\mathbf{z})) + \rho(\mathbf{x}^*(\mathbf{z}) - \mathbf{z}) \right]. \tag{4.18}$$

From the Hoffman's bound in Lemma A.3, we have

$$\mathrm{dist}(\boldsymbol{\lambda}, \Lambda^*(\mathbf{z}))^2$$

$$= \min_{\boldsymbol{\lambda}(\mathbf{z}) \text{ satisfying } (4.18)} \|\boldsymbol{\lambda} - \boldsymbol{\lambda}(\mathbf{z})\|^2$$

$$\leq \theta^2 \|\mathbf{A}^T \boldsymbol{\lambda} + \left[ \nabla f(\mathbf{x}^*(\mathbf{z})) + \rho(\mathbf{x}^*(\mathbf{z}) - \mathbf{z}) \right] \|^2$$

$$= \theta^2 \left\| \left[ \nabla f(\mathbf{x}(\mathbf{z}, \boldsymbol{\lambda})) + \rho(\mathbf{x}(\mathbf{z}, \boldsymbol{\lambda}) - \mathbf{z}) + \beta \mathbf{A}^T (\mathbf{A}\mathbf{x}(\mathbf{z}, \boldsymbol{\lambda}) - \mathbf{b}) \right] \right.$$

$$\left. - \left[ \nabla f(\mathbf{x}^*(\mathbf{z})) + \rho(\mathbf{x}^*(\mathbf{z}) - \mathbf{z}) \right] \right\|^2$$

$$\overset{a}{\leq} \theta^2 (L + \rho + \beta\widetilde{\sigma}^2)^2 \|\mathbf{x}(\mathbf{z}, \boldsymbol{\lambda}) - \mathbf{x}^*(\mathbf{z})\|^2$$

$$\overset{b}{\leq} \frac{\theta^2 (L + \rho + \beta\widetilde{\sigma}^2)^2}{\rho - L}$$

$$\times \left\langle \nabla f(\mathbf{x}(\mathbf{z}, \boldsymbol{\lambda})) + \rho(\mathbf{x}(\mathbf{z}, \boldsymbol{\lambda}) - \mathbf{z}) - \nabla f(\mathbf{x}^*(\mathbf{z})) - \rho(\mathbf{x}^*(\mathbf{z}) - \mathbf{z}), \right.$$

$$\left. \mathbf{x}(\mathbf{z}, \boldsymbol{\lambda}) - \mathbf{x}^*(\mathbf{z}) \right\rangle$$

$$= -\frac{\theta^2 (L + \rho + \beta\widetilde{\sigma}^2)^2}{\rho - L}$$

$$\times \left\langle \mathbf{A}^T \boldsymbol{\lambda} + \beta \mathbf{A}^T (\mathbf{A}\mathbf{x}(\mathbf{z}, \boldsymbol{\lambda}) - \mathbf{b}) - \mathbf{A}^T \boldsymbol{\lambda}^*(\mathbf{z}), \mathbf{x}(\mathbf{z}, \boldsymbol{\lambda}) - \mathbf{x}^*(\mathbf{z}) \right\rangle$$

$$= -\frac{\theta^2 (L + \rho + \beta\widetilde{\sigma}^2)^2}{\rho - L} \left( \langle \boldsymbol{\lambda} - \boldsymbol{\lambda}^*(\mathbf{z}), \mathbf{A}\mathbf{x}(\mathbf{z}, \boldsymbol{\lambda}) - \mathbf{b} \rangle + \beta \|\mathbf{A}\mathbf{x}(\mathbf{z}, \boldsymbol{\lambda}) - \mathbf{b}\|^2 \right)$$

$$\leq \frac{\theta^2 (L + \rho + \beta\widetilde{\sigma}^2)^2}{\rho - L} \|\boldsymbol{\lambda} - \boldsymbol{\lambda}^*(\mathbf{z})\| \|\mathbf{A}\mathbf{x}(\mathbf{z}, \boldsymbol{\lambda}) - \mathbf{b}\|,$$

where $\overset{a}{\leq}$ uses the $(L + \rho + \beta\tilde{\sigma}^2)$-smoothness of $P(\mathbf{x}, \mathbf{z}, \boldsymbol{\lambda})$ with respect to $\mathbf{x}$ and $\overset{b}{\leq}$ uses the $(\rho - L)$-strong convexity of $f(\mathbf{x}) + \frac{\rho}{2}\|\mathbf{x} - \mathbf{z}\|^2$ and (A.8). Choosing $\boldsymbol{\lambda}^*(\mathbf{z})$ to be the point such that $\|\boldsymbol{\lambda} - \boldsymbol{\lambda}^*(\mathbf{z})\| = \text{dist}(\boldsymbol{\lambda}, \Lambda^*(\mathbf{z}))$, we have (4.17). $\qquad\square$

Next, we give the estimates on the changes in $P$, $d$, and $M$.

**Lemma 4.4** *Assume that $f$ is $L$-smooth with respect to $\mathbf{x}$. Let*

$$0 < \alpha_1 \leq \frac{1}{L + \beta\sigma^2 + \rho}, \qquad \alpha_2 > 0, \qquad 0 < \alpha_3 \leq 1, \quad and \quad \rho > L.$$

*For Algorithm 4.2, we have*

$$P(\mathbf{x}^k, \mathbf{z}^k, \boldsymbol{\lambda}^{k-1}) - P(\mathbf{x}^{k+1}, \mathbf{z}^{k+1}, \boldsymbol{\lambda}^k)$$

$$\geq \frac{1}{2\alpha_1}\|\mathbf{x}^{k+1} - \mathbf{x}^k\|^2 + \frac{\rho}{2\alpha_3}\|\mathbf{z}^{k+1} - \mathbf{z}^k\|^2 - \alpha_2\|A\mathbf{x}^k - \mathbf{b}\|^2, \tag{4.19}$$

$$d(\mathbf{z}^{k+1}, \boldsymbol{\lambda}^k) - d(\mathbf{z}^k, \boldsymbol{\lambda}^{k-1})$$

$$\geq \alpha_2 \left\langle A\mathbf{x}^k - \mathbf{b}, A\mathbf{x}(\mathbf{z}^k, \boldsymbol{\lambda}^k) - \mathbf{b} \right\rangle$$

$$+ \frac{\rho}{2}\left\langle \mathbf{z}^{k+1} - \mathbf{z}^k, \mathbf{z}^{k+1} + \mathbf{z}^k - 2\mathbf{x}(\mathbf{z}^{k+1}, \boldsymbol{\lambda}^k) \right\rangle, \tag{4.20}$$

$$M(\mathbf{z}^{k+1}) - M(\mathbf{z}^k)$$

$$\leq \rho\left\langle \mathbf{z}^{k+1} - \mathbf{z}^k, \mathbf{z}^k - \mathbf{x}^*(\mathbf{z}^k) \right\rangle + \frac{\rho}{2}\left(\frac{\rho}{\rho - L} + 1\right)\|\mathbf{z}^{k+1} - \mathbf{z}^k\|^2. \tag{4.21}$$

***Proof*** From the update of $\boldsymbol{\lambda}$, we have

$$P(\mathbf{x}^k, \mathbf{z}^k, \boldsymbol{\lambda}^{k-1}) - P(\mathbf{x}^k, \mathbf{z}^k, \boldsymbol{\lambda}^k) = -\alpha_2\|A\mathbf{x}^k - \mathbf{b}\|^2.$$

Since $P$ is $(L + \beta\sigma^2 + \rho)$-smooth for each $\mathbf{x}_j$, by (A.4) and considering that updating $\mathbf{x}_j$ is a standard coordinate gradient descent, we have

$$P\left(\mathbf{x}_1^{k+1}, \cdots, \mathbf{x}_{j-1}^{k+1}, \mathbf{x}_j^k, \mathbf{x}_{j+1}^k, \cdots, \mathbf{x}_m^k, \mathbf{z}^k, \boldsymbol{\lambda}^k\right)$$

$$- P\left(\mathbf{x}_1^{k+1}, \cdots, \mathbf{x}_{j-1}^{k+1}, \mathbf{x}_j^{k+1}, \mathbf{x}_{j+1}^k, \cdots, \mathbf{x}_m^k, \mathbf{z}^k, \boldsymbol{\lambda}^k\right)$$

$$\geq -\left\langle \nabla_j P\left(\mathbf{x}_1^{k+1}, \cdots, \mathbf{x}_{j-1}^{k+1}, \mathbf{x}_j^k, \mathbf{x}_{j+1}^k, \cdots, \mathbf{x}_m^k, \mathbf{z}^k, \boldsymbol{\lambda}^k\right), \mathbf{x}_j^{k+1} - \mathbf{x}_j^k \right\rangle$$

$$- \frac{L + \beta\sigma^2 + \rho}{2}\left\|\mathbf{x}_j^{k+1} - \mathbf{x}_j^k\right\|^2$$

$$= \frac{1}{\alpha_1} \left\| \mathbf{x}_j^{k+1} - \mathbf{x}_j^k \right\|^2 - \frac{L + \beta\sigma^2 + \rho}{2} \left\| \mathbf{x}_j^{k+1} - \mathbf{x}_j^k \right\|^2$$

$$\geq \frac{1}{2\alpha_1} \left\| \mathbf{x}_j^{k+1} - \mathbf{x}_j^k \right\|^2.$$

Summing up for $i = 1, \cdots, m$, we have

$$P(\mathbf{x}^k, \mathbf{z}^k, \boldsymbol{\lambda}^k) - P(\mathbf{x}^{k+1}, \mathbf{z}^k, \boldsymbol{\lambda}^k) \geq \frac{1}{2\alpha_1} \|\mathbf{x}^{k+1} - \mathbf{x}^k\|^2.$$

From the update of $\mathbf{z}$, we have

$$P(\mathbf{x}^{k+1}, \mathbf{z}^k, \boldsymbol{\lambda}^k) - P(\mathbf{x}^{k+1}, \mathbf{z}^{k+1}, \boldsymbol{\lambda}^k)$$

$$= \frac{\rho}{2} \left( \|\mathbf{x}^{k+1} - \mathbf{z}^k\|^2 - \|\mathbf{x}^{k+1} - \mathbf{z}^{k+1}\|^2 \right)$$

$$= \frac{\rho}{2} \left\langle \mathbf{z}^{k+1} - \mathbf{z}^k, 2\mathbf{x}^{k+1} - \mathbf{z}^k - \mathbf{z}^{k+1} \right\rangle$$

$$= \frac{\rho}{2} \left( \frac{2}{\alpha_3} - 1 \right) \|\mathbf{z}^{k+1} - \mathbf{z}^k\|^2$$

$$\geq \frac{\rho}{2\alpha_3} \|\mathbf{z}^{k+1} - \mathbf{z}^k\|^2$$

for $\alpha_3 \leq 1$. Summing up the above three inequalities, we have (4.19).
   Similarly, we have

$$d(\mathbf{z}^k, \boldsymbol{\lambda}^k) - d(\mathbf{z}^k, \boldsymbol{\lambda}^{k-1})$$

$$= P(\mathbf{x}(\mathbf{z}^k, \boldsymbol{\lambda}^k), \mathbf{z}^k, \boldsymbol{\lambda}^k) - P(\mathbf{x}(\mathbf{z}^k, \boldsymbol{\lambda}^{k-1}), \mathbf{z}^k, \boldsymbol{\lambda}^{k-1})$$

$$\geq P(\mathbf{x}(\mathbf{z}^k, \boldsymbol{\lambda}^k), \mathbf{z}^k, \boldsymbol{\lambda}^k) - P(\mathbf{x}(\mathbf{z}^k, \boldsymbol{\lambda}^k), \mathbf{z}^k, \boldsymbol{\lambda}^{k-1})$$

$$= \left\langle \boldsymbol{\lambda}^k - \boldsymbol{\lambda}^{k-1}, \mathbf{A}\mathbf{x}(\mathbf{z}^k, \boldsymbol{\lambda}^k) - \mathbf{b} \right\rangle$$

$$= \alpha_2 \left\langle \mathbf{A}\mathbf{x}^k - \mathbf{b}, \mathbf{A}\mathbf{x}(\mathbf{z}^k, \boldsymbol{\lambda}^k) - \mathbf{b} \right\rangle$$

and

$$d(\mathbf{z}^{k+1}, \boldsymbol{\lambda}^k) - d(\mathbf{z}^k, \boldsymbol{\lambda}^k)$$

$$= P(\mathbf{x}(\mathbf{z}^{k+1}, \boldsymbol{\lambda}^k), \mathbf{z}^{k+1}, \boldsymbol{\lambda}^k) - P(\mathbf{x}(\mathbf{z}^k, \boldsymbol{\lambda}^k), \mathbf{z}^k, \boldsymbol{\lambda}^k)$$

$$\geq P(\mathbf{x}(\mathbf{z}^{k+1}, \boldsymbol{\lambda}^k), \mathbf{z}^{k+1}, \boldsymbol{\lambda}^k) - P(\mathbf{x}(\mathbf{z}^{k+1}, \boldsymbol{\lambda}^k), \mathbf{z}^k, \boldsymbol{\lambda}^k)$$

$$= \frac{\rho}{2} \left( \|\mathbf{x}(\mathbf{z}^{k+1}, \boldsymbol{\lambda}^k) - \mathbf{z}^{k+1}\|^2 - \|\mathbf{x}(\mathbf{z}^{k+1}, \boldsymbol{\lambda}^k) - \mathbf{z}^k\|^2 \right)$$

$$= \frac{\rho}{2} \left\langle \mathbf{z}^{k+1} - \mathbf{z}^k, \mathbf{z}^{k+1} + \mathbf{z}^k - 2\mathbf{x}(\mathbf{z}^{k+1}, \boldsymbol{\lambda}^k) \right\rangle.$$

Adding them, we have (4.20).

To prove (4.21), we first prove

$$\nabla M(\mathbf{z}) = \rho \left( \mathbf{z} - \mathbf{x}^*(\mathbf{z}) \right). \tag{4.22}$$

Indeed, define

$$\phi(\mathbf{x}, \mathbf{z}) = \frac{\rho}{2} \|\mathbf{z}\|^2 - f(\mathbf{x}) - \frac{\rho}{2} \|\mathbf{x} - \mathbf{z}\|^2 \text{ and}$$

$$\psi(\mathbf{z}) = \max_{\mathbf{A}\mathbf{x}=\mathbf{b}} \phi(\mathbf{x}, \mathbf{z}) = \frac{\rho}{2} \|\mathbf{z}\|^2 - M(\mathbf{z}).$$

Then $\phi(\mathbf{x}, \cdot)$ is convex for each $\mathbf{x} \in \{\mathbf{x} | \mathbf{A}\mathbf{x} = \mathbf{b}\}$, and

$$\mathbf{x}^*(\mathbf{z}) = \operatorname*{argmax}_{\mathbf{A}\mathbf{x}=\mathbf{b}} \phi(\mathbf{x}, \mathbf{z})$$

is a singleton. So by Danskin's theorem (Theorem A.1) and Proposition A.7, we have that $\psi(\mathbf{z})$ is differentiable and

$$\nabla \psi(\mathbf{z}) = \nabla_{\mathbf{z}} \phi\left(\mathbf{x}^*(\mathbf{z}), \mathbf{z}\right) = \rho \mathbf{x}^*(\mathbf{z}).$$

Thus $\nabla M(\mathbf{z}) = \rho \mathbf{z} - \nabla \psi(\mathbf{z}) = \rho (\mathbf{z} - \mathbf{x}^*(\mathbf{z}))$.

From (4.22) and (4.12), we have

$$\|\nabla M(\mathbf{z}^k) - \nabla M(\mathbf{z}^{k+1})\| \leq \rho \left( \frac{\rho}{\rho - L} + 1 \right) \|\mathbf{z}^k - \mathbf{z}^{k+1}\|.$$

So $M(\mathbf{z})$ is $\rho \left( \frac{\rho}{\rho-L} + 1 \right)$-smooth. Thus by (A.4), we have (4.21). $\qquad \square$

Now we are ready to prove that $\Phi^k$ decreases sufficiently.

**Lemma 4.5** *Assume that $f$ is $L$-smooth with respect to $\mathbf{x}$. Choosing $\alpha_1$, $\alpha_2$, and $\alpha_3$ appropriately and letting $\rho > L$, then for Algorithm 4.2 we have*

$$\Phi^k - \Phi^{k+1} \geq \delta \left( \|\mathbf{x}^{k+1} - \mathbf{x}^k\|^2 + \|\mathbf{z}^{k+1} - \mathbf{z}^k\|^2 + \|\mathbf{A}\mathbf{x}(\mathbf{z}^k, \boldsymbol{\lambda}^k) - \mathbf{b}\|^2 \right) \tag{4.23}$$

*for some positive constant $\delta$.*

**Proof** From (4.19), (4.20), and (4.21), we have

$$\Phi^k - \Phi^{k+1}$$

$$\geq \frac{1}{2\alpha_1}\|\mathbf{x}^{k+1} - \mathbf{x}^k\|^2 + \frac{\rho}{2\alpha_3}\|\mathbf{z}^{k+1} - \mathbf{z}^k\|^2 - \alpha_2\|\mathbf{A}\mathbf{x}^k - \mathbf{b}\|^2$$

$$+ 2\alpha_2\left\langle \mathbf{A}\mathbf{x}^k - \mathbf{b}, \mathbf{A}\mathbf{x}(\mathbf{z}^k, \boldsymbol{\lambda}^k) - \mathbf{b}\right\rangle$$

$$+ \rho\left\langle \mathbf{z}^{k+1} - \mathbf{z}^k, \mathbf{z}^{k+1} + \mathbf{z}^k - 2\mathbf{x}(\mathbf{z}^{k+1}, \boldsymbol{\lambda}^k)\right\rangle$$

$$+ 2\rho\left\langle \mathbf{z}^{k+1} - \mathbf{z}^k, \mathbf{x}^*(\mathbf{z}^k) - \mathbf{z}^k\right\rangle - \rho\left(\frac{\rho}{\rho - L} + 1\right)\|\mathbf{z}^{k+1} - \mathbf{z}^k\|^2$$

$$= \frac{1}{2\alpha_1}\|\mathbf{x}^{k+1} - \mathbf{x}^k\|^2 + \left[\frac{\rho}{2\alpha_3} - \rho\left(\frac{\rho}{\rho - L} + 1\right)\right]\|\mathbf{z}^{k+1} - \mathbf{z}^k\|^2$$

$$- \alpha_2\|\mathbf{A}\mathbf{x}^k - \mathbf{b}\|^2 + 2\alpha_2\left\langle \mathbf{A}\mathbf{x}^k - \mathbf{b}, \mathbf{A}\mathbf{x}(\mathbf{z}^k, \boldsymbol{\lambda}^k) - \mathbf{b}\right\rangle + \rho\Big\langle \mathbf{z}^{k+1} - \mathbf{z}^k,$$

$$\mathbf{z}^{k+1} - \mathbf{z}^k - 2\left(\mathbf{x}(\mathbf{z}^{k+1}, \boldsymbol{\lambda}^k) - \mathbf{x}(\mathbf{z}^k, \boldsymbol{\lambda}^k)\right) - 2\left(\mathbf{x}(\mathbf{z}^k, \boldsymbol{\lambda}^k) - \mathbf{x}^*(\mathbf{z}^k)\right)\Big\rangle.$$

Since

$$-2\rho\left\langle \mathbf{z}^{k+1} - \mathbf{z}^k, \mathbf{x}(\mathbf{z}^k, \boldsymbol{\lambda}^k) - \mathbf{x}^*(\mathbf{z}^k)\right\rangle$$

$$\geq -\frac{\rho}{c_1}\|\mathbf{z}^{k+1} - \mathbf{z}^k\|^2 - \rho c_1\|\mathbf{x}(\mathbf{z}^k, \boldsymbol{\lambda}^k) - \mathbf{x}^*(\mathbf{z}^k)\|^2$$

$$\overset{a}{\geq} -\frac{\rho}{c_1}\|\mathbf{z}^{k+1} - \mathbf{z}^k\|^2 - \frac{\rho c_1\widetilde{\sigma}^2}{(\rho - L)^2}\|\boldsymbol{\lambda}^k - \boldsymbol{\lambda}^*(\mathbf{z}^k)\|^2$$

$$\overset{b}{\geq} -\frac{\rho}{c_1}\|\mathbf{z}^{k+1} - \mathbf{z}^k\|^2 - \frac{\rho c_1\widetilde{\sigma}^2}{(\rho - L)^2}\frac{\theta^4(L + \rho + \beta\widetilde{\sigma}^2)^4}{(\rho - L)^2}\|\mathbf{A}\mathbf{x}(\mathbf{z}^k, \boldsymbol{\lambda}^k) - \mathbf{b}\|^2,$$

$$-2\rho\left\langle \mathbf{z}^{k+1} - \mathbf{z}^k, \mathbf{x}(\mathbf{z}^{k+1}, \boldsymbol{\lambda}^k) - \mathbf{x}(\mathbf{z}^k, \boldsymbol{\lambda}^k)\right\rangle$$

$$\geq -2\rho\|\mathbf{z}^{k+1} - \mathbf{z}^k\|\|\mathbf{x}(\mathbf{z}^{k+1}, \boldsymbol{\lambda}^k) - \mathbf{x}(\mathbf{z}^k, \boldsymbol{\lambda}^k)\|$$

$$\overset{c}{\geq} -\frac{2\rho^2}{\rho - L}\|\mathbf{z}^k - \mathbf{z}^{k+1}\|^2,$$

and

$$-\alpha_2\|\mathbf{A}\mathbf{x}^k - \mathbf{b}\|^2 + 2\alpha_2\left\langle \mathbf{A}\mathbf{x}^k - \mathbf{b}, \mathbf{A}\mathbf{x}(\mathbf{z}^k, \boldsymbol{\lambda}^k) - \mathbf{b}\right\rangle$$

$$= \alpha_2\|\mathbf{A}\mathbf{x}(\mathbf{z}^k, \boldsymbol{\lambda}^k) - \mathbf{b}\|^2 - \alpha_2\|\mathbf{A}\mathbf{x}(\mathbf{z}^k, \boldsymbol{\lambda}^k) - \mathbf{A}\mathbf{x}^k\|^2$$

$$\geq \alpha_2\|\mathbf{A}\mathbf{x}(\mathbf{z}^k, \boldsymbol{\lambda}^k) - \mathbf{b}\|^2 - \alpha_2\widetilde{\sigma}^2\|\mathbf{x}(\mathbf{z}^k, \boldsymbol{\lambda}^k) - \mathbf{x}^k\|^2$$

$$\overset{d}{\geq} \alpha_2 \|\mathbf{Ax}(\mathbf{z}^k, \boldsymbol{\lambda}^k) - \mathbf{b}\|^2$$

$$- \alpha_2 \widetilde{\sigma}^2 \frac{2 + 2\alpha_1^2 \left[2mL^2 + (m-1)m\beta^2\sigma^4\right]}{\alpha_1^2(\rho - L)^2} \|\mathbf{x}^{k+1} - \mathbf{x}^k\|^2,$$

where we use (4.14) in $\overset{a}{\geq}$ and (4.17) in $\overset{b}{\geq}$ with $\boldsymbol{\lambda}^*(\mathbf{z}^k)$ chosen as the point such that $\|\boldsymbol{\lambda}^k - \boldsymbol{\lambda}^*(\mathbf{z}^k)\| = \text{dist}(\boldsymbol{\lambda}^k, \Lambda^*(\mathbf{z}^k))$, (4.13) in $\overset{c}{\geq}$, and (4.15) in $\overset{d}{\geq}$, and $c_1 > 0$ will be chosen later, we have

$$\Phi^k - \Phi^{k+1}$$

$$\geq \left\{\frac{1}{2\alpha_1} - \frac{2\alpha_2\widetilde{\sigma}^2}{\alpha_1^2(\rho - L)^2} - \frac{2\alpha_2\widetilde{\sigma}^2 \left[2mL^2 + (m-1)m\beta^2\sigma^4\right]}{(\rho - L)^2}\right\} \|\mathbf{x}^{k+1} - \mathbf{x}^k\|^2$$

$$+ \left(\frac{\rho}{2\alpha_3} - \frac{\rho}{c_1} - \frac{3\rho^2}{\rho - L}\right) \|\mathbf{z}^{k+1} - \mathbf{z}^k\|^2$$

$$+ \left[\alpha_2 - \frac{\rho c_1 \widetilde{\sigma}^2 \theta^4 (L + \rho + \beta\widetilde{\sigma}^2)^4}{(\rho - L)^4}\right] \|\mathbf{Ax}(\mathbf{z}^k, \boldsymbol{\lambda}^k) - \mathbf{b}\|^2.$$

We choose

$$\delta \leq \min\left\{\frac{(\rho - L)^2}{12\widetilde{\sigma}^2(L + \beta\sigma^2 + \rho)}, \frac{L + \beta\sigma^2 + \rho}{6\left\{1 + \frac{4\widetilde{\sigma}^2[2mL^2 + (m-1)m\beta^2\sigma^4]}{(\rho - L)^2}\right\}}\right\},$$

$$\alpha_3 \leq \min\left\{\frac{\rho}{3\left(\delta + \frac{3\rho^2}{\rho - L}\right)}, \frac{(\rho - L)^4 \delta}{6\rho\widetilde{\sigma}^2\theta^4(L + \rho + \beta\widetilde{\sigma}^2)^4}, 1\right\},$$

$$\alpha_2 = \frac{6\rho\alpha_3\widetilde{\sigma}^2\theta^4(L + \rho + \beta\widetilde{\sigma}^2)^4}{(\rho - L)^4} + \delta,$$

$$\alpha_1 \in \left[\frac{12\widetilde{\sigma}^2\delta}{(\rho - L)^2}, \frac{1}{L + \beta\sigma^2 + \rho}\right],$$

where we also take the requirements on $\alpha_i (i = 1, 2, 3)$ in Lemma 4.4 into consideration. Then letting $c_1 = 6\alpha_3$, we can check that

$$\frac{1}{2\alpha_1} - \frac{2\alpha_2\widetilde{\sigma}^2}{\alpha_1^2(\rho - L)^2} - \frac{2\alpha_2\widetilde{\sigma}^2 \left[2mL^2 + (m-1)m\beta^2\sigma^4\right]}{(\rho - L)^2}$$

$$\overset{a}{\geq} \frac{1}{6\alpha_1} - \frac{2\alpha_2\widetilde{\sigma}^2 \left[2mL^2 + (m-1)m\beta^2\sigma^4\right]}{(\rho - L)^2}$$

$$\overset{b}{\geq} \frac{L + \beta\sigma^2 + \rho}{6} - \frac{4\delta\widetilde{\sigma}^2 \left[2mL^2 + (m-1)m\beta^2\sigma^4\right]}{(\rho - L)^2} \geq \delta,$$

$$\frac{\rho}{2\alpha_3} - \frac{\rho}{c_1} - \frac{3\rho^2}{\rho - L} = \frac{\rho}{3\alpha_3} - \frac{3\rho^2}{\rho - L} \geq \delta, \quad \text{and}$$

$$\alpha_2 - \frac{\rho c_1 \widetilde{\sigma}^2 \theta^4 (L + \rho + \beta\widetilde{\sigma}^2)^4}{(\rho - L)^4} = \alpha_2 - \frac{6\rho\alpha_3 \widetilde{\sigma}^2 \theta^4 (L + \rho + \beta\widetilde{\sigma}^2)^4}{(\rho - L)^4} = \delta,$$

where $\overset{a}{\geq}$ uses $\alpha_2 \leq 2\delta \leq \frac{\alpha_1(\rho - L)^2}{6\widetilde{\sigma}^2}$ and $\overset{b}{\geq}$ uses $\alpha_1 \leq \frac{1}{L + \beta\sigma^2 + \rho}$ and $\alpha_2 \leq 2\delta$. So (4.23) is proven.                                                                                          $\square$

Then we have the convergence rate result for Algorithm 4.2.

**Theorem 4.3** *Assume that $f$ is $L$-smooth with respect to $\mathbf{x}$. Choosing $\alpha_1$, $\alpha_2$, and $\alpha_3$ appropriately and letting $\rho > L$, then Algorithm 4.2 needs $O(\frac{1}{\epsilon^2})$ iterations to find an $\epsilon$-approximate KKT point $(\mathbf{x}, \boldsymbol{\lambda})$. Namely,*

$$\|\mathbf{A}\mathbf{x} - \mathbf{b}\| \leq O(\epsilon) \quad \text{and} \quad \|\nabla f(\mathbf{x}) + \mathbf{A}^T \boldsymbol{\lambda}\| \leq O(\epsilon).$$

**Proof** Denote $f^* = \min_{\mathbf{A}\mathbf{x}=\mathbf{b}} f(\mathbf{x})$, we have $M(\mathbf{z}) \geq f^*$. So for any $k$ we have

$$\Phi^k = \left(P(\mathbf{x}^k, \mathbf{z}^k, \boldsymbol{\lambda}^{k-1}) - d(\mathbf{z}^k, \boldsymbol{\lambda}^{k-1})\right) + \left(M(\mathbf{z}^k) - d(\mathbf{z}^k, \boldsymbol{\lambda}^{k-1})\right) + M(\mathbf{z}^k)$$

$$\geq M(\mathbf{z}^k) \geq f^*.$$

Summing (4.23) over $k = 0, \cdots, K$, we have

$$\min_{k=0,\cdots,K} \left\{ \|\mathbf{x}^{k+1} - \mathbf{x}^k\|^2 + \|\mathbf{z}^{k+1} - \mathbf{z}^k\|^2 + \|\mathbf{A}\mathbf{x}(\mathbf{z}^k, \boldsymbol{\lambda}^k) - \mathbf{b}\|^2 \right\}$$

$$\leq \frac{\Phi_0 - f^*}{\delta(K + 1)}.$$

Thus, we have

$$\min_{k=0,\cdots,K} \left\{ \|\mathbf{x}^{k+1} - \mathbf{x}^k\| + \|\mathbf{z}^{k+1} - \mathbf{z}^k\| + \|\mathbf{A}\mathbf{x}(\mathbf{z}^k, \boldsymbol{\lambda}^k) - \mathbf{b}\| \right\} \leq \epsilon$$

after $O(\frac{1}{\epsilon^2})$ iterations.

From (4.15), we have

$$\|\mathbf{A}\mathbf{x}^k - \mathbf{b}\|$$
$$\leq \|\mathbf{A}\mathbf{x}(\mathbf{z}^k, \boldsymbol{\lambda}^k) - \mathbf{b}\| + \widetilde{\sigma}\|\mathbf{x}(\mathbf{z}^k, \boldsymbol{\lambda}^k) - \mathbf{x}^k\|$$
$$\leq \|\mathbf{A}\mathbf{x}(\mathbf{z}^k, \boldsymbol{\lambda}^k) - \mathbf{b}\| + \frac{\widetilde{\sigma}\left[1 + \alpha_1\sqrt{2mL^2 + (m-1)m\beta^2\sigma^4}\right]}{\alpha_1(\rho - L)}\|\mathbf{x}^{k+1} - \mathbf{x}^k\|$$
$$= O(\epsilon).$$

Since

$$\nabla_{\mathbf{x}}P(\mathbf{x}^k, \mathbf{z}^k, \boldsymbol{\lambda}^k) = \nabla f(\mathbf{x}^k) + \mathbf{A}^T\boldsymbol{\lambda}^k + \rho(\mathbf{x}^k - \mathbf{z}^k) + \beta\mathbf{A}^T(\mathbf{A}\mathbf{x}^k - \mathbf{b}),$$

from (4.16) and the update of $\mathbf{z}$, we have

$$\|\nabla f(\mathbf{x}^k) + \mathbf{A}^T\boldsymbol{\lambda}^k\|$$
$$\leq \|\nabla_{\mathbf{x}}P(\mathbf{x}^k, \mathbf{z}^k, \boldsymbol{\lambda}^k)\| + \rho\|\mathbf{x}^k - \mathbf{z}^k\| + \beta\widetilde{\sigma}\|\mathbf{A}\mathbf{x}^k - \mathbf{b}\|$$
$$\leq \left[\frac{1}{\alpha_1} + \sqrt{2mL^2 + (m-1)m\beta^2\sigma^4}\right]\|\mathbf{x}^{k+1} - \mathbf{x}^k\|$$
$$+ \rho\left(\frac{1}{\alpha_3}\|\mathbf{z}^{k+1} - \mathbf{z}^k\| + \|\mathbf{x}^{k+1} - \mathbf{x}^k\|\right) + \beta\widetilde{\sigma}\|\mathbf{A}\mathbf{x}^k - \mathbf{b}\|$$
$$\leq O(\epsilon).$$

$\square$

## 4.3  ADMM for Multilinearly Constrained Optimization

In this section, we introduce how to use ADMM to solve problems with multilinear constraints in the form of $\mathbf{X}\mathbf{Y} = \mathbf{Z}$, where multilinear means that $\mathbf{X}\mathbf{Y} = \mathbf{Z}$ is linear with respect to the individual variables of $\mathbf{X}$ and $\mathbf{Y}$, but nonconvex for $\mathbf{X}$ and $\mathbf{Y}$ jointly. There have been a wide range of problems which can be modeled with multilinear constraints, for example, non-negative matrix factorization [2, 5], RPCA [3], and the training of neural networks [9, 10]. Gao et al. [4] gave a unified proof for a family of nonconvex problems with multilinear constraints. To simplify the description, we only introduce the convergence proof for RPCA to illustrate the main proof techniques.

As introduced in Sect. 1.1, we consider the following model of RPCA:

$$\min_{\mathbf{U},\mathbf{V},\mathbf{Z},\mathbf{Y}} \left[ \frac{1}{2} \left( \|\mathbf{U}\|^2 + \|\mathbf{V}\|^2 \right) + \tau \|\mathbf{Z}\|_1 + \frac{\mu}{2} \|\mathbf{Y}\|^2 \right],$$

$$s.t. \quad \mathbf{UV} + \mathbf{Z} - \mathbf{M} = \mathbf{Y}.$$

The augmented Lagrangian function is

$$L_\beta(\mathbf{U}, \mathbf{V}, \mathbf{Z}, \mathbf{Y}, \boldsymbol{\Lambda}) = \frac{1}{2} \left( \|\mathbf{U}\|^2 + \|\mathbf{V}\|^2 \right) + \tau \|\mathbf{Z}\|_1 + \frac{\mu}{2} \|\mathbf{Y}\|^2$$

$$+ \langle \boldsymbol{\Lambda}, \mathbf{UV} + \mathbf{Z} - \mathbf{M} - \mathbf{Y} \rangle + \frac{\beta}{2} \|\mathbf{UV} + \mathbf{Z} - \mathbf{M} - \mathbf{Y}\|^2.$$

We can use ADMM to solve the above problem by alternating between updating $(\mathbf{V}, \mathbf{Z})$ and $(\mathbf{U}, \mathbf{Y})$:

$$(\mathbf{V}^{k+1}, \mathbf{Z}^{k+1}) = \operatorname*{argmin}_{\mathbf{V}, \mathbf{Z}} \left( \frac{1}{2} \|\mathbf{V}\|^2 + \tau \|\mathbf{Z}\|_1 \right.$$

$$\left. + \frac{\beta}{2} \left\| \mathbf{U}^k \mathbf{V} + \mathbf{Z} - \mathbf{M} - \mathbf{Y}^k + \frac{1}{\beta} \boldsymbol{\Lambda}^k \right\|^2 \right), \tag{4.24a}$$

$$(\mathbf{U}^{k+1}, \mathbf{Y}^{k+1}) = \operatorname*{argmin}_{\mathbf{U}, \mathbf{Y}} \left( \frac{1}{2} \|\mathbf{U}\|^2 + \frac{\mu}{2} \|\mathbf{Y}\|^2 \right.$$

$$\left. + \frac{\beta}{2} \left\| \mathbf{U}\mathbf{V}^{k+1} + \mathbf{Z}^{k+1} - \mathbf{M} - \mathbf{Y} + \frac{1}{\beta} \boldsymbol{\Lambda}^k \right\|^2 \right), \tag{4.24b}$$

$$\boldsymbol{\Lambda}^{k+1} = \boldsymbol{\Lambda}^k + \beta (\mathbf{U}^{k+1}\mathbf{V}^{k+1} + \mathbf{Z}^{k+1} - \mathbf{M} - \mathbf{Y}^{k+1}). \tag{4.24c}$$

The algorithm is presented in Algorithm 4.3.

---

**Algorithm 4.3** ADMM for RPCA

---

Initialize $\mathbf{U}^0$, $\mathbf{Y}^0$, and $\boldsymbol{\Lambda}^0$.
**for** $k = 0, 1, 2, 3, \cdots$ **do**
   Update $\mathbf{X}^{k+1}$, $\mathbf{Y}^{k+1}$, $\mathbf{U}^{k+1}$, $\mathbf{V}^{k+1}$, and $\boldsymbol{\Lambda}^{k+1}$ by (4.24a)–(4.24c), respectively.
**end for**

---

We can prove the $O\left(\frac{1}{\epsilon^2}\right)$ iteration complexity of Algorithm 4.3.

**Lemma 4.6**  *For Algorithm 4.3, we have*

$$L_\beta(\mathbf{U}^{k+1}, \mathbf{V}^{k+1}, \mathbf{Z}^{k+1}, \mathbf{Y}^{k+1}, \mathbf{\Lambda}^{k+1}) - L_\beta(\mathbf{U}^k, \mathbf{V}^k, \mathbf{Z}^k, \mathbf{Y}^k, \mathbf{\Lambda}^k)$$

$$\leq -\frac{1}{2}\|\mathbf{V}^{k+1} - \mathbf{V}^k\|^2 - \frac{1}{2}\|\mathbf{U}^{k+1} - \mathbf{U}^k\|^2 - \left(\frac{\mu}{2} - \frac{\mu^2}{\beta}\right)\|\mathbf{Y}^{k+1} - \mathbf{Y}^k\|^2.$$

(4.25)

***Proof***  From the optimality conditions, we have

$$\mathbf{0} = \mathbf{V}^{k+1} + \beta(\mathbf{U}^k)^T\left(\mathbf{U}^k\mathbf{V}^{k+1} + \mathbf{Z}^{k+1} - \mathbf{M} - \mathbf{Y}^k + \frac{1}{\beta}\mathbf{\Lambda}^k\right),$$

(4.26)

$$\mathbf{0} \in \partial\|\mathbf{Z}^{k+1}\|_1 + \frac{\beta}{\tau}\left(\mathbf{U}^k\mathbf{V}^{k+1} + \mathbf{Z}^{k+1} - \mathbf{M} - \mathbf{Y}^k + \frac{1}{\beta}\mathbf{\Lambda}^k\right),$$

(4.27)

$$\mathbf{0} = \mathbf{U}^{k+1} + \beta\left(\mathbf{U}^{k+1}\mathbf{V}^{k+1} + \mathbf{Z}^{k+1} - \mathbf{M} - \mathbf{Y}^{k+1} + \frac{1}{\beta}\mathbf{\Lambda}^k\right)(\mathbf{V}^{k+1})^T$$

$$= \mathbf{U}^{k+1} + \mathbf{\Lambda}^{k+1}(\mathbf{V}^{k+1})^T,$$

(4.28)

$$\mathbf{0} = \mu\mathbf{Y}^{k+1} - \beta\left(\mathbf{U}^{k+1}\mathbf{V}^{k+1} + \mathbf{Z}^{k+1} - \mathbf{M} - \mathbf{Y}^{k+1} + \frac{1}{\beta}\mathbf{\Lambda}^k\right)$$

$$= \mu\mathbf{Y}^{k+1} - \mathbf{\Lambda}^{k+1}.$$

(4.29)

We can check that

$$L_\beta(\mathbf{U}^k, \mathbf{V}^{k+1}, \mathbf{Z}^{k+1}, \mathbf{Y}^k, \mathbf{\Lambda}^k) - L_\beta(\mathbf{U}^k, \mathbf{V}^k, \mathbf{Z}^k, \mathbf{Y}^k, \mathbf{\Lambda}^k)$$

$$= \frac{1}{2}\|\mathbf{V}^{k+1}\|^2 - \frac{1}{2}\|\mathbf{V}^k\|^2 + \tau\|\mathbf{Z}^{k+1}\|_1 - \tau\|\mathbf{Z}^k\|_1$$

$$\quad + \left\langle \mathbf{\Lambda}^k, \mathbf{U}^k(\mathbf{V}^{k+1} - \mathbf{V}^k) + \mathbf{Z}^{k+1} - \mathbf{Z}^k \right\rangle$$

$$\quad + \frac{\beta}{2}\|\mathbf{U}^k\mathbf{V}^{k+1} + \mathbf{Z}^{k+1} - \mathbf{M} - \mathbf{Y}^k\|^2 - \frac{\beta}{2}\|\mathbf{U}^k\mathbf{V}^k + \mathbf{Z}^k - \mathbf{M} - \mathbf{Y}^k\|^2$$

$$= -\frac{1}{2}\|\mathbf{V}^{k+1} - \mathbf{V}^k\|^2 - \left\langle \mathbf{V}^{k+1}, \mathbf{V}^k \right\rangle + \left\langle \mathbf{V}^{k+1}, \mathbf{V}^{k+1} \right\rangle$$

$$\quad + \tau\|\mathbf{Z}^{k+1}\|_1 - \tau\|\mathbf{Z}^k\|_1 + \left\langle \mathbf{\Lambda}^k, \mathbf{U}^k(\mathbf{V}^{k+1} - \mathbf{V}^k) + \mathbf{Z}^{k+1} - \mathbf{Z}^k \right\rangle$$

$$\quad - \frac{\beta}{2}\|\mathbf{U}^k(\mathbf{V}^{k+1} - \mathbf{V}^k) + \mathbf{Z}^{k+1} - \mathbf{Z}^k\|^2$$

$$\quad - \beta\left\langle \mathbf{U}^k\mathbf{V}^{k+1} + \mathbf{Z}^{k+1} - \mathbf{M} - \mathbf{Y}^k, \mathbf{U}^k\mathbf{V}^k + \mathbf{Z}^k - \mathbf{M} - \mathbf{Y}^k \right\rangle$$

$$\quad + \beta\left\langle \mathbf{U}^k\mathbf{V}^{k+1} + \mathbf{Z}^{k+1} - \mathbf{M} - \mathbf{Y}^k, \mathbf{U}^k\mathbf{V}^{k+1} + \mathbf{Z}^{k+1} - \mathbf{M} - \mathbf{Y}^k \right\rangle$$

$$= -\frac{1}{2}\|\mathbf{V}^{k+1} - \mathbf{V}^k\|^2 + \left\langle \mathbf{V}^{k+1} - \mathbf{V}^k, \mathbf{V}^{k+1} \right\rangle + \tau \|\mathbf{Z}^{k+1}\|_1 - \tau \|\mathbf{Z}^k\|_1$$

$$- \frac{\beta}{2}\|\mathbf{U}^k(\mathbf{V}^{k+1} - \mathbf{V}^k) + \mathbf{Z}^{k+1} - \mathbf{Z}^k\|^2$$

$$+ \beta \left\langle \frac{1}{\beta}\mathbf{\Lambda}^k + \mathbf{U}^k\mathbf{V}^{k+1} + \mathbf{Z}^{k+1} - \mathbf{M} - \mathbf{Y}^k, \mathbf{U}^k(\mathbf{V}^{k+1} - \mathbf{V}^k) + \mathbf{Z}^{k+1} - \mathbf{Z}^k \right\rangle$$

$$\overset{a}{\leq} -\frac{1}{2}\|\mathbf{V}^{k+1} - \mathbf{V}^k\|^2 \tag{4.30}$$

and

$$L_\beta(\mathbf{U}^{k+1}, \mathbf{V}^{k+1}, \mathbf{Z}^{k+1}, \mathbf{Y}^{k+1}, \mathbf{\Lambda}^k) - L_\beta(\mathbf{U}^k, \mathbf{V}^{k+1}, \mathbf{Z}^{k+1}, \mathbf{Y}^k, \mathbf{\Lambda}^k)$$

$$= \frac{1}{2}\|\mathbf{U}^{k+1}\|^2 - \frac{1}{2}\|\mathbf{U}^k\|^2 + \frac{\mu}{2}\|\mathbf{Y}^{k+1}\|^2 - \frac{\mu}{2}\|\mathbf{Y}^k\|^2$$

$$+ \left\langle \mathbf{\Lambda}^k, (\mathbf{U}^{k+1} - \mathbf{U}^k)\mathbf{V}^{k+1} - \mathbf{Y}^{k+1} + \mathbf{Y}^k \right\rangle$$

$$+ \frac{\beta}{2}\|\mathbf{U}^{k+1}\mathbf{V}^{k+1} + \mathbf{Z}^{k+1} - \mathbf{M} - \mathbf{Y}^{k+1}\|^2 - \frac{\beta}{2}\|\mathbf{U}^k\mathbf{V}^{k+1} + \mathbf{Z}^{k+1} - \mathbf{M} - \mathbf{Y}^k\|^2$$

$$= -\frac{1}{2}\|\mathbf{U}^{k+1} - \mathbf{U}^k\|^2 - \left\langle \mathbf{U}^{k+1}, \mathbf{U}^k \right\rangle + \left\langle \mathbf{U}^{k+1}, \mathbf{U}^{k+1} \right\rangle$$

$$- \frac{\mu}{2}\|\mathbf{Y}^{k+1} - \mathbf{Y}^k\|^2 - \mu \left\langle \mathbf{Y}^{k+1}, \mathbf{Y}^k \right\rangle + \mu \left\langle \mathbf{Y}^{k+1}, \mathbf{Y}^{k+1} \right\rangle$$

$$+ \left\langle \mathbf{\Lambda}^k, (\mathbf{U}^{k+1} - \mathbf{U}^k)\mathbf{V}^{k+1} - \mathbf{Y}^{k+1} + \mathbf{Y}^k \right\rangle$$

$$- \frac{\beta}{2}\|(\mathbf{U}^{k+1} - \mathbf{U}^k)\mathbf{V}^{k+1} - \mathbf{Y}^{k+1} + \mathbf{Y}^k\|^2$$

$$- \beta \left\langle \mathbf{U}^{k+1}\mathbf{V}^{k+1} + \mathbf{Z}^{k+1} - \mathbf{M} - \mathbf{Y}^{k+1}, \mathbf{U}^k\mathbf{V}^{k+1} + \mathbf{Z}^{k+1} - \mathbf{M} - \mathbf{Y}^k \right\rangle$$

$$+ \beta \left\langle \mathbf{U}^{k+1}\mathbf{V}^{k+1} + \mathbf{Z}^{k+1} - \mathbf{M} - \mathbf{Y}^{k+1}, \mathbf{U}^{k+1}\mathbf{V}^{k+1} + \mathbf{Z}^{k+1} - \mathbf{M} - \mathbf{Y}^{k+1} \right\rangle$$

$$= -\frac{1}{2}\|\mathbf{U}^{k+1} - \mathbf{U}^k\|^2 + \left\langle \mathbf{U}^{k+1} - \mathbf{U}^k, \mathbf{U}^{k+1} \right\rangle - \frac{\mu}{2}\|\mathbf{Y}^{k+1} - \mathbf{Y}^k\|^2$$

$$+ \mu \left\langle \mathbf{Y}^{k+1} - \mathbf{Y}^k, \mathbf{Y}^{k+1} \right\rangle - \frac{\beta}{2}\|(\mathbf{U}^{k+1} - \mathbf{U}^k)\mathbf{V}^{k+1} - \mathbf{Y}^{k+1} + \mathbf{Y}^k\|^2$$

$$+ \beta \left\langle \frac{1}{\beta}\mathbf{\Lambda}^k + \mathbf{U}^{k+1}\mathbf{V}^{k+1} + \mathbf{Z}^{k+1} - \mathbf{M} - \mathbf{Y}^{k+1}, (\mathbf{U}^{k+1} - \mathbf{U}^k)\mathbf{V}^{k+1} - \mathbf{Y}^{k+1} + \mathbf{Y}^k \right\rangle$$

$$= -\frac{1}{2}\|\mathbf{U}^{k+1} - \mathbf{U}^k\|^2 + \left\langle \mathbf{U}^{k+1} - \mathbf{U}^k, \mathbf{U}^{k+1} \right\rangle - \frac{\mu}{2}\|\mathbf{Y}^{k+1} - \mathbf{Y}^k\|^2$$

$$+ \mu \left\langle \mathbf{Y}^{k+1} - \mathbf{Y}^k, \mathbf{Y}^{k+1} \right\rangle - \frac{\beta}{2}\|(\mathbf{U}^{k+1} - \mathbf{U}^k)\mathbf{V}^{k+1} - \mathbf{Y}^{k+1} + \mathbf{Y}^k\|^2$$

$$+ \left\langle \boldsymbol{\Lambda}^{k+1}, (\mathbf{U}^{k+1} - \mathbf{U}^k)\mathbf{V}^{k+1} - \mathbf{Y}^{k+1} + \mathbf{Y}^k \right\rangle$$

$$\overset{b}{\leq} -\frac{1}{2}\|\mathbf{U}^{k+1} - \mathbf{U}^k\|^2 - \frac{\mu}{2}\|\mathbf{Y}^{k+1} - \mathbf{Y}^k\|^2, \tag{4.31}$$

where we use (4.26) and (4.27) in $\overset{a}{\leq}$ and (4.28) and (4.29) in $\overset{b}{\leq}$. We also have

$$L_\beta(\mathbf{U}^{k+1}, \mathbf{V}^{k+1}, \mathbf{Z}^{k+1}, \mathbf{Y}^{k+1}, \boldsymbol{\Lambda}^{k+1}) - L_\beta(\mathbf{U}^{k+1}, \mathbf{V}^{k+1}, \mathbf{Z}^{k+1}, \mathbf{Y}^{k+1}, \boldsymbol{\Lambda}^k)$$

$$= \frac{1}{\beta}\|\boldsymbol{\Lambda}^{k+1} - \boldsymbol{\Lambda}^k\|^2 = \frac{\mu^2}{\beta}\|\mathbf{Y}^{k+1} - \mathbf{Y}^k\|^2. \tag{4.32}$$

By adding (4.30)–(4.32) we have

$$L_\beta(\mathbf{U}^{k+1}, \mathbf{V}^{k+1}, \mathbf{Z}^{k+1}, \mathbf{Y}^{k+1}, \boldsymbol{\Lambda}^{k+1}) - L_\beta(\mathbf{U}^k, \mathbf{V}^k, \mathbf{Z}^k, \mathbf{Y}^k, \boldsymbol{\Lambda}^k)$$

$$\leq -\frac{1}{2}\|\mathbf{V}^{k+1} - \mathbf{V}^k\|^2 - \frac{1}{2}\|\mathbf{U}^{k+1} - \mathbf{U}^k\|^2 - \left(\frac{\mu}{2} - \frac{\mu^2}{\beta}\right)\|\mathbf{Y}^{k+1} - \mathbf{Y}^k\|^2.$$

$\square$

**Theorem 4.4** *Let $\beta > 2\mu$. Suppose that the sequence $\{\mathbf{U}^k, \mathbf{V}^k, \mathbf{Z}^k, \mathbf{Y}^k, \boldsymbol{\Lambda}^k\}_k$ is bounded. Then Algorithm 4.3 needs $O(\frac{1}{\epsilon^2})$ iterations to find an $\epsilon$-approximate KKT point, i.e., $(\mathbf{U}^{k+1}, \mathbf{V}^{k+1}, \mathbf{Z}^{k+1}, \mathbf{Y}^{k+1}, \boldsymbol{\Lambda}^{k+1})$ with $k = O(\epsilon^{-2})$, such that*

$$\mathbf{U}^{k+1} + \boldsymbol{\Lambda}^{k+1}(\mathbf{V}^{k+1})^T = 0, \quad \|\mathbf{V}^{k+1} + (\mathbf{U}^{k+1})^T\boldsymbol{\Lambda}^{k+1}\| \leq O(\epsilon),$$

$$\mu\mathbf{Y}^{k+1} - \boldsymbol{\Lambda}^{k+1} = 0, \quad \text{dist}\left(-\frac{1}{\tau}\boldsymbol{\Lambda}^{k+1}, \partial\|\mathbf{Z}^{k+1}\|_1\right) \leq O(\epsilon), \text{ and}$$

$$\|\mathbf{U}^{k+1}\mathbf{V}^{k+1} + \mathbf{Z}^{k+1} - \mathbf{M} - \mathbf{Y}^{k+1}\| \leq O(\epsilon).$$

*Proof* We first prove $L_\beta(\mathbf{U}^k, \mathbf{V}^k, \mathbf{Z}^k, \mathbf{Y}^k, \boldsymbol{\Lambda}^k) \geq 0$. From (4.29), we have

$$L_\beta(\mathbf{U}^k, \mathbf{V}^k, \mathbf{Z}^k, \mathbf{Y}^k, \boldsymbol{\Lambda}^k)$$

$$= \frac{1}{2}\left(\|\mathbf{U}^k\|^2 + \|\mathbf{V}^k\|^2\right) + \tau\|\mathbf{Z}^k\|_1 + \frac{\mu}{2}\|\mathbf{Y}^k\|^2$$

$$+ \mu\left\langle \mathbf{Y}^k, \mathbf{U}^k\mathbf{V}^k + \mathbf{Z}^k - \mathbf{M} - \mathbf{Y}^k \right\rangle + \frac{\beta}{2}\|\mathbf{U}^k\mathbf{V}^k + \mathbf{Z}^k - \mathbf{M} - \mathbf{Y}^k\|^2$$

$$= \frac{1}{2}\left(\|\mathbf{U}^k\|^2 + \|\mathbf{V}^k\|^2\right) + \tau\|\mathbf{Z}^k\|_1 + \frac{\mu}{2}\|\mathbf{Y}^k + \mathbf{U}^k\mathbf{V}^k + \mathbf{Z}^k - \mathbf{M} - \mathbf{Y}^k\|^2$$

$$+ \frac{\beta - \mu}{2}\|\mathbf{U}^k\mathbf{V}^k + \mathbf{Z}^k - \mathbf{M} - \mathbf{Y}^k\|^2$$

$$\geq 0.$$

Summing (4.25) over $k = 0, 1, \cdots, K$, we have

$$
\min_{0 \le k \le K} \left\{ \frac{1}{2} \|\mathbf{V}^{k+1} - \mathbf{V}^k\|^2 + \frac{1}{2} \|\mathbf{U}^{k+1} - \mathbf{U}^k\|^2 + \left( \frac{\mu}{2} - \frac{\mu^2}{\beta} \right) \|\mathbf{Y}^{k+1} - \mathbf{Y}^k\|^2 \right\}
$$
$$
\le \frac{L_\beta(\mathbf{U}^0, \mathbf{V}^0, \mathbf{Z}^0, \mathbf{Y}^0, \boldsymbol{\Lambda}^0)}{K + 1}.
$$

Thus, we know that the algorithm needs $O(\frac{1}{\epsilon^2})$ iterations to find $(\mathbf{U}^k, \mathbf{V}^k, \mathbf{Z}^k, \mathbf{Y}^k, \boldsymbol{\Lambda}^k)$ such that

$$
\|\mathbf{V}^{k+1} - \mathbf{V}^k\| \le \epsilon, \quad \|\mathbf{U}^{k+1} - \mathbf{U}^k\| \le \epsilon, \quad \|\mathbf{Y}^{k+1} - \mathbf{Y}^k\| \le \epsilon.
$$

From (4.24c) and (4.29), we have

$$
\|\mathbf{U}^{k+1}\mathbf{V}^{k+1} + \mathbf{Z}^{k+1} - \mathbf{M} - \mathbf{Y}^{k+1}\|
$$
$$
= \frac{1}{\beta} \|\boldsymbol{\Lambda}^{k+1} - \boldsymbol{\Lambda}^k\| = \frac{\mu}{\beta} \|\mathbf{Y}^{k+1} - \mathbf{Y}^k\| \le O(\epsilon).
$$

From (4.28) and (4.29), we have

$$
\mathbf{U}^{k+1} + \boldsymbol{\Lambda}^{k+1}(\mathbf{V}^{k+1})^T = \mathbf{0} \quad \text{and} \quad \mu \mathbf{Y}^{k+1} - \boldsymbol{\Lambda}^{k+1} = \mathbf{0},
$$

respectively. From (4.26), we have

$$
\mathbf{V}^{k+1} + (\mathbf{U}^{k+1})^T \boldsymbol{\Lambda}^{k+1}
$$
$$
= \mathbf{V}^{k+1} + \beta(\mathbf{U}^{k+1})^T \left( \mathbf{U}^{k+1}\mathbf{V}^{k+1} + \mathbf{Z}^{k+1} - \mathbf{M} - \mathbf{Y}^{k+1} + \frac{1}{\beta}\boldsymbol{\Lambda}^k \right)
$$
$$
= \mathbf{V}^{k+1} + \beta(\mathbf{U}^k)^T \left( \mathbf{U}^k\mathbf{V}^{k+1} + \mathbf{Z}^{k+1} - \mathbf{M} - \mathbf{Y}^k + \frac{1}{\beta}\boldsymbol{\Lambda}^k \right)
$$
$$
+ \beta(\mathbf{U}^k)^T \left[ (\mathbf{U}^{k+1} - \mathbf{U}^k)\mathbf{V}^{k+1} - (\mathbf{Y}^{k+1} - \mathbf{Y}^k) \right] + (\mathbf{U}^{k+1} - \mathbf{U}^k)^T \boldsymbol{\Lambda}^{k+1}
$$
$$
= \beta(\mathbf{U}^k)^T \left[ (\mathbf{U}^{k+1} - \mathbf{U}^k)\mathbf{V}^{k+1} - (\mathbf{Y}^{k+1} - \mathbf{Y}^k) \right] + (\mathbf{U}^{k+1} - \mathbf{U}^k)^T \boldsymbol{\Lambda}^{k+1}.
$$

So by the boundedness of $\{\mathbf{U}^k, \mathbf{V}^k, \mathbf{Z}^k, \mathbf{Y}^k, \boldsymbol{\Lambda}^k\}_k$ we have

$$
\|\mathbf{V}^{k+1} + (\mathbf{U}^{k+1})^T \boldsymbol{\Lambda}^{k+1}\| \le O(\epsilon).
$$

From (4.27), we have

$$
\frac{\beta}{\tau} \left[ (\mathbf{U}^{k+1} - \mathbf{U}^k)\mathbf{V}^{k+1} - (\mathbf{Y}^{k+1} - \mathbf{Y}^k) \right] \in \partial \|\mathbf{Z}^{k+1}\|_1 + \frac{1}{\tau}\boldsymbol{\Lambda}^{k+1}
$$

and thus

$$\mathrm{dist}\left(-\frac{1}{\tau}\mathbf{\Lambda}^{k+1}, \partial\|\mathbf{Z}^{k+1}\|_1\right) \leq O(\epsilon),$$

due to the boundedness of $\{\mathbf{U}^k, \mathbf{V}^k, \mathbf{Z}^k, \mathbf{Y}^k, \mathbf{\Lambda}^k\}_k$. □

# References

1. R.I. Bot, D.-K. Nguyen, The proximal alternating direction method of multipliers in the nonconvex setting: Convergence analysis and rates. Math. Oper. Res. **45**(2), 682–712 (2020)
2. S. Boyd, L. Vandenberghe, *Convex Optimization* (Cambridge University Press, Cambridge, 2004)
3. E.J. Candes, X. Li, Y. Ma, J. Wright, Robust principal component analysis? J. ACM **58**(3), 1–37 (2011)
4. W. Gao, D. Goldfarb, F.E. Curtis, ADMM for multiaffine constrained optimization. Optim. Methods Softw. **35**(2), 257–303 (2020)
5. D. Hajinezhad, T.-H. Chang, X. Wang, Q. Shi, M. Hong, Nonnegative matrix factorization using ADMM: algorithm and convergence analysis, in *IEEE International Conference on Acoustics, Speech, and Signal Processing* (2016), pp. 4742–4746
6. M. Hong, Z.-Q. Luo, M. Razaviyayn, Convergence analysis of alternating direction method of multipliers for a family of nonconvex problems. SIAM J. Optim. **26**(1), 337–364 (2016)
7. B. Jiang, T. Lin, S. Ma, S. Zhang, Structured nonconvex and nonsmooth optimization: algorithms and iteration complexity analysis. Comput. Optim. Appl. **72**(1), 115–157 (2019)
8. G. Li, T.K. Pong, Global convergence of splitting methods for nonconvex composite optimization. SIAM J. Optim. **25**(4), 2434–2460 (2015)
9. J. Li, M. Xiao, C. Fang, Y. Dai, C. Xu, Z. Lin, Training deep neural networks by lifted proximal operator machines. IEEE Trans. Pattern Anal. Mach. Intell. **44**(6), 3334–3348 (2022)
10. G. Taylor, R. Burmeister, Z. Xu, B. Singh, A. Patel, T. Goldstein, Training neural networks without gradients: a scalable ADMM approach, in *International Conference on Machine Learning* (2016), pp. 2722–2731
11. F. Wang, W. Cao, Z. Xu, Convergence of multi-block Bregman ADMM for nonconvex composite problems. Sci. China Inf. Sci. **61**(12), 1–12 (2018)
12. Y. Wang, W. Yin, J. Zeng, Global convergence of ADMM in nonconvex nonsmooth optimization. J. Sci. Comput. **78**(1), 29–63 (2020)
13. J. Zhang, Z.-Q. Luo, A proximal alternating direction method of multiplier for linearly constrained nonconvex minimization. SIAM J. Optim. **30**(3), 2272–2302 (2020)

# Chapter 5
# Stochastic ADMM

Consider the following linearly constrained separable optimization problem:

$$\min_{\mathbf{x}_1, \mathbf{x}_2} \left( f_1(\mathbf{x}_1) + f_2(\mathbf{x}_2) \right),$$

$$s.t. \ \mathbf{A}_1\mathbf{x}_1 + \mathbf{A}_2\mathbf{x}_2 = \mathbf{b}. \tag{5.1}$$

For lots of machine learning problems, $f_1(\mathbf{x}_1)$ is typically a loss over the data and $f_2(\mathbf{x}_2)$ is a regularizer that controls the complexity of the model or provides the prior information of the solution. For such problems, we may assume that $f_1$ has a structure as follows:

$$f_1(\mathbf{x}_1) \equiv \mathbb{E}F(\mathbf{x}_1; \xi), \tag{5.2}$$

where $F(\mathbf{x}_1; \xi)$ is a stochastic component indexed by a random number $\xi$. For traditional machine learning, the data are often finitely sampled. If we denote each component function as $F_i(\mathbf{x})$, we can rewrite $f_1(\mathbf{x})$ as below:

$$f_1(\mathbf{x}_1) \equiv \frac{1}{n} \sum_{i=1}^{n} F_i(\mathbf{x}_1), \tag{5.3}$$

where $n$ is the number of individual functions. When $n$ is finite, (5.3) is an offline problem, with examples including empirical risk minimization. $n$ can also go to infinity, which is a general case. In the rest of this chapter, when we study the finite-sum (offline) problem, we shall use the formula (5.3); otherwise, we use (5.2).

When $n$ is large, accessing the exact function value of $f_1(\mathbf{x}_1)$ or its gradient may be very expensive and even impossible when $n = \infty$. To deal with such large-scale problems, the standard way is to estimate the full gradient via one or several randomly sampled counterparts from individual functions. We call algorithms using this technique as stochastic algorithms.

© The Author(s), under exclusive license to Springer Nature Singapore Pte Ltd. 2022
Z. Lin et al., *Alternating Direction Method of Multipliers for Machine Learning*,
https://doi.org/10.1007/978-981-16-9840-8_5

In this chapter, we will introduce a variety of stochastic ADMM methods to solve Problem (5.1) with $f_1$ in the form of (5.2) or (5.3). In practice, the stochastic algorithms are often much faster than the deterministic ones. To begin with, in Sect. 5.1 we introduce the naive stochastic ADMM algorithm, which randomly samples an individual function and uses its gradient as an estimator of the full gradient. Recall that the deterministic ADMM converges in $O(1/K)$ when $f_1$ and $f_2$ are convex functions, where $K$ is the iteration number. In comparison, we will show that the naive stochastic ADMM can only achieve an $O\left(1/\sqrt{K}\right)$ rate under similar conditions. The slower convergence rate is caused by the variance of the noisy gradient. Because the variance will not converge to zero through the updates, we have to choose a decreasing stepsize. We will introduce variance reduction (VR) technique in Sect. 5.2, which can reduce the adverse effect of the noise, especially for the offline case. We show that the convergence rate can be improved to $O(1/K)$ when the stochastic algorithm is equipped with VR. Later in Sect. 5.3, we consider fusing VR and momentum techniques, and show that a non-ergodic $O(1/K)$ rate is achievable under mild conditions. Finally, in Sect. 5.4, we extend our analysis to the nonconvex setting.

## 5.1   Stochastic ADMM

We consider Problem (5.1) with $f_1(\mathbf{x}_1) \equiv \mathbb{E}_\xi F(\mathbf{x}_1, \xi)$. In each iteration, we independently sample a stochastic index $\xi$ and compute the stochastic gradient $\nabla F(\mathbf{x}_1, \xi)$. For the simplicity of our expressions, we denote $\nabla F(\mathbf{x}_1, \xi)$ by $\tilde{\nabla} f_1(\mathbf{x}_1)$. The algorithm of Stochastic ADMM (SADMM) was designed in [9] and is shown in Algorithm 5.1, in which the approximated augmented Lagrangian function $\hat{L}_\beta^k$ is given as

$$\hat{L}_\beta^k(\mathbf{x}_1, \mathbf{x}_2, \boldsymbol{\lambda}) = f_1\left(\mathbf{x}_1^k\right) + \left\langle \tilde{\nabla} f_1\left(\mathbf{x}_1^k\right), \mathbf{x}_1 - \mathbf{x}_1^k \right\rangle + f_2(\mathbf{x}_2)$$

$$+ \frac{\beta}{2}\left\|\mathbf{A}_1\mathbf{x}_1 + \mathbf{A}_2\mathbf{x}_2 - \mathbf{b} + \frac{1}{\beta}\boldsymbol{\lambda}\right\|^2 + \frac{1}{2\eta_{k+1}}\left\|\mathbf{x}_1 - \mathbf{x}_1^k\right\|^2. \quad (5.4)$$

Note that for the simplicity of our analysis, we only linearize the objective function $f_1$ in Algorithm 5.1. In Sects. 5.2 and 5.3, we will consider also linearizing the augmented term $\frac{\beta}{2}\left\|\mathbf{A}_1\mathbf{x}_1 + \mathbf{A}_2\mathbf{x}_2 - \mathbf{b} + \frac{1}{\beta}\boldsymbol{\lambda}\right\|^2$. Our result can also be extended to the case where $f_2$ has the expectation structure. An example is shown in Sect. 5.3.

Now we begin the analysis. The proof is taken from [9]. We first prove a useful lemma.

---

**Algorithm 5.1** Stochastic ADMM (SADMM)

---

**Input:** $\mathbf{x}_1^0$, $\mathbf{x}_2^0$, and $\boldsymbol{\lambda}^0 = \mathbf{0}$.
  **for** $k = 0, 1, 2, \cdots$ **do**
  1    $\mathbf{x}_1^{k+1} = \mathrm{argmin}_{\mathbf{x}_1} \hat{L}_\beta^k(\mathbf{x}_1, \mathbf{x}_2^k, \boldsymbol{\lambda}^k)$, with $\hat{L}_\beta^k(\mathbf{x}_1, \mathbf{x}_2, \boldsymbol{\lambda})$ given in (5.4).
  2    $\mathbf{x}_2^{k+1} = \mathrm{argmin}_{\mathbf{x}_2} \hat{L}_\beta^k(\mathbf{x}_1^{k+1}, \mathbf{x}_2, \boldsymbol{\lambda}^k)$.
  3    $\boldsymbol{\lambda}^{k+1} = \boldsymbol{\lambda}^k + \beta(\mathbf{A}_1\mathbf{x}_1^{k+1} + \mathbf{A}_2\mathbf{x}_2^{k+1} - \mathbf{b})$.
  **end for** $k$

---

**Lemma 5.1** *Assume that $f_1$ is $\mu$-strongly convex and $L$-smooth and $f_2$ is convex. For $k \geq 0$, if the stepsize $\eta_{k+1} \leq 1/(2L)$, then for any $\tilde{\boldsymbol{\lambda}}$, we have*

$$f_1\left(\mathbf{x}_1^{k+1}\right) + f_2\left(\mathbf{x}_2^{k+1}\right) - f_1\left(\mathbf{x}_1^*\right) - f_2\left(\mathbf{x}_2^*\right) + \left\langle \tilde{\boldsymbol{\lambda}}, \mathbf{A}_1\mathbf{x}_1^{k+1} + \mathbf{A}_2\mathbf{x}_2^{k+1} - \mathbf{b} \right\rangle$$

$$\leq \eta_{k+1} \left\| \tilde{\nabla} f_1(\mathbf{x}_1^k) - \nabla f_1(\mathbf{x}_1^k) \right\|^2$$

$$+ \left( \frac{1}{2\eta_{k+1}} - \frac{\mu}{2} \right) \left\| \mathbf{x}_1^k - \mathbf{x}_1^* \right\|^2 - \frac{1}{2\eta_{k+1}} \left\| \mathbf{x}_1^{k+1} - \mathbf{x}_1^* \right\|^2$$

$$+ \left\langle \nabla f_1\left(\mathbf{x}_1^k\right) - \tilde{\nabla} f_1\left(\mathbf{x}_1^k\right), \mathbf{x}_1^k - \mathbf{x}_1^* \right\rangle + \frac{1}{2\beta} \left( \|\tilde{\boldsymbol{\lambda}} - \boldsymbol{\lambda}^k\|^2 - \|\tilde{\boldsymbol{\lambda}} - \boldsymbol{\lambda}^{k+1}\|^2 \right)$$

$$+ \frac{\beta}{2} \left( \left\| \mathbf{A}_2\mathbf{x}_2^k - \mathbf{A}_2\mathbf{x}_2^* \right\|^2 - \left\| \mathbf{A}_2\mathbf{x}_2^{k+1} - \mathbf{A}_2\mathbf{x}_2^* \right\|^2 \right), \tag{5.5}$$

*where $(\mathbf{x}_1^*, \mathbf{x}_2^*)$ is any optimal solution to Problem (5.1).*

**Proof** Because $\mathbf{x}_1^{k+1} = \mathrm{argmin}_{\mathbf{x}_1} \hat{L}_\beta^k(\mathbf{x}_1, \mathbf{x}_2^k, \boldsymbol{\lambda}^k)$, by the first-order optimality, we have

$$\tilde{\nabla} f_1\left(\mathbf{x}_1^k\right) + \beta\mathbf{A}_1^T\left(\mathbf{A}_1\mathbf{x}_1^{k+1} + \mathbf{A}_2\mathbf{x}_2^k - \mathbf{b}\right) + \mathbf{A}_1^T\boldsymbol{\lambda}^k + \frac{1}{\eta_{k+1}}\left(\mathbf{x}_1^{k+1} - \mathbf{x}_1^k\right) = \mathbf{0}. \tag{5.6}$$

Because $f_1$ is $L$-smooth, we have

$$f_1\left(\mathbf{x}_1^{k+1}\right) \leq f_1\left(\mathbf{x}_1^k\right) + \left\langle \nabla f_1\left(\mathbf{x}_1^k\right), \mathbf{x}_1^{k+1} - \mathbf{x}_1^k \right\rangle + \frac{L}{2}\left\| \mathbf{x}_1^{k+1} - \mathbf{x}_1^k \right\|^2. \tag{5.7}$$

Because $f_1$ is $\mu$-strongly convex, we have

$$f_1\left(\mathbf{x}_1^k\right) \leq f_1\left(\mathbf{x}_1^*\right) + \left\langle \nabla f_1\left(\mathbf{x}_1^k\right), \mathbf{x}_1^k - \mathbf{x}_1^* \right\rangle - \frac{\mu}{2}\left\| \mathbf{x}_1^k - \mathbf{x}_1^* \right\|^2. \tag{5.8}$$

Therefore, combining (5.7) with (5.8), we have

$$
\begin{aligned}
f_1\left(\mathbf{x}_1^{k+1}\right) &\le f_1\left(\mathbf{x}_1^*\right) + \left\langle \nabla f_1\left(\mathbf{x}_1^k\right), \mathbf{x}_1^{k+1} - \mathbf{x}_1^* \right\rangle + \frac{L}{2}\left\|\mathbf{x}_1^{k+1} - \mathbf{x}_1^k\right\|^2 - \frac{\mu}{2}\left\|\mathbf{x}_1^k - \mathbf{x}_1^*\right\|^2 \\
&= f_1\left(\mathbf{x}_1^*\right) + \left\langle \nabla f_1\left(\mathbf{x}_1^k\right) - \tilde{\nabla} f_1\left(\mathbf{x}_1^k\right), \mathbf{x}_1^k - \mathbf{x}_1^* \right\rangle + \frac{L}{2}\left\|\mathbf{x}_1^{k+1} - \mathbf{x}_1^k\right\|^2 \\
&\quad - \frac{\mu}{2}\left\|\mathbf{x}_1^k - \mathbf{x}_1^*\right\|^2 + \underbrace{\left\langle \tilde{\nabla} f_1\left(\mathbf{x}_1^k\right) - \nabla f_1\left(\mathbf{x}_1^k\right), \mathbf{x}_1^k - \mathbf{x}_1^{k+1} \right\rangle}_{I_1} \\
&\quad + \underbrace{\left\langle \tilde{\nabla} f_1\left(\mathbf{x}_1^k\right), \mathbf{x}_1^{k+1} - \mathbf{x}_1^* \right\rangle}_{I_2}.
\end{aligned}
$$

By the Cauchy–Schwartz inequality, we have for $I_1$ that

$$
I_1 \le \eta_{k+1}\left\|\tilde{\nabla} f_1\left(\mathbf{x}_1^k\right) - \nabla f_1\left(\mathbf{x}_1^k\right)\right\|^2 + \frac{1}{4\eta_{k+1}}\left\|\mathbf{x}_1^k - \mathbf{x}_1^{k+1}\right\|^2. \tag{5.9}
$$

Plugging (5.9) and (5.6) into $I_1$ and $I_2$, respectively, we have

$$
\begin{aligned}
f_1&\left(\mathbf{x}_1^{k+1}\right) - f_1\left(\mathbf{x}_1^*\right) \\
&\le \left\langle \beta\left(\mathbf{A}_1\mathbf{x}_1^{k+1} + \mathbf{A}_2\mathbf{x}_2^k - \mathbf{b}\right) + \boldsymbol{\lambda}^k, \mathbf{A}_1\mathbf{x}_1^* - \mathbf{A}_1\mathbf{x}_1^{k+1} \right\rangle \\
&\quad + \left\langle \nabla f_1\left(\mathbf{x}_1^k\right) - \tilde{\nabla} f_1\left(\mathbf{x}_1^k\right), \mathbf{x}_1^k - \mathbf{x}_1^* \right\rangle \\
&\quad + \frac{1}{\eta_{k+1}}\left\langle \mathbf{x}_1^{k+1} - \mathbf{x}_1^k, \mathbf{x}_1^* - \mathbf{x}_1^{k+1} \right\rangle + \eta_{k+1}\left\|\tilde{\nabla} f_1\left(\mathbf{x}_1^k\right) - \nabla f_1\left(\mathbf{x}_1^k\right)\right\|^2 \\
&\quad + \left(\frac{1}{4\eta_{k+1}} + \frac{L}{2}\right)\left\|\mathbf{x}_1^k - \mathbf{x}_1^{k+1}\right\|^2 - \frac{\mu}{2}\left\|\mathbf{x}_1^k - \mathbf{x}_1^*\right\|^2. \tag{5.10}
\end{aligned}
$$

On the other hand, because

$$
\mathbf{x}_2^{k+1} = \underset{\mathbf{x}_2}{\operatorname{argmin}} \, \hat{L}_\beta^k\left(\mathbf{x}_1^{k+1}, \mathbf{x}_2, \boldsymbol{\lambda}^k\right),
$$

by the first-order optimality, we have

$$
-\left[\beta \mathbf{A}_2^T\left(\mathbf{A}_1\mathbf{x}_1^{k+1} + \mathbf{A}_2\mathbf{x}_2^{k+1} - \mathbf{b}\right) + \mathbf{A}_2^T\boldsymbol{\lambda}^k\right] \in \partial f_2\left(\mathbf{x}_2^{k+1}\right). \tag{5.11}
$$

So by the convexity of $f_2$, we have

$$
f_2 \left( \mathbf{x}_2^{k+1} \right) - f_2 \left( \mathbf{x}_2^* \right)
$$

$$
\overset{a}{\leq} \left\langle \beta \left( \mathbf{A}_1 \mathbf{x}_1^{k+1} + \mathbf{A}_2 \mathbf{x}_2^{k+1} - \mathbf{b} \right) + \boldsymbol{\lambda}^k, \mathbf{A}_2 \mathbf{x}_2^* - \mathbf{A}_2 \mathbf{x}_2^{k+1} \right\rangle, \tag{5.12}
$$

where $\overset{a}{\leq}$ uses (5.11). Adding (5.12) and (5.10) and using $\eta_{k+1} \leq 1/(2L)$ and the fact that

$$
\left\langle \mathbf{x}_1^{k+1} - \mathbf{x}_1^k, \mathbf{x}_1^* - \mathbf{x}_1^{k+1} \right\rangle = \frac{1}{2} \left\| \mathbf{x}_1^k - \mathbf{x}_1^* \right\|^2 - \frac{1}{2} \left\| \mathbf{x}_1^{k+1} - \mathbf{x}_1^* \right\|^2 - \frac{1}{2} \left\| \mathbf{x}_1^{k+1} - \mathbf{x}_1^k \right\|^2
$$

from (A.2), we obtain

$$
f_1 \left( \mathbf{x}_1^{k+1} \right) - f_1 \left( \mathbf{x}_1^* \right) + f_2 \left( \mathbf{x}_2^{k+1} \right) - f_2 \left( \mathbf{x}_2^* \right)
$$

$$
\leq \left( \frac{1}{2\eta_{k+1}} - \frac{\mu}{2} \right) \left\| \mathbf{x}_1^k - \mathbf{x}_1^* \right\|^2 - \frac{1}{2\eta_{k+1}} \left\| \mathbf{x}_1^{k+1} - \mathbf{x}_1^* \right\|^2
$$

$$
+ \left\langle \nabla f_1 \left( \mathbf{x}_1^k \right) - \tilde{\nabla} f_1 \left( \mathbf{x}_1^k \right), \mathbf{x}_1^k - \mathbf{x}_1^* \right\rangle
$$

$$
+ \left\langle \beta \left( \mathbf{A}_1 \mathbf{x}_1^{k+1} + \mathbf{A}_2 \mathbf{x}_2^{k+1} - \mathbf{b} \right) + \boldsymbol{\lambda}^k, \mathbf{A}_1 \mathbf{x}_1^* - \mathbf{A}_1 \mathbf{x}_1^{k+1} + \mathbf{A}_2 \mathbf{x}_2^* - \mathbf{A}_2 \mathbf{x}_2^{k+1} \right\rangle
$$

$$
+ \beta \left\langle \mathbf{A}_2 \mathbf{x}_2^k - \mathbf{A}_2 \mathbf{x}_2^{k+1}, \mathbf{A}_1 \mathbf{x}_1^* - \mathbf{A}_1 \mathbf{x}_1^{k+1} \right\rangle
$$

$$
+ \eta_{k+1} \left\| \tilde{\nabla} f_1 \left( \mathbf{x}_1^k \right) - \nabla f_1 \left( \mathbf{x}_1^k \right) \right\|^2. \tag{5.13}
$$

Observing that

$$
\mathbf{A}_1 \mathbf{x}_1^* + \mathbf{A}_2 \mathbf{x}_2^* = \mathbf{b} \quad \text{and} \quad \boldsymbol{\lambda}^{k+1} = \boldsymbol{\lambda}^k + \beta \left( \mathbf{A}_1 \mathbf{x}_1^{k+1} + \mathbf{A}_2 \mathbf{x}_2^{k+1} - \mathbf{b} \right)
$$

from line 3 in Algorithm 5.1, we have

$$
\left\langle \left[ \beta \left( \mathbf{A}_1 \mathbf{x}_1^{k+1} + \mathbf{A}_2 \mathbf{x}_2^{k+1} - \mathbf{b} \right) + \boldsymbol{\lambda}^k \right] - \tilde{\boldsymbol{\lambda}}, \mathbf{A}_1 \mathbf{x}_1^* - \mathbf{A}_1 \mathbf{x}_1^{k+1} + \mathbf{A}_2 \mathbf{x}_2^* - \mathbf{A}_2 \mathbf{x}_2^{k+1} \right\rangle
$$

$$
= \frac{1}{\beta} \left\langle \tilde{\boldsymbol{\lambda}} - \boldsymbol{\lambda}^{k+1}, \boldsymbol{\lambda}^{k+1} - \boldsymbol{\lambda}^k \right\rangle
$$

$$
\overset{a}{=} -\frac{1}{2\beta} \left\| \tilde{\boldsymbol{\lambda}} - \boldsymbol{\lambda}^{k+1} \right\|^2 - \frac{1}{2\beta} \left\| \boldsymbol{\lambda}^k - \boldsymbol{\lambda}^{k+1} \right\|^2 + \frac{1}{2\beta} \left\| \tilde{\boldsymbol{\lambda}} - \boldsymbol{\lambda}^k \right\|^2, \tag{5.14}
$$

where $\overset{a}{=}$ uses (A.2). Moreover,

$$\beta \left\langle \mathbf{A}_2 \mathbf{x}_2^k - \mathbf{A}_2 \mathbf{x}_2^{k+1}, \mathbf{A}_1 \mathbf{x}_1^* - \mathbf{A}_1 \mathbf{x}_1^{k+1} \right\rangle$$

$$= \beta \left\langle \left( \mathbf{A}_2 \mathbf{x}_2^k - \mathbf{A}_2 \mathbf{x}_2^* \right) - \left( \mathbf{A}_2 \mathbf{x}_2^{k+1} - \mathbf{A}_2 \mathbf{x}_2^* \right), -\left( \mathbf{A}_1 \mathbf{x}_1^{k+1} - \mathbf{A}_1 \mathbf{x}_1^* \right) - \mathbf{0} \right\rangle$$

$$\overset{a}{=} \frac{\beta}{2} \left\| \mathbf{A}_2 \mathbf{x}_2^k - \mathbf{A}_2 \mathbf{x}_2^* \right\|^2 - \frac{\beta}{2} \left\| \mathbf{A}_2 \mathbf{x}_2^{k+1} - \mathbf{A}_2 \mathbf{x}_2^* \right\|^2 - \frac{\beta}{2} \left\| \mathbf{A}_2 \mathbf{x}_2^k + \mathbf{A}_1 \mathbf{x}_1^{k+1} - \mathbf{b} \right\|^2$$

$$+ \frac{\beta}{2} \left\| \mathbf{A}_2 \mathbf{x}_2^{k+1} + \mathbf{A}_1 \mathbf{x}_1^{k+1} - \mathbf{b} \right\|^2$$

$$\leq \frac{\beta}{2} \left\| \mathbf{A}_2 \mathbf{x}_2^k - \mathbf{A}_2 \mathbf{x}_2^* \right\|^2 - \frac{\beta}{2} \left\| \mathbf{A}_2 \mathbf{x}_2^{k+1} - \mathbf{A}_2 \mathbf{x}_2^* \right\|^2 + \frac{\beta}{2} \left\| \mathbf{A}_2 \mathbf{x}_2^{k+1} + \mathbf{A}_1 \mathbf{x}_1^{k+1} - \mathbf{b} \right\|^2$$

$$= \frac{\beta}{2} \left\| \mathbf{A}_2 \mathbf{x}_2^k - \mathbf{A}_2 \mathbf{x}_2^* \right\|^2 - \frac{\beta}{2} \left\| \mathbf{A}_2 \mathbf{x}_2^{k+1} - \mathbf{A}_2 \mathbf{x}_2^* \right\|^2 + \frac{1}{2\beta} \left\| \boldsymbol{\lambda}^k - \boldsymbol{\lambda}^{k+1} \right\|^2,$$

$$\tag{5.15}$$

where $\overset{a}{=}$ follows (A.3) and uses $\mathbf{A}_1 \mathbf{x}_1^* + \mathbf{A}_2 \mathbf{x}_2^* = \mathbf{b}$. Adding $\left\langle \tilde{\boldsymbol{\lambda}}, \mathbf{A}_1 \mathbf{x}_1^{k+1} + \mathbf{A}_2 \mathbf{x}_2^{k+1} - \mathbf{b} \right\rangle$ to both sides of (5.13), and plugging in (5.14) and (5.15), we obtain (5.5).  □

We now provide the convergence result for Algorithm 5.1.

**Theorem 5.1** *Under the assumptions of Lemma 5.1, assume that the variance of $f_1$'s gradient is uniformly bounded by $\sigma^2$, i.e.,*

$$\mathbb{E}_\xi \| \nabla F_1(\mathbf{x}_1, \xi) - \nabla f_1(\mathbf{x}_1) \|^2 \leq \sigma^2 \text{ for all given } \mathbf{x}_1.$$

*Define*

$$D_1 = \| \mathbf{x}_1^0 - \mathbf{x}_1^* \| \quad and \quad D_2 = \| \mathbf{A}_2 \mathbf{x}_2^0 - \mathbf{A}_2 \mathbf{x}_2^* \|.$$

*For the generally convex case, i.e., $\mu = 0$, set the stepsize $\eta_k = 1/(2L + \sqrt{k}\sigma/D_1)$,*

$$\bar{\mathbf{x}}_1^K = \frac{1}{\sum_{k=1}^K \eta_k} \sum_{k=1}^K \eta_k \mathbf{x}_1^k \quad and \quad \bar{\mathbf{x}}_2^K = \frac{1}{\sum_{k=1}^K \eta_k} \sum_{k=1}^K \eta_k \mathbf{x}_2^k,$$

*then for any $\rho > 0$ and sufficiently large $K$, we have*

$$\mathbb{E} f_1 \left( \bar{\mathbf{x}}_1^K \right) + \mathbb{E} f_2 \left( \bar{\mathbf{x}}_2^K \right) - f_1 \left( \mathbf{x}_1^* \right) - f_2 \left( \mathbf{x}_2^* \right) + \rho \mathbb{E} \| \mathbf{A}_1 \bar{\mathbf{x}}_1^K + \mathbf{A}_2 \bar{\mathbf{x}}_2^K - \mathbf{b} \|$$

$$\leq \frac{2 D_1 \sigma \log K}{\sqrt{K}} + \frac{\sigma}{\sqrt{K}} \left[ \frac{D_1}{2} + \frac{\rho^2}{2\beta(2L D_1 + \sigma)} + \frac{\beta D_2^2}{2(2L D_1 + \sigma)} \right].$$

*For the strongly convex case, i.e., $\mu > 0$, set the stepsize $\eta_k = 1/(2L + k\mu)$,*

$$\bar{\mathbf{x}}_1^K = \frac{1}{K} \sum_{k=1}^{K} \mathbf{x}_1^k \quad \text{and} \quad \bar{\mathbf{x}}_2^K = \frac{1}{K} \sum_{k=1}^{K} \mathbf{x}_2^k,$$

*then for any $\rho > 0$, we have*

$$\mathbb{E} f_1\left(\bar{\mathbf{x}}_1^K\right) + \mathbb{E} f_2\left(\bar{\mathbf{x}}_2^K\right) - f_1\left(\mathbf{x}_1^*\right) - f_2\left(\mathbf{x}_2^*\right) + \rho \mathbb{E}\left\|\mathbf{A}_1\bar{\mathbf{x}}_1^K + \mathbf{A}_2\bar{\mathbf{x}}_2^K - \mathbf{b}\right\|$$

$$\leq \frac{\sigma^2(\log K + 1)}{\mu K} + \frac{1}{K}\left(LD_1^2 + \frac{\rho^2}{2\beta} + \frac{\beta D_2^2}{2}\right).$$

**Proof** For the case $\mu = 0$, (5.5) reduces to

$$f_1\left(\mathbf{x}_1^{k+1}\right) + f_2\left(\mathbf{x}_2^{k+1}\right) - f_1\left(\mathbf{x}_1^*\right) - f_2\left(\mathbf{x}_2^*\right) + \left\langle\tilde{\boldsymbol{\lambda}}, \mathbf{A}_1\mathbf{x}_1^{k+1} + \mathbf{A}_2\mathbf{x}_2^{k+1} - \mathbf{b}\right\rangle$$

$$\leq \eta_{k+1}\left\|\tilde{\nabla} f_1\left(\mathbf{x}_1^k\right) - \nabla f_1\left(\mathbf{x}_1^k\right)\right\|^2 + \frac{1}{2\eta_{k+1}}\left\|\mathbf{x}_1^k - \mathbf{x}_1^*\right\|^2 - \frac{1}{2\eta_{k+1}}\left\|\mathbf{x}_1^{k+1} - \mathbf{x}_1^*\right\|^2$$

$$+ \left\langle\nabla f_1\left(\mathbf{x}_1^k\right) - \tilde{\nabla} f_1\left(\mathbf{x}_1^k\right), \mathbf{x}_1^k - \mathbf{x}_1^*\right\rangle + \frac{1}{2\beta}\left(\left\|\tilde{\boldsymbol{\lambda}} - \boldsymbol{\lambda}^k\right\|^2 - \left\|\tilde{\boldsymbol{\lambda}} - \boldsymbol{\lambda}^{k+1}\right\|^2\right)$$

$$+ \frac{\beta}{2}\left(\left\|\mathbf{A}_2\mathbf{x}_2^k - \mathbf{A}_2\mathbf{x}_2^*\right\|^2 - \left\|\mathbf{A}_2\mathbf{x}_2^{k+1} - \mathbf{A}_2\mathbf{x}_2^*\right\|^2\right).$$

Multiplying both sides with $\eta_{k+1}$, summing the above result from $k = 0$ to $K - 1$, using $\eta_{k+1} < \eta_k$ to drop some non-positive terms, and finally dividing both sides with $\sum_{k=1}^{K} \eta_k$, we have

$$\frac{1}{\sum_{k=1}^{K} \eta_k} \sum_{k=1}^{K} \eta_k \left(f_1\left(\mathbf{x}_1^k\right) + f_2\left(\mathbf{x}_2^k\right)\right) - f_1\left(\mathbf{x}_1^*\right) - f_2\left(\mathbf{x}_2^*\right)$$

$$+ \left\langle\tilde{\boldsymbol{\lambda}}, \frac{1}{\sum_{k=1}^{K} \eta_k} \sum_{k=1}^{K} \eta_k \left(\mathbf{A}_1\mathbf{x}_1^k + \mathbf{A}_2\mathbf{x}_2^k\right) - \mathbf{b}\right\rangle$$

$$\leq \frac{1}{\sum_{k=1}^{K} \eta_k} \left[\sum_{k=0}^{K-1} \eta_{k+1}^2 \left\|\tilde{\nabla} f_1\left(\mathbf{x}_1^k\right) - \nabla f_1\left(\mathbf{x}_1^k\right)\right\|^2 + \frac{1}{2}\left\|\mathbf{x}_1^0 - \mathbf{x}_1^*\right\|^2\right]$$

$$+ \sum_{k=0}^{K-1} \eta_{k+1} \left\langle \nabla f_1 \left( \mathbf{x}_1^k \right) - \tilde{\nabla} f_1 \left( \mathbf{x}_1^k \right), \mathbf{x}_1^k - \mathbf{x}_1^* \right\rangle$$

$$+ \frac{\eta_1}{2\beta} \left\| \tilde{\boldsymbol{\lambda}} - \boldsymbol{\lambda}^0 \right\|^2 + \frac{\beta \eta_1}{2} \left\| \mathbf{A}_2 \mathbf{x}_2^0 - \mathbf{A}_2 \mathbf{x}_2^* \right\|^2 \right]. \tag{5.16}$$

By the convexity of $f_1$ and $f_2$, we have

$$f_1 \left( \bar{\mathbf{x}}_1^K \right) + f_2 \left( \bar{\mathbf{x}}_2^K \right) \leq \frac{1}{\sum_{k=1}^K \eta_k} \sum_{k=1}^K \eta_k \left( f_1 \left( \mathbf{x}_1^k \right) + f_2 \left( \mathbf{x}_2^k \right) \right). \tag{5.17}$$

Letting

$$\tilde{\boldsymbol{\lambda}} = \frac{\rho \left( \mathbf{A}_1 \bar{\mathbf{x}}_1^K + \mathbf{A}_2 \bar{\mathbf{x}}_2^K - \mathbf{b} \right)}{\left\| \mathbf{A}_1 \bar{\mathbf{x}}_1^K + \mathbf{A}_2 \bar{\mathbf{x}}_2^K - \mathbf{b} \right\|},$$

recalling $\boldsymbol{\lambda}^0 = \mathbf{0}$ in Algorithm 5.1, taking full expectation on (5.16), and using (5.17), we have

$$\mathbb{E} f_1 \left( \bar{\mathbf{x}}_1^K \right) + \mathbb{E} f_2 \left( \bar{\mathbf{x}}_2^K \right) - f_1 \left( \mathbf{x}_1^* \right) - f_2 \left( \mathbf{x}_2^* \right) + \rho \mathbb{E} \left\| \mathbf{A}_1 \bar{\mathbf{x}}_1^K + \mathbf{A}_2 \bar{\mathbf{x}}_2^K - \mathbf{b} \right\|$$

$$\leq \frac{1}{\sum_{k=1}^K \eta_k} \left[ \sum_{k=0}^{K-1} \eta_{k+1}^2 \mathbb{E} \left\| \tilde{\nabla} f_1 \left( \mathbf{x}_1^k \right) - \nabla f_1 \left( \mathbf{x}_1^k \right) \right\|^2 \right.$$

$$+ \sum_{k=0}^{K-1} \eta_{k+1} \mathbb{E} \left\langle \nabla f_1 \left( \mathbf{x}_1^k \right) - \tilde{\nabla} f_1 \left( \mathbf{x}_1^k \right), \mathbf{x}_1^k - \mathbf{x}_1^* \right\rangle$$

$$\left. + \frac{D_1^2}{2} + \frac{\eta_1 \rho^2}{2\beta} + \frac{\beta \eta_1 D_2^2}{2} \right].$$

For each $k \geq 1$, we have

$$\mathbb{E} \left\| \tilde{\nabla} f_1 \left( \mathbf{x}_1^k \right) - \nabla f_1 \left( \mathbf{x}_1^k \right) \right\|^2$$

$$= \mathbb{E}_1 \mathbb{E}_2 \cdots \mathbb{E}_{k-1} \left[ \mathbb{E}_k \left\| \tilde{\nabla} f_1 \left( \mathbf{x}_1^k \right) - \nabla f_1 \left( \mathbf{x}_1^k \right) \right\|^2 \right]$$

$$\leq \mathbb{E}_1 \mathbb{E}_2 \cdots \mathbb{E}_{k-1} \left[ \sigma^2 \right] = \sigma^2, \tag{5.18}$$

where $\mathbb{E}_k$ denotes the expectation taken only on the random number at iteration $k$ given the previous updates. We also have

$$\mathbb{E}\left\langle \nabla f_1\left(\mathbf{x}_1^k\right) - \tilde{\nabla} f_1\left(\mathbf{x}_1^k\right), \mathbf{x}_1^k - \mathbf{x}_1^*\right\rangle$$

$$= \mathbb{E}_1 \mathbb{E}_2 \cdots \mathbb{E}_{k-1}\left[\mathbb{E}_k\left\langle \nabla f_1\left(\mathbf{x}_1^k\right) - \tilde{\nabla} f_1\left(\mathbf{x}_1^k\right), \mathbf{x}_1^k - \mathbf{x}_1^*\right\rangle\right]$$

$$= \mathbb{E}_1 \mathbb{E}_2 \cdots \mathbb{E}_{k-1}[0] = 0. \tag{5.19}$$

Moreover, when $K$ is sufficiently large we have

$$\sum_{k=1}^{K} \eta_k^2 \leq \int_0^K \left(2L + \frac{\sigma}{D_1}\sqrt{x}\right)^{-2} dx$$

$$= 2\left(\frac{D_1}{\sigma}\right)^2 \left[\log\left(1 + \frac{\sigma}{2LD_1}\sqrt{K}\right) - \left(1 + \frac{2LD_1}{\sigma\sqrt{K}}\right)^{-1}\right]$$

$$\leq 2\left(\frac{D_1}{\sigma}\right)^2 \log K$$

and

$$\sum_{k=1}^{K} \eta_k \geq \int_1^{K+1} \left(2L + \frac{\sigma}{D_1}\sqrt{x}\right)^{-1} dx$$

$$= 4L\left(\frac{D_1}{\sigma}\right)^2 \left[\frac{\sigma}{2LD_1}\left(\sqrt{K+1} - 1\right)\right.$$

$$\left. - \log\left(1 + \frac{\sigma}{2LD_1}\sqrt{K+1}\right) + \log\left(1 + \frac{\sigma}{2LD_1}\right)\right]$$

$$\geq \frac{D_1}{\sigma}\sqrt{K}.$$

Putting the pieces together and using $\eta_1 = 1/(2L + \sigma/D_1)$, we obtain

$$\mathbb{E}f_1\left(\bar{\mathbf{x}}_1^K\right) + \mathbb{E}f_2\left(\bar{\mathbf{x}}_2^K\right) - f_1\left(\mathbf{x}_1^*\right) - f_2\left(\mathbf{x}_2^*\right) + \rho\mathbb{E}\left\|\mathbf{A}_1\bar{\mathbf{x}}_1^K + \mathbf{A}_2\bar{\mathbf{x}}_2^K - \mathbf{b}\right\|$$

$$\leq \frac{1}{\sqrt{K}}\left[2D_1\sigma \log K + \frac{D_1\sigma}{2} + \frac{\sigma\rho^2}{2\beta(2LD_1 + \sigma)} + \frac{\beta\sigma D_2^2}{2(2LD_1 + \sigma)}\right].$$

For the $\mu$-strongly convex case, using $\eta_k = 1/(2L + k\mu)$, (5.5) becomes

$$f_1\left(\mathbf{x}_1^{k+1}\right) + f_2\left(\mathbf{x}_2^{k+1}\right) - f_1\left(\mathbf{x}_1^*\right) - f_2\left(\mathbf{x}_2^*\right) + \left\langle \tilde{\boldsymbol{\lambda}}, \mathbf{A}_1\mathbf{x}_1^{k+1} + \mathbf{A}_2\mathbf{x}_2^{k+1} - \mathbf{b}\right\rangle$$

$$\leq \eta_{k+1}\left\|\tilde{\nabla}f_1\left(\mathbf{x}_1^k\right) - \nabla f_1\left(\mathbf{x}_1^k\right)\right\|^2 + \frac{1}{2\eta_k}\left\|\mathbf{x}_1^k - \mathbf{x}_1^*\right\|^2 - \frac{1}{2\eta_{k+1}}\left\|\mathbf{x}_1^{k+1} - \mathbf{x}_1^*\right\|^2$$

$$+ \left\langle \nabla f_1\left(\mathbf{x}_1^k\right) - \tilde{\nabla}f_1\left(\mathbf{x}_1^k\right), \mathbf{x}_1^k - \mathbf{x}_1^*\right\rangle + \frac{1}{2\beta}\left(\left\|\tilde{\boldsymbol{\lambda}} - \boldsymbol{\lambda}^k\right\|^2 - \left\|\tilde{\boldsymbol{\lambda}} - \boldsymbol{\lambda}^{k+1}\right\|^2\right)$$

$$+ \frac{\beta}{2}\left(\left\|\mathbf{A}_2\mathbf{x}_2^k - \mathbf{A}_2\mathbf{x}_2^*\right\|^2 - \left\|\mathbf{A}_2\mathbf{x}_2^{k+1} - \mathbf{A}_2\mathbf{x}_2^*\right\|^2\right).$$

Summing the above result from $k = 0$ to $K - 1$, dropping some non-positive terms, dividing both sides with $K$, again using the convexity of $f_1$ and $f_2$, letting

$$\tilde{\boldsymbol{\lambda}} = \frac{\rho\left(\mathbf{A}_1\bar{\mathbf{x}}_1^K + \mathbf{A}_2\bar{\mathbf{x}}_2^K - \mathbf{b}\right)}{\left\|\mathbf{A}_1\bar{\mathbf{x}}_1^K + \mathbf{A}_2\bar{\mathbf{x}}_2^K - \mathbf{b}\right\|},$$

taking full expectation, and finally applying (5.18) and (5.19), we have

$$\mathbb{E}f_1\left(\bar{\mathbf{x}}_1^K\right) + \mathbb{E}f_2\left(\bar{\mathbf{x}}_2^K\right) - f_1\left(\mathbf{x}_1^*\right) - f_2\left(\mathbf{x}_2^*\right) + \rho\mathbb{E}\|\mathbf{A}_1\bar{\mathbf{x}}_1^K + \mathbf{A}_2\bar{\mathbf{x}}_2^K - \mathbf{b}\|$$

$$\leq \frac{\sigma^2}{K}\sum_{k=1}^{K}\frac{1}{2L + k\mu} + \frac{L}{K}\left\|\mathbf{x}_1^0 - \mathbf{x}_1^*\right\|^2 + \frac{1}{2\beta K}\left\|\tilde{\boldsymbol{\lambda}} - \boldsymbol{\lambda}^0\right\|^2 + \frac{\beta}{2K}\left\|\mathbf{A}_2\mathbf{x}_2^0 - \mathbf{A}_2\mathbf{x}_2^*\right\|^2$$

$$\overset{a}{\leq} \frac{\sigma^2(\log K + 1)}{\mu K} + \frac{LD_1^2}{K} + \frac{\rho^2}{2\beta K} + \frac{\beta D_2^2}{2K},$$

where $\overset{a}{\leq}$ uses

$$\sum_{k=1}^{K}\frac{1}{2L + k\mu} < \sum_{k=1}^{K}\frac{1}{k\mu} < \int_1^K \frac{1}{\mu x}\mathrm{d}x + \frac{1}{\mu} = \frac{\log K + 1}{\mu}.$$

$\square$

## 5.2   Variance Reduction

The Variance Reduction (VR) technique is initially designed to solve the problem

$$\min_{\mathbf{x}\in\mathbb{R}^d} \frac{1}{n}\sum_{i=1}^{n} F_i(\mathbf{x}).$$

**Algorithm 5.2** SVRG-ADMM

---

**Input** $\mathbf{x}_{0,1}^0, \mathbf{x}_{0,2}^0, \boldsymbol{\lambda}_0^0$. Set epoch length $m$, $\tilde{\mathbf{x}}_{0,1} = \mathbf{x}_{0,1}^0$, and stepsize $\eta$.

1 **for** $s = 0$ **to** $S - 1$ **do**

2     **for** $k = 0$ **to** $m - 1$ **do**

3        Randomly sample $i_{k,s}$ from $[n]$.

4        $\tilde{\nabla} f_1(\mathbf{x}_{s,1}^k) = \nabla F_{i_{k,s}}(\mathbf{x}_{s,1}^k) - \nabla F_{i_{k,s}}(\tilde{\mathbf{x}}_{s,1}) + \frac{1}{n} \sum_{i=1}^n \nabla F_i(\tilde{\mathbf{x}}_{s,1})$.

5        Update $\mathbf{x}_{s,1}^{k+1}$ by (5.21).

6        Update $\mathbf{x}_{s,2}^{k+1}$ by (5.22).

7        Update dual variable: $\boldsymbol{\lambda}_s^{k+1} = \boldsymbol{\lambda}_s^k + \beta \left( \mathbf{A}_1 \mathbf{x}_{s,1}^{k+1} + \mathbf{A}_2 \mathbf{x}_{s,2}^{k+1} - \mathbf{b} \right)$.

8     **end for** $k$

9     $\tilde{\mathbf{x}}_{s+1,i} = \frac{1}{m} \sum_{k=1}^m \mathbf{x}_{s,i}^k$, $\mathbf{x}_{s+1,i}^0 = \mathbf{x}_{s,i}^m$, for $i = 1, 2$, and $\boldsymbol{\lambda}_{s+1}^0 = \boldsymbol{\lambda}_s^m$.

   **end for** $s$

---

It is known that the standard Stochastic Gradient Descent (SGD) will enjoy a sublinear convergence rate when each $F_i$ is strongly convex and smooth. Surprisingly, the VR technique can accelerate stochastic algorithms to a linear convergence rate. The first VR method may be SAG [10], which uses the sum of the latest individual gradients as an estimator. The method requires $O(nd)$ memory storage and the estimated gradient is a biased gradient estimator. Later, lots of VR methods have been designed, e.g., SDCA [11], MISO [7], SVRG [6], and SAGA [2].

In this section, we introduce the application of VR to ADMM methods. We show that for the offline problems, VR improves the convergence rate to $O(1/K)$ for the generally convex case. We use a classical VR method called SVRG [6]. Its main technique is to frequently pre-store a snapshot vector and to control the variance via the snapshot vector and the latest iterate.

Specifically, consider the problem:

$$\min_{\mathbf{x}_1, \mathbf{x}_2} \left( f_1(\mathbf{x}_1) + f_2(\mathbf{x}_2) \right),$$

$$s.t. \ \mathbf{A}_1 \mathbf{x}_1 + \mathbf{A}_2 \mathbf{x}_2 = \mathbf{b}, \tag{5.20}$$

where $f_1(\mathbf{x}_1) = \frac{1}{n} \sum_{i=1}^n F_i(\mathbf{x}_1)$. We introduce SVRG-ADMM proposed by Zheng and Kwok [12]. The algorithm is shown in Algorithm 5.2. In the process of solving the primal variable, we linearize both $f_1(\mathbf{x}_1)$ and the augmented term $\frac{\beta}{2} \left\| \mathbf{A}_1 \mathbf{x}_1 + \mathbf{A}_2 \mathbf{x}_2 - \mathbf{b} + \frac{1}{\beta} \boldsymbol{\lambda} \right\|^2$. So

$$\mathbf{x}_{s,1}^{k+1} = \operatorname*{argmin}_{\mathbf{x}_1} \left( \left\langle \tilde{\nabla} f_1 \left( \mathbf{x}_{s,1}^k \right), \mathbf{x}_1 - \mathbf{x}_{s,1}^k \right\rangle \right.$$

$$+ \left\langle \beta \left( \mathbf{A}_1 \mathbf{x}_{s,1}^k + \mathbf{A}_2 \mathbf{x}_{s,2}^k - \mathbf{b} \right) + \boldsymbol{\lambda}_s^k, \mathbf{A}_1 \left( \mathbf{x}_1 - \mathbf{x}_{s,1}^k \right) \right\rangle$$

$$\left. + \frac{1}{2\eta_1} \left\| \mathbf{x}_1 - \mathbf{x}_{s,1}^k \right\|^2 \right) \tag{5.21}$$

and

$$\mathbf{x}_{s,2}^{k+1} = \underset{\mathbf{x}_2}{\operatorname{argmin}} \left( f_2(\mathbf{x}_2) + \left\langle \beta \left( \mathbf{A}_1 \mathbf{x}_{s,1}^{k+1} + \mathbf{A}_2 \mathbf{x}_{s,2}^k - \mathbf{b} \right) + \lambda_s^k, \mathbf{A}_2 \left( \mathbf{x}_2 - \mathbf{x}_{s,2}^k \right) \right\rangle \right.$$
$$\left. + \frac{1}{2\eta_2} \left\| \mathbf{x}_2 - \mathbf{x}_{s,2}^k \right\|^2 \right), \tag{5.22}$$

where

$$\eta_1 = 1 / \left( 9L + \beta \|\mathbf{A}_1\|_2^2 \right) \quad \text{and} \quad \eta_2 = 1 / \left( \beta \|\mathbf{A}_2\|_2^2 \right).$$

The main step to reduce the variance is line 4 of Algorithm 5.2, where the gradient is estimated by

$$\tilde{\nabla} f_1 \left( \mathbf{x}_{s,1}^k \right) = \nabla F_{i_{k,s}} \left( \mathbf{x}_{s,1}^k \right) - \nabla F_{i_{k,s}} (\tilde{\mathbf{x}}_{s,1}) + \frac{1}{n} \sum_{i=1}^n \nabla F_i (\tilde{\mathbf{x}}_{s,1}),$$

in which $\tilde{\mathbf{x}}_{s,1}$ is the snapshot vector and $\frac{1}{n} \sum_{i=1}^n \nabla F_i (\tilde{\mathbf{x}}_{s,1})$ is re-computed at the beginning of the outer loop.

The following analysis is taken from [12]. We first show that the variance of this specially estimated gradient can be controlled by the lemma below.

**Lemma 5.2** *Assume that $F_i$ is convex and $L$-smooth for all $i \in [n]$. Let $\mathbb{E}_k$ denote the expectation taken only on the random number $i_{k,s}$ conditioned on $\mathbf{x}_{s,1}^k$. Then we have*

$$\mathbb{E}_k \tilde{\nabla} f_1 \left( \mathbf{x}_{s,1}^k \right) = \nabla f_1 \left( \mathbf{x}_{s,1}^k \right).$$

*Let $(\mathbf{x}_1^*, \mathbf{x}_2^*, \boldsymbol{\lambda}^*)$ be any optimal solution of (5.20). We have*

$$\mathbb{E}_k \left\| \tilde{\nabla} f_1 \left( \mathbf{x}_{s,1}^k \right) - \nabla f_1 \left( \mathbf{x}_{s,1}^k \right) \right\|^2 \le 4L \left[ H_1 \left( \mathbf{x}_{s,1}^k \right) + H_1(\tilde{\mathbf{x}}_{s,1}) \right], \tag{5.23}$$

*where $H_1(\mathbf{x}_1) = f_1(\mathbf{x}_1) - f_1(\mathbf{x}_1^*) - \langle \nabla f_1(\mathbf{x}_1^*), \mathbf{x}_1 - \mathbf{x}_1^* \rangle$.*

**Proof** First, we have

$$\mathbb{E}_k \tilde{\nabla} f_1 \left( \mathbf{x}_{s,1}^k \right) = \mathbb{E}_k \nabla F_{i_{k,s}} \left( \mathbf{x}_{s,1}^k \right) - \mathbb{E}_k \left( \nabla F_{i_{k,s}} (\tilde{\mathbf{x}}_{s,1}) - \frac{1}{n} \sum_{i=1}^n \nabla F_i (\tilde{\mathbf{x}}_{s,1}) \right)$$
$$= \nabla f_1 \left( \mathbf{x}_{s,1}^k \right).$$

Thus $\tilde{\nabla} f_1(\mathbf{x}_{s,1}^k)$ is an unbiased estimator of $\nabla f_1(\mathbf{x}_{s,1}^k)$. Then

$$
\mathbb{E}_k \left\| \tilde{\nabla} f_1\left(\mathbf{x}_{s,1}^k\right) - \nabla f_1\left(\mathbf{x}_{s,1}^k\right) \right\|^2
$$

$$
= \mathbb{E}_k \left\| \nabla F_{i_{k,s}}\left(\mathbf{x}_{s,1}^k\right) - \nabla F_{i_{k,s}}(\tilde{\mathbf{x}}_{s,1}) + \nabla f_1(\tilde{\mathbf{x}}_{s,1}) - \nabla f_1\left(\mathbf{x}_{s,1}^k\right) \right\|^2
$$

$$
\overset{a}{\leq} \mathbb{E}_k \left\| \nabla F_{i_{k,s}}\left(\mathbf{x}_{s,1}^k\right) - \nabla F_{i_{k,s}}(\tilde{\mathbf{x}}_{s,1}) \right\|^2
$$

$$
\overset{b}{\leq} 2\mathbb{E}_k \left\| \nabla F_{i_{k,s}}\left(\mathbf{x}_{s,1}^k\right) - \nabla F_{i_{k,s}}\left(\mathbf{x}_1^*\right) \right\|^2
$$

$$
+ 2\mathbb{E}_k \left\| \nabla F_{i_{k,s}}(\tilde{\mathbf{x}}_{s,1}) - \nabla F_{i_{k,s}}\left(\mathbf{x}_1^*\right) \right\|^2 , \tag{5.24}
$$

where $\overset{a}{\leq}$ uses

$$
\mathbb{E}_k \left( \nabla F_{i_{k,s}}\left(\mathbf{x}_{s,1}^k\right) - \nabla F_{i_{k,s}}(\tilde{\mathbf{x}}_{s,1}) \right) = \nabla f_1\left(\mathbf{x}_{s,1}^k\right) - \nabla f_1(\tilde{\mathbf{x}}_{s,1})
$$

and $\mathbb{E}\|\boldsymbol{\xi} - \mathbb{E}\boldsymbol{\xi}\|^2 \leq \mathbb{E}\|\boldsymbol{\xi}\|^2$ for random vector $\boldsymbol{\xi}$ (Proposition A.3) and $\overset{b}{\leq}$ uses $\|\mathbf{a} + \mathbf{b}\|^2 \leq 2\|\mathbf{a}\|^2 + 2\|\mathbf{b}\|^2$. Since $f_i(\mathbf{x})$ is convex and $L$-smooth, it follows from (A.5) that

$$
\|\nabla F_i(\mathbf{x}) - \nabla F_i(\mathbf{y})\|^2 \leq 2L \left( F_i(\mathbf{x}) - F_i(\mathbf{y}) + \langle \nabla F_i(\mathbf{y}), \mathbf{y} - \mathbf{x} \rangle \right). \tag{5.25}
$$

Letting $\mathbf{x} = \mathbf{x}_{s,1}^k$ and $\mathbf{y} = \mathbf{x}_1^*$ in (5.25) and summing the result with $i = 1$ to $n$, we have

$$
\mathbb{E}_k \left\| \nabla F_{i_{k,s}}\left(\mathbf{x}_{s,1}^k\right) - \nabla F_{i_{k,s}}\left(\mathbf{x}_1^*\right) \right\|^2 \leq 2L H_1\left(\mathbf{x}_{s,1}^k\right). \tag{5.26}
$$

In the same way, we have

$$
\mathbb{E}_k \left\| \nabla F_{i_{k,s}}(\tilde{\mathbf{x}}_{s,1}) - \nabla F_{i_{k,s}}\left(\mathbf{x}_1^*\right) \right\|^2 \leq 2L H_1(\tilde{\mathbf{x}}_{s,1}). \tag{5.27}
$$

Plugging (5.26) and (5.27) into (5.24), we have (5.23).                           $\square$

We then study the inner loop. For the sake of simplicity, we drop the subscript $s$ in the analysis of inner loop, since it is clear from the context.

**Lemma 5.3** *Assume that $F_i$ is convex and $L$-smooth for $i \in [n]$ and $f_2$ is convex. Then for $k \geq 0$,*

$$
\mathbb{E}_k f_1 \left( \mathbf{x}_1^{k+1} \right) - f_1 \left( \mathbf{x}_1^* \right) + \mathbb{E}_k f_2 \left( \mathbf{x}_2^{k+1} \right) - f_2 \left( \mathbf{x}_2^* \right) + \mathbb{E}_k \left\langle \boldsymbol{\lambda}^*, \mathbf{A}_1 \mathbf{x}_1^{k+1} + \mathbf{A}_2 \mathbf{x}_2^{k+1} - \mathbf{b} \right\rangle
$$

$$
\leq \frac{1}{4} \left( H_1 \left( \mathbf{x}_1^k \right) + H_1 (\tilde{\mathbf{x}}_1) \right) + \left\| \mathbf{x}_1^k - \mathbf{x}_1^* \right\|_{\mathbf{G}_1}^2 - \mathbb{E}_k \left\| \mathbf{x}_1^{k+1} - \mathbf{x}_1^* \right\|_{\mathbf{G}_1}^2 + \left\| \mathbf{x}_2^k - \mathbf{x}_2^* \right\|_{\mathbf{G}_2}^2
$$

$$
- \mathbb{E}_k \left\| \mathbf{x}_2^{k+1} - \mathbf{x}_2^* \right\|_{\mathbf{G}_2}^2 + \frac{1}{2\beta} \left\| \boldsymbol{\lambda}^* - \boldsymbol{\lambda}^k \right\|^2 - \frac{1}{2\beta} \mathbb{E}_k \left\| \boldsymbol{\lambda}^* - \boldsymbol{\lambda}^{k+1} \right\|^2, \tag{5.28}
$$

*where $\mathbf{G}_1 = \frac{1}{2} \left[ \left( \beta \|\mathbf{A}_1\|_2^2 + 9L \right) \mathbf{I} - \beta \mathbf{A}_1^T \mathbf{A}_1 \right]$ and $\mathbf{G}_2 = \frac{\beta}{2} \|\mathbf{A}_2\|_2^2 \mathbf{I}$.*

***Proof*** By the optimality of solution $\mathbf{x}_1^{k+1}$ in (5.21), we have

$$
\tilde{\nabla} f_1 \left( \mathbf{x}_1^k \right) + \beta \mathbf{A}_1^T \left( \mathbf{A}_1 \mathbf{x}_1^k + \mathbf{A}_2 \mathbf{x}_2^k - \mathbf{b} \right) + \mathbf{A}_1^T \boldsymbol{\lambda}^k + \frac{1}{\eta_1} \left( \mathbf{x}_1^{k+1} - \mathbf{x}_1^k \right) = \mathbf{0}. \tag{5.29}
$$

By the $L$-smoothness and the convexity of $f_1$, we have

$$
\begin{aligned}
f_1 \left( \mathbf{x}_1^{k+1} \right) &\leq f_1 \left( \mathbf{x}_1^k \right) + \left\langle \nabla f_1 \left( \mathbf{x}_1^k \right), \mathbf{x}_1^{k+1} - \mathbf{x}_1^k \right\rangle + \frac{L}{2} \left\| \mathbf{x}_1^{k+1} - \mathbf{x}_1^k \right\|^2 \\
&\leq f_1 \left( \mathbf{x}_1^* \right) + \left\langle \nabla f_1 \left( \mathbf{x}_1^k \right), \mathbf{x}_1^{k+1} - \mathbf{x}_1^* \right\rangle + \frac{L}{2} \left\| \mathbf{x}_1^{k+1} - \mathbf{x}_1^k \right\|^2 \\
&= f_1 \left( \mathbf{x}_1^* \right) + \left\langle \nabla f_1 \left( \mathbf{x}_1^k \right) - \tilde{\nabla} f_1 \left( \mathbf{x}_1^k \right), \mathbf{x}_1^k - \mathbf{x}_1^* \right\rangle + \frac{L}{2} \left\| \mathbf{x}_1^{k+1} - \mathbf{x}_1^k \right\|^2 \\
&\quad + \underbrace{\left\langle \tilde{\nabla} f_1 \left( \mathbf{x}_1^k \right) - \nabla f_1 \left( \mathbf{x}_1^k \right), \mathbf{x}_1^k - \mathbf{x}_1^{k+1} \right\rangle}_{I_1} + \underbrace{\left\langle \tilde{\nabla} f_1 \left( \mathbf{x}_1^k \right), \mathbf{x}_1^{k+1} - \mathbf{x}_1^* \right\rangle}_{I_2}.
\end{aligned}
\tag{5.30}
$$

By the Cauchy–Schwartz inequality, we have for $I_1$ that

$$
\begin{aligned}
\mathbb{E}_k I_1 &\leq \frac{1}{16L} \mathbb{E}_k \left\| \tilde{\nabla} f_1 \left( \mathbf{x}_1^k \right) - \nabla f_1 \left( \mathbf{x}_1^k \right) \right\|^2 + 4L \mathbb{E}_k \left\| \mathbf{x}_1^k - \mathbf{x}_1^{k+1} \right\|^2 \\
&\overset{a}{\leq} \frac{1}{4} \left( H_1 \left( \mathbf{x}_1^k \right) + H_1 \left( \tilde{\mathbf{x}}_1 \right) \right) + 4L \mathbb{E}_k \left\| \mathbf{x}_1^k - \mathbf{x}_1^{k+1} \right\|^2, \tag{5.31}
\end{aligned}
$$

where in $\overset{a}{\leq}$ we apply Lemma 5.2. On the other hand, we have

$$
\mathbb{E}_k \left\langle \nabla f_1 \left( \mathbf{x}_1^k \right) - \tilde{\nabla} f_1 \left( \mathbf{x}_1^k \right), \mathbf{x}_1^k - \mathbf{x}_1^* \right\rangle = 0.
$$

Thus taking conditional expectation $\mathbb{E}_k$ on (5.30), bounding $I_1$ by (5.31), and plugging (5.29) into $I_2$, we have

$$
\mathbb{E}_k f_1\left(\mathbf{x}_1^{k+1}\right) - f_1\left(\mathbf{x}_1^*\right)
$$

$$
\leq \frac{1}{4}\left(H_1\left(\mathbf{x}_1^k\right) + H_1(\tilde{\mathbf{x}}_1)\right) + \frac{9L}{2}\mathbb{E}_k\left\|\mathbf{x}_1^k - \mathbf{x}_1^{k+1}\right\|^2
$$

$$
+ \mathbb{E}_k\left\langle \beta\left(\mathbf{A}_1\mathbf{x}_1^k + \mathbf{A}_2\mathbf{x}_2^k - \mathbf{b}\right) + \lambda^k, \mathbf{A}_1\mathbf{x}_1^* - \mathbf{A}_1\mathbf{x}_1^{k+1}\right\rangle
$$

$$
+ \frac{1}{\eta_1}\mathbb{E}_k\left\langle \mathbf{x}_1^{k+1} - \mathbf{x}_1^k, \mathbf{x}_1^* - \mathbf{x}_1^{k+1}\right\rangle. \tag{5.32}
$$

By the optimality of solution $\mathbf{x}_2^{k+1}$ in (5.22), we have

$$
-\left[\beta\mathbf{A}_2^T\left(\mathbf{A}_1\mathbf{x}_1^{k+1} + \mathbf{A}_2\mathbf{x}_2^k - \mathbf{b}\right) + \mathbf{A}_2^T\lambda^k + \frac{1}{\eta_2}\left(\mathbf{x}_2^{k+1} - \mathbf{x}_2^k\right)\right] \in \partial f_2\left(\mathbf{x}_2^{k+1}\right). \tag{5.33}
$$

So by the convexity of $f_2$, we have

$$
f_2\left(\mathbf{x}_2^{k+1}\right) - f_2\left(\mathbf{x}_2^*\right)
$$

$$
\overset{a}{\leq} \left\langle \beta\left(\mathbf{A}_1\mathbf{x}_1^{k+1} + \mathbf{A}_2\mathbf{x}_2^k - \mathbf{b}\right) + \lambda^k, \mathbf{A}_2\mathbf{x}_2^* - \mathbf{A}_2\mathbf{x}_2^{k+1}\right\rangle
$$

$$
+ \frac{1}{\eta_2}\left\langle \mathbf{x}_2^{k+1} - \mathbf{x}_2^k, \mathbf{x}_2^* - \mathbf{x}_2^{k+1}\right\rangle, \tag{5.34}
$$

where $\overset{a}{\leq}$ uses (5.33). Taking conditional expectation on (5.34), adding it with (5.32), then adding $\mathbb{E}_k\left\langle \lambda^*, \mathbf{A}_1\mathbf{x}_1^{k+1} + \mathbf{A}_2\mathbf{x}_2^{k+1} - \mathbf{b}\right\rangle$ on both sides, and finally using the fact that

$$
\left\langle \mathbf{x}_i^{k+1} - \mathbf{x}_i^k, \mathbf{x}_i^* - \mathbf{x}_i^{k+1}\right\rangle = \frac{1}{2}\left\|\mathbf{x}_i^k - \mathbf{x}_i^*\right\|^2 - \frac{1}{2}\left\|\mathbf{x}_i^{k+1} - \mathbf{x}_i^*\right\|^2 - \frac{1}{2}\left\|\mathbf{x}_i^{k+1} - \mathbf{x}_i^k\right\|^2
$$

for $i = 1, 2$ from (A.2), we have

$$
\mathbb{E}_k f_1\left(\mathbf{x}_1^{k+1}\right) - f_1\left(\mathbf{x}_1^*\right) + \mathbb{E}_k f_2\left(\mathbf{x}_2^{k+1}\right) - f_2\left(\mathbf{x}_2^*\right) + \mathbb{E}_k\left\langle \lambda^*, \mathbf{A}_1\mathbf{x}_1^{k+1} + \mathbf{A}_2\mathbf{x}_2^{k+1} - \mathbf{b}\right\rangle
$$

$$
\leq \frac{1}{4}\left(H_1\left(\mathbf{x}_1^k\right) + H_1(\tilde{\mathbf{x}}_1)\right) + \frac{1}{2\eta_1}\left\|\mathbf{x}_1^k - \mathbf{x}_1^*\right\|^2 - \frac{1}{2\eta_1}\mathbb{E}_k\left\|\mathbf{x}_1^{k+1} - \mathbf{x}_1^*\right\|^2
$$

$$
+ \frac{1}{2\eta_2}\left\|\mathbf{x}_2^k - \mathbf{x}_2^*\right\|^2 - \frac{1}{2\eta_2}\mathbb{E}_k\left\|\mathbf{x}_2^{k+1} - \mathbf{x}_2^*\right\|^2 - \frac{1}{2\eta_2}\mathbb{E}_k\left\|\mathbf{x}_2^{k+1} - \mathbf{x}_2^k\right\|^2
$$

$$
+ \mathbb{E}_k\left\langle \beta\left(\mathbf{A}_1\mathbf{x}_1^{k+1} + \mathbf{A}_2\mathbf{x}_2^k - \mathbf{b}\right) + \lambda^k - \lambda^*, \mathbf{A}_1\mathbf{x}_1^* - \mathbf{A}_1\mathbf{x}_1^{k+1} + \mathbf{A}_2\mathbf{x}_2^* - \mathbf{A}_2\mathbf{x}_2^{k+1}\right\rangle
$$

$$+ \beta \mathbb{E}_k \left\langle \mathbf{A}_1 \mathbf{x}_1^k - \mathbf{A}_1 \mathbf{x}_1^{k+1}, \mathbf{A}_1 \mathbf{x}_1^* - \mathbf{A}_1 \mathbf{x}_1^{k+1} \right\rangle$$

$$- \left( \frac{1}{2\eta_1} - \frac{9L}{2} \right) \mathbb{E}_k \left\| \mathbf{x}_1^{k+1} - \mathbf{x}_1^k \right\|^2 . \tag{5.35}$$

For the fourth line of (5.35), we have

$$\left\langle \beta \left( \mathbf{A}_1 \mathbf{x}_1^{k+1} + \mathbf{A}_2 \mathbf{x}_2^k - \mathbf{b} \right) + \lambda^k - \lambda^*, \mathbf{A}_1 \mathbf{x}_1^* - \mathbf{A}_1 \mathbf{x}_1^{k+1} + \mathbf{A}_2 \mathbf{x}_2^* - \mathbf{A}_2 \mathbf{x}_2^{k+1} \right\rangle$$

$$= \underbrace{\left\langle \beta \left( \mathbf{A}_1 \mathbf{x}_1^{k+1} + \mathbf{A}_2 \mathbf{x}_2^{k+1} - \mathbf{b} \right) + \lambda^k - \lambda^*, \mathbf{A}_1 \mathbf{x}_1^* - \mathbf{A}_1 \mathbf{x}_1^{k+1} + \mathbf{A}_2 \mathbf{x}_2^* - \mathbf{A}_2 \mathbf{x}_2^{k+1} \right\rangle}_{I_3}$$

$$+ \underbrace{\beta \left\langle \mathbf{A}_2 \mathbf{x}_2^k - \mathbf{A}_2 \mathbf{x}_2^{k+1}, \mathbf{A}_1 \mathbf{x}_1^* - \mathbf{A}_1 \mathbf{x}_1^{k+1} \right\rangle}_{I_4}$$

$$+ \beta \left\langle \mathbf{A}_2 \mathbf{x}_2^k - \mathbf{A}_2 \mathbf{x}_2^{k+1}, \mathbf{A}_2 \mathbf{x}_2^* - \mathbf{A}_2 \mathbf{x}_2^{k+1} \right\rangle . \tag{5.36}$$

For $I_3$, since

$$\lambda^{k+1} = \lambda^k + \beta \left( \mathbf{A}_1 \mathbf{x}_1^{k+1} + \mathbf{A}_2 \mathbf{x}_2^{k+1} - \mathbf{b} \right),$$

using the same argument as (5.14), we have

$$I_3 = -\frac{1}{2\beta} \left\| \lambda^* - \lambda^{k+1} \right\|^2 - \frac{1}{2\beta} \left\| \lambda^k - \lambda^{k+1} \right\|^2 + \frac{1}{2\beta} \left\| \lambda^* - \lambda^k \right\|^2 . \tag{5.37}$$

For $I_4$, using the same argument as (5.15), we have

$$I_4 \leq \frac{\beta}{2} \left\| \mathbf{A}_2 \mathbf{x}_2^k - \mathbf{A}_2 \mathbf{x}_2^* \right\|^2 - \frac{\beta}{2} \left\| \mathbf{A}_2 \mathbf{x}_2^{k+1} - \mathbf{A}_2 \mathbf{x}_2^* \right\|^2 + \frac{1}{2\beta} \left\| \lambda^k - \lambda^{k+1} \right\|^2 . \tag{5.38}$$

Plugging (5.36), (5.37), and (5.38) into (5.35), and using (by (A.1))

$$\left\langle \mathbf{A}_i \mathbf{x}_i^k - \mathbf{A}_i \mathbf{x}_i^{k+1}, \mathbf{A}_i \mathbf{x}_i^* - \mathbf{A}_i \mathbf{x}_i^{k+1} \right\rangle - \frac{1}{2} \left\| \mathbf{A}_i^T \mathbf{A}_i \right\|_2 \left\| \mathbf{x}_i^{k+1} - \mathbf{x}_i^k \right\|^2$$

$$= \frac{1}{2} \left\| \mathbf{A}_i \mathbf{x}_i^{k+1} - \mathbf{A}_i \mathbf{x}_i^* \right\|^2 + \frac{1}{2} \left\| \mathbf{A}_i \mathbf{x}_i^{k+1} - \mathbf{A}_i \mathbf{x}_i^k \right\|^2 - \frac{1}{2} \left\| \mathbf{A}_i \mathbf{x}_i^k - \mathbf{A}_i \mathbf{x}_i^* \right\|^2$$

$$- \frac{1}{2} \left\| \mathbf{A}_i^T \mathbf{A}_i \right\|_2 \left\| \mathbf{x}_i^{k+1} - \mathbf{x}_i^k \right\|^2$$

$$\leq \frac{1}{2} \left\| \mathbf{A}_i \left( \mathbf{x}_i^{k+1} - \mathbf{x}_i^* \right) \right\|^2 - \frac{1}{2} \left\| \mathbf{A}_i \left( \mathbf{x}_i^k - \mathbf{x}_i^* \right) \right\|^2$$

for $i = 1, 2$, we can obtain (5.28).                                                     $\square$

Now we give the convergence result for Algorithm 5.2.

**Theorem 5.2**  *Under the assumptions of Lemma 5.3, letting*

$$D_\lambda = \left\| \lambda^* - \lambda_0^0 \right\|,$$

$$D_i = \left\| \mathbf{x}_{0,i}^0 - \mathbf{x}_i^* \right\|_{\mathbf{G}_i}, \quad i = 1, 2,$$

$$D_f = f_1 \left( \mathbf{x}_{0,1}^0 \right) - f_1 \left( \mathbf{x}_1^* \right) - \left\langle \nabla f_1 \left( \mathbf{x}_1^* \right), \mathbf{x}_{0,1}^0 - \mathbf{x}_1^* \right\rangle, \text{ and}$$

$$\bar{\mathbf{x}}_i^S = \frac{1}{S} \sum_{s=1}^S \tilde{\mathbf{x}}_{s,i}, \quad i = 1, 2,$$

*we have*

$$\mathbb{E} \left( f_1 \left( \bar{\mathbf{x}}_1^S \right) + f_2 \left( \bar{\mathbf{x}}_2^S \right) - f_1 \left( \mathbf{x}_1^* \right) - f_2 \left( \mathbf{x}_2^* \right) + \left\langle \lambda^*, \mathbf{A}_1 \bar{\mathbf{x}}_1^S + \mathbf{A}_2 \bar{\mathbf{x}}_2^S - \mathbf{b} \right\rangle \right)$$

$$\leq \frac{(m+1)D_f}{2Sm} + \frac{D_\lambda^2}{\beta m S} + \frac{2 \left( D_1^2 + D_2^2 \right)}{mS}, \tag{5.39}$$

$$\mathbb{E} \left\| \mathbf{A}_1 \bar{\mathbf{x}}_1^S + \mathbf{A}_2 \bar{\mathbf{x}}_2^S - \mathbf{b} \right\| \leq \frac{D_\lambda}{m\beta S} + \frac{\sqrt{D_\lambda^2 + 2\beta \left( D_1^2 + D_2^2 \right) + \frac{\beta(m+1)}{2} D_f}}{m\beta S}. \tag{5.40}$$

***Proof***  Because $(\mathbf{x}_1^*, \mathbf{x}_2^*, \lambda^*)$ is a KKT point of Problem (5.20), we have

$$\mathbf{A}_1^T \lambda^* + \nabla f_1 \left( \mathbf{x}_1^* \right) = \mathbf{0} \quad \text{and} \quad - \mathbf{A}_2^T \lambda^* \in \partial f_2 \left( \mathbf{x}_2^* \right).$$

In Lemma 5.3, plugging $\nabla f_1 \left( \mathbf{x}_1^* \right) = -\mathbf{A}_1^T \lambda^*$ into the definition of $H_1 \left( \mathbf{x}_1^k \right) + H_1(\tilde{\mathbf{x}}_1)$, we obtain from (5.28) that

$$\mathbb{E}_k f_1 \left( \mathbf{x}_1^{k+1} \right) - f_1 \left( \mathbf{x}_1^* \right) + \mathbb{E}_k f_2 \left( \mathbf{x}_2^{k+1} \right) - f_2 \left( \mathbf{x}_2^* \right) + \mathbb{E}_k \left\langle \lambda^*, \mathbf{A}_1 \mathbf{x}_1^{k+1} + \mathbf{A}_2 \mathbf{x}_2^{k+1} - \mathbf{b} \right\rangle$$

$$\leq \frac{1}{4} \left( f_1 \left( \mathbf{x}_1^k \right) - f_1 \left( \mathbf{x}_1^* \right) + \left\langle \lambda^*, \mathbf{A}_1 \left( \mathbf{x}_1^k - \mathbf{x}_1^* \right) \right\rangle \right)$$

$$+ \frac{1}{4} \left( f_1(\tilde{\mathbf{x}}_1) - f_1 \left( \mathbf{x}_1^* \right) + \left\langle \lambda^*, \mathbf{A}_1 \left( \tilde{\mathbf{x}}_1 - \mathbf{x}_1^* \right) \right\rangle \right)$$

$$+ \left\| \mathbf{x}_1^k - \mathbf{x}_1^* \right\|_{\mathbf{G}_1}^2 - \mathbb{E}_k \left\| \mathbf{x}_1^{k+1} - \mathbf{x}_1^* \right\|_{\mathbf{G}_1}^2 + \left\| \mathbf{x}_2^k - \mathbf{x}_2^* \right\|_{\mathbf{G}_2}^2 - \mathbb{E}_k \left\| \mathbf{x}_2^{k+1} - \mathbf{x}_2^* \right\|_{\mathbf{G}_2}^2$$

$$+ \frac{1}{2\beta} \left\| \lambda^* - \lambda^k \right\|^2 - \frac{1}{2\beta} \mathbb{E}_k \left\| \lambda^* - \lambda^{k+1} \right\|^2. \tag{5.41}$$

By taking full expectation on (5.41), using $\mathbf{b} = \mathbf{A}_1 \mathbf{x}_1^* + \mathbf{A}_2 \mathbf{x}_2^*$, arranging terms, and adding back the subscript $s$, we have

$$
\frac{3}{4} \mathbb{E} \left( f_1 \left( \mathbf{x}_{s,1}^{k+1} \right) - f_1 \left( \mathbf{x}_1^* \right) + \left\langle \boldsymbol{\lambda}^*, \mathbf{A}_1 \left( \mathbf{x}_{s,1}^{k+1} - \mathbf{x}_1^* \right) \right\rangle \right)
$$

$$
+ \mathbb{E} \left( f_2 \left( \mathbf{x}_{s,2}^{k+1} \right) - f_2 \left( \mathbf{x}_2^* \right) + \left\langle \boldsymbol{\lambda}^*, \mathbf{A}_2 \left( \mathbf{x}_{s,2}^{k+1} - \mathbf{x}_2^* \right) \right\rangle \right)
$$

$$
\leq \frac{1}{4} \mathbb{E} \left( f_1 \left( \mathbf{x}_{s,1}^{k} \right) - f_1 \left( \mathbf{x}_1^* \right) + \left\langle \boldsymbol{\lambda}^*, \mathbf{A}_1 \left( \mathbf{x}_{s,1}^{k} - \mathbf{x}_1^* \right) \right\rangle \right)
$$

$$
- \frac{1}{4} \mathbb{E} \left( f_1 \left( \mathbf{x}_{s,1}^{k+1} \right) - f_1 \left( \mathbf{x}_1^* \right) + \left\langle \boldsymbol{\lambda}^*, \mathbf{A}_1 \left( \mathbf{x}_{s,1}^{k+1} - \mathbf{x}_1^* \right) \right\rangle \right)
$$

$$
+ \frac{1}{4} \mathbb{E} \left( f_1 \left( \tilde{\mathbf{x}}_{s,1} \right) - f_1 \left( \mathbf{x}_1^* \right) + \left\langle \boldsymbol{\lambda}^*, \mathbf{A}_1 \left( \tilde{\mathbf{x}}_{s,1} - \mathbf{x}_1^* \right) \right\rangle \right)
$$

$$
+ \frac{1}{2\beta} \mathbb{E} \left\| \boldsymbol{\lambda}^* - \boldsymbol{\lambda}_s^k \right\|^2 - \frac{1}{2\beta} \mathbb{E} \left\| \boldsymbol{\lambda}^* - \boldsymbol{\lambda}_s^{k+1} \right\|^2
$$

$$
+ \mathbb{E} \left\| \mathbf{x}_{s,1}^{k} - \mathbf{x}_1^* \right\|_{\mathbf{G}_1}^2 - \mathbb{E} \left\| \mathbf{x}_{s,1}^{k+1} - \mathbf{x}_1^* \right\|_{\mathbf{G}_1}^2
$$

$$
+ \mathbb{E} \left\| \mathbf{x}_{s,2}^{k} - \mathbf{x}_2^* \right\|_{\mathbf{G}_2}^2 - \mathbb{E} \left\| \mathbf{x}_{s,2}^{k+1} - \mathbf{x}_2^* \right\|_{\mathbf{G}_2}^2 .
$$

Summing the above inequality from $k = 0$ to $m - 1$, we have

$$
\frac{3}{4} \sum_{k=1}^{m} \mathbb{E} \left( f_1 \left( \mathbf{x}_{s,1}^{k} \right) - f_1 \left( \mathbf{x}_1^* \right) + \left\langle \boldsymbol{\lambda}^*, \mathbf{A}_1 \left( \mathbf{x}_{s,1}^{k} - \mathbf{x}_1^* \right) \right\rangle \right)
$$

$$
+ \sum_{k=1}^{m} \mathbb{E} \left( f_2 \left( \mathbf{x}_{s,2}^{k} \right) - f_2 \left( \mathbf{x}_2^* \right) + \left\langle \boldsymbol{\lambda}^*, \mathbf{A}_2 \left( \mathbf{x}_{s,2}^{k} - \mathbf{x}_2^* \right) \right\rangle \right)
$$

$$
\leq \frac{1}{4} \mathbb{E} \left( f_1 \left( \mathbf{x}_{s,1}^{0} \right) - f_1 \left( \mathbf{x}_1^* \right) + \left\langle \boldsymbol{\lambda}^*, \mathbf{A}_1 \left( \mathbf{x}_{s,1}^{0} - \mathbf{x}_1^* \right) \right\rangle \right)
$$

$$
- \frac{1}{4} \mathbb{E} \left( f_1 \left( \mathbf{x}_{s,1}^{m} \right) - f_1 \left( \mathbf{x}_1^* \right) + \left\langle \boldsymbol{\lambda}^*, \mathbf{A}_1 \left( \mathbf{x}_{s,1}^{m} - \mathbf{x}_1^* \right) \right\rangle \right)
$$

$$
+ \frac{m}{4} \mathbb{E} \left( f_1 (\tilde{\mathbf{x}}_{s,1}) - f_1 \left( \mathbf{x}_1^* \right) + \left\langle \boldsymbol{\lambda}^*, \mathbf{A}_1 \left( \tilde{\mathbf{x}}_{s,1} - \mathbf{x}_1^* \right) \right\rangle \right)
$$

$$
+ \frac{1}{2\beta} \mathbb{E} \left\| \boldsymbol{\lambda}^* - \boldsymbol{\lambda}_s^0 \right\|^2 - \frac{1}{2\beta} \mathbb{E} \left\| \boldsymbol{\lambda}^* - \boldsymbol{\lambda}_s^m \right\|^2
$$

$$
+ \mathbb{E} \left\| \mathbf{x}_{s,1}^{0} - \mathbf{x}_1^* \right\|_{\mathbf{G}_1}^2 - \mathbb{E} \left\| \mathbf{x}_{s,1}^{m} - \mathbf{x}_1^* \right\|_{\mathbf{G}_1}^2
$$

$$
+ \mathbb{E} \left\| \mathbf{x}_{s,2}^{0} - \mathbf{x}_2^* \right\|_{\mathbf{G}_2}^2 - \mathbb{E} \left\| \mathbf{x}_{s,2}^{m} - \mathbf{x}_2^* \right\|_{\mathbf{G}_2}^2 .
$$

Then using

$$\tilde{\mathbf{x}}_{s+1,i} = \frac{1}{m} \sum_{k=1}^{m} \mathbf{x}_{s,1}^k, \quad \mathbf{x}_{s+1,i}^0 = \mathbf{x}_{s,i}^m, \quad \text{for } i = 1, 2, \quad \text{and} \quad \lambda_{s+1}^0 = \lambda_s^m,$$

by the convexity of $f_1$ and $f_2$ and

$$f_2(\mathbf{x}_2) - f_2\left(\mathbf{x}_2^*\right) + \left\langle \lambda^*, \mathbf{A}_2\left(\mathbf{x}_2 - \mathbf{x}_2^*\right)\right\rangle \geq 0,$$

we obtain

$$\frac{m}{2}\mathbb{E}\left(f_1(\tilde{\mathbf{x}}_{s+1,1}) - f_1\left(\mathbf{x}_1^*\right) + \left\langle \lambda^*, \mathbf{A}_1\left(\tilde{\mathbf{x}}_{s+1,1} - \mathbf{x}_1^*\right)\right\rangle\right)$$

$$+ \frac{m}{2}\mathbb{E}\left(f_2(\tilde{\mathbf{x}}_{s+1,2}) - f_2\left(\mathbf{x}_2^*\right) + \left\langle \lambda^*, \mathbf{A}_2\left(\tilde{\mathbf{x}}_{s+1,2} - \mathbf{x}_2^*\right)\right\rangle\right)$$

$$\leq \frac{1}{4}\mathbb{E}\left(f_1\left(\mathbf{x}_{s,1}^0\right) - f_1\left(\mathbf{x}_1^*\right) + \left\langle \lambda^*, \mathbf{A}_1\left(\mathbf{x}_{s,1}^0 - \mathbf{x}_1^*\right)\right\rangle\right)$$

$$- \frac{1}{4}\mathbb{E}\left(f_1\left(\mathbf{x}_{s+1,1}^0\right) - f_1\left(\mathbf{x}_1^*\right) + \left\langle \lambda^*, \mathbf{A}_1\left(\mathbf{x}_{s+1,1}^0 - \mathbf{x}_1^*\right)\right\rangle\right)$$

$$+ \frac{m}{4}\mathbb{E}\left(f_1(\tilde{\mathbf{x}}_{s,1}) - f_1\left(\mathbf{x}_1^*\right) + \left\langle \lambda^*, \mathbf{A}_1\left(\tilde{\mathbf{x}}_{s,1} - \mathbf{x}_1^*\right)\right\rangle\right)$$

$$- \frac{m}{4}\mathbb{E}\left(f_1(\tilde{\mathbf{x}}_{s+1,1}) - f_1\left(\mathbf{x}_1^*\right) + \left\langle \lambda^*, \mathbf{A}_1\left(\tilde{\mathbf{x}}_{s+1,1} - \mathbf{x}_1^*\right)\right\rangle\right)$$

$$+ \frac{1}{2\beta}\mathbb{E}\left\|\lambda^* - \lambda_s^0\right\|^2 - \frac{1}{2\beta}\mathbb{E}\left\|\lambda^* - \lambda_{s+1}^0\right\|^2$$

$$+ \mathbb{E}\left\|\mathbf{x}_{s,1}^0 - \mathbf{x}_1^*\right\|_{\mathbf{G}_1}^2 - \mathbb{E}\left\|\mathbf{x}_{s+1,1}^0 - \mathbf{x}_1^*\right\|_{\mathbf{G}_1}^2$$

$$+ \mathbb{E}\left\|\mathbf{x}_{s,2}^0 - \mathbf{x}_2^*\right\|_{\mathbf{G}_2}^2 - \mathbb{E}\left\|\mathbf{x}_{s+1,2}^0 - \mathbf{x}_2^*\right\|_{\mathbf{G}_2}^2. \tag{5.42}$$

Summing (5.42) from $s = 0$ to $S - 1$, recalling $\bar{\mathbf{x}}_i^S = \frac{1}{S}\sum_{s=1}^S \tilde{\mathbf{x}}_{s,i}$ for $i = 1, 2$, by the convexity of $f_1$ and $f_2$ and

$$f_1(\mathbf{x}_1) - f_1\left(\mathbf{x}_1^*\right) + \left\langle \lambda^*, \mathbf{A}_1\left(\mathbf{x}_1 - \mathbf{x}_1^*\right)\right\rangle \geq 0,$$

we have

$$\frac{mS}{2}\mathbb{E}\left(f_1\left(\bar{\mathbf{x}}_1^S\right) - f_1\left(\mathbf{x}_1^*\right) + \left\langle \lambda^*, \mathbf{A}_1\left(\bar{\mathbf{x}}_1^S - \mathbf{x}_1^*\right)\right\rangle\right)$$

$$+ \frac{mS}{2}\mathbb{E}\left(f_2\left(\bar{\mathbf{x}}_2^S\right) - f_2\left(\mathbf{x}_2^*\right) + \left\langle \lambda^*, \mathbf{A}_2\left(\bar{\mathbf{x}}_2^S - \mathbf{x}_2^*\right)\right\rangle\right)$$

$$\leq \frac{1}{4}\mathbb{E}\left(f_1\left(\mathbf{x}_{0,1}^0\right) - f_1\left(\mathbf{x}_1^*\right) + \left\langle \lambda^*, \mathbf{A}_1\left(\mathbf{x}_{0,1}^0 - \mathbf{x}_1^*\right)\right\rangle\right)$$

$$+ \frac{m}{4} \mathbb{E} \left( f_1(\tilde{\mathbf{x}}_{0,1}) - f_1\left(\mathbf{x}_1^*\right) + \left\langle \boldsymbol{\lambda}^*, \mathbf{A}_1\left(\tilde{\mathbf{x}}_{0,1} - \mathbf{x}_1^*\right) \right\rangle \right)$$

$$+ \frac{1}{2\beta} \left\| \boldsymbol{\lambda}^* - \boldsymbol{\lambda}_0^0 \right\|^2 - \frac{1}{2\beta} \mathbb{E} \left\| \boldsymbol{\lambda}^* - \boldsymbol{\lambda}_S^0 \right\|^2$$

$$+ \left\| \mathbf{x}_{0,1}^0 - \mathbf{x}_1^* \right\|_{\mathbf{G}_1}^2 + \left\| \mathbf{x}_{0,2}^0 - \mathbf{x}_2^* \right\|_{\mathbf{G}_2}^2. \tag{5.43}$$

For (5.43), using

$$\frac{1}{2\beta} \mathbb{E} \left\| \boldsymbol{\lambda}^* - \boldsymbol{\lambda}_S^0 \right\|^2 \geq 0, \quad \mathbf{b} = \mathbf{A}_1 \mathbf{x}_1^* + \mathbf{A}_2 \mathbf{x}_2^*, \quad \tilde{\mathbf{x}}_{0,1} = \mathbf{x}_{0,1}^0,$$

and $\mathbf{A}_1^T \boldsymbol{\lambda}^* = -\nabla f_1(\mathbf{x}_1^*)$ in the KKT conditions, we can obtain (5.39).
  On the other hand, using

$$f_i\left(\bar{\mathbf{x}}_i^S\right) - f_i\left(\mathbf{x}_i^*\right) + \left\langle \boldsymbol{\lambda}^*, \mathbf{A}_i\left(\bar{\mathbf{x}}_i^S - \mathbf{x}_i^*\right) \right\rangle \geq 0, \text{ for } i = 1, 2,$$

from (5.43) we have

$$\mathbb{E} \left\| \boldsymbol{\lambda}^* - \boldsymbol{\lambda}_S^0 \right\|^2 \leq D_\lambda^2 + 2\beta \left( D_1^2 + D_2^2 \right) + \frac{\beta(m+1)}{2} D_f.$$

Then it follows by the Jensen's inequality (Proposition A.4) that

$$\mathbb{E} \left\| \boldsymbol{\lambda}^* - \boldsymbol{\lambda}_S^0 \right\| \leq \sqrt{D_\lambda^2 + 2\beta \left( D_1^2 + D_2^2 \right) + \frac{\beta(m+1)}{2} D_f}. \tag{5.44}$$

Therefore, we obtain (5.40) via

$$\mathbb{E} \left\| \mathbf{A}_1 \bar{\mathbf{x}}_1^S + \mathbf{A}_2 \bar{\mathbf{x}}_2^S - \mathbf{b} \right\|$$

$$= \mathbb{E} \left\| \frac{1}{mS} \sum_{s=1}^{S} \sum_{k=1}^{m} \left( \mathbf{A}_1 \mathbf{x}_{s-1,1}^k + \mathbf{A}_2 \mathbf{x}_{s-1,2}^k - \mathbf{b} \right) \right\|$$

$$= \frac{1}{m\beta S} \mathbb{E} \left\| \boldsymbol{\lambda}_S^0 - \boldsymbol{\lambda}_0^0 \right\|$$

$$\leq \frac{1}{m\beta S} \left( \mathbb{E} \left\| \boldsymbol{\lambda}_S^0 - \boldsymbol{\lambda}^* \right\| + \left\| \boldsymbol{\lambda}_0^0 - \boldsymbol{\lambda}^* \right\| \right)$$

$$\overset{a}{\leq} \frac{1}{m\beta S} \left( \sqrt{D_\lambda^2 + 2\beta \left( D_1^2 + D_2^2 \right) + \frac{\beta(m+1)}{2} D_f} + D_\lambda \right),$$

where $\overset{a}{\leq}$ uses (5.44).                                         $\square$

## 5.3   Momentum Acceleration

When applying the VR technique, the algorithms are transformed to act like a deterministic algorithm. So it is possible to fuse the momentum technique. In this section, we fuse the VR with the momentum techniques for the ADMM algorithm. As an example, we provide an ergodic (actually only averaging the last several iterates) $O(1/K)$ stochastic ADMM in the convex setting. Our method can also be generalized to design algorithms that ensure the bounds related to the objective functions decrease in the same way as shown in Theorem 3.10.

We consider the convex finite-sum problem with linear constraints in the general setting:

$$\min_{\mathbf{x}_1, \mathbf{x}_2} \left( h_1(\mathbf{x}_1) + f_1(\mathbf{x}_1) + h_2(\mathbf{x}_2) + \frac{1}{n} \sum_{i=1}^{n} F_{2,i}(\mathbf{x}_2) \right),$$

$$s.t. \ \mathbf{A}_1 \mathbf{x}_1 + \mathbf{A}_2 \mathbf{x}_2 = \mathbf{b}, \tag{5.45}$$

where $f_1(\mathbf{x}_1)$ and $F_{2,i}(\mathbf{x}_2)$ with $i \in [n]$ are convex and $L_1$-smooth and $L_2$-smooth, respectively, and $h_1(\mathbf{x}_1)$ and $h_2(\mathbf{x}_2)$ are also convex and their proximal mappings can be solved efficiently. We define

$$f_2(\mathbf{x}_2) = \frac{1}{n} \sum_{i=1}^{n} F_{2,i}(\mathbf{x}_2),$$

$$J_1(\mathbf{x}_1) = h_1(\mathbf{x}_1) + f_1(\mathbf{x}_1), \quad J_2(\mathbf{x}_2) = h_2(\mathbf{x}_2) + f_2(\mathbf{x}_2),$$

$$\mathbf{x} = \left( \mathbf{x}_1^T, \mathbf{x}_2^T \right)^T, \quad \mathbf{A} = [\mathbf{A}_1, \mathbf{A}_2], \quad \text{and} \quad J(\mathbf{x}) = J_1(\mathbf{x}_1) + J_2(\mathbf{x}_2).$$

To begin with, we list the notations and variables in Table 5.1. The algorithm is designed in [3], which has double loops: In the inner loop, we update primal variables $\mathbf{x}_{s,1}^k$ and $\mathbf{x}_{s,2}^k$ through extrapolation terms $\mathbf{y}_{s,1}^k$ and $\mathbf{y}_{s,2}^k$ and the dual variable $\lambda_s^k$; in the outer loop, we maintain snapshot vectors $\tilde{\mathbf{x}}_{s+1,1}$, $\tilde{\mathbf{x}}_{s+1,2}$, and $\tilde{\mathbf{b}}_{s+1}$, and

**Table 5.1** Notations and variables

| Notation | Meaning | Variable | Meaning |
|---|---|---|---|
| $\langle \mathbf{x}, \mathbf{y} \rangle_\mathbf{G}, \|\mathbf{x}\|_\mathbf{G}$ | $\mathbf{x}^T \mathbf{G} \mathbf{y}, \sqrt{\mathbf{x}^T \mathbf{G} \mathbf{x}}$ | $\mathbf{y}_{s,1}^k, \mathbf{y}_{s,2}^k$ | extrapolation variables |
| $J_i(\mathbf{x}_i)$ | $h_i(\mathbf{x}_i) + f_i(\mathbf{x}_i)$ | $\mathbf{x}_{s,1}^k, \mathbf{x}_{s,2}^k$ | primal variables |
| $\mathbf{x}$ | $\left( \mathbf{x}_1^T, \mathbf{x}_2^T \right)^T$ | $\lambda_s^k, \lambda_s^k, \hat{\lambda}^k$ | dual and temporary variables |
| $\mathbf{y}$ | $\left( \mathbf{y}_1^T, \mathbf{y}_2^T \right)^T$ | $\tilde{\mathbf{x}}_{s,1}, \tilde{\mathbf{x}}_{s,2}, \tilde{\mathbf{b}}_s$ | snapshot vectors |
| $J(\mathbf{x})$ | $J_1(\mathbf{x}_1) + J_2(\mathbf{x}_2)$ | | used for VR |
| $\mathbf{A}$ | $[\mathbf{A}_1, \mathbf{A}_2]$ | $(\mathbf{x}_1^*, \mathbf{x}_2^*, \lambda^*)$ | KKT point of (5.45) |
| $\mathcal{I}_{k,s}$ | mini-batch indices | $b$ | batch size |

---

**Algorithm 5.3** Inner loop of Acc-SADMM

**for** $k = 0$ to $m - 1$ **do**

    Update dual variable: $\lambda_s^k = \tilde{\lambda}_s^k + \frac{\beta\theta_2}{\theta_{1,s}}\left(\mathbf{A}_1\mathbf{x}_{s,1}^k + \mathbf{A}_2\mathbf{x}_{s,2}^k - \tilde{\mathbf{b}}_s\right)$.

    Update $\mathbf{x}_{s,1}^{k+1}$ by (5.46).

    Update $\mathbf{x}_{s,2}^{k+1}$ by (5.47).

    Update dual variable: $\tilde{\lambda}_s^{k+1} = \lambda_s^k + \beta\left(\mathbf{A}_1\mathbf{x}_{s,1}^{k+1} + \mathbf{A}_2\mathbf{x}_{s,2}^{k+1} - \mathbf{b}\right)$.

    Update $\mathbf{y}_s^{k+1}$ by $\mathbf{y}_s^{k+1} = \mathbf{x}_s^{k+1} + (1 - \theta_{1,s} - \theta_2)\left(\mathbf{x}_s^{k+1} - \mathbf{x}_s^k\right)$.

**end for** $k$

---

then assign the initial value to the extrapolation terms $\mathbf{y}_{s+1,1}^0$ and $\mathbf{y}_{s+1,2}^0$. The whole algorithm is shown in Algorithm 5.4. In the process of solving primal variables, we linearize both the smooth term $f_i(\mathbf{x}_i)$ and the augmented term $\frac{\beta}{2}\|\mathbf{A}_1\mathbf{x}_1 + \mathbf{A}_2\mathbf{x}_2 - \mathbf{b} + \frac{1}{\beta}\lambda\|^2$. The update rules of $\mathbf{x}_1$ and $\mathbf{x}_2$ can be written as

$$
\begin{aligned}
\mathbf{x}_{s,1}^{k+1} = \operatorname*{argmin}_{\mathbf{x}_1} &\left[ h_1(\mathbf{x}_1) + \left\langle \nabla f_1\left(\mathbf{y}_{s,1}^k\right), \mathbf{x}_1 \right\rangle \right. \\
&+ \left\langle \frac{\beta}{\theta_{1,s}}\left(\mathbf{A}_1\mathbf{y}_{s,1}^k + \mathbf{A}_2\mathbf{y}_{s,2}^k - \mathbf{b}\right) + \lambda_s^k, \mathbf{A}_1\mathbf{x}_1 \right\rangle \\
&+ \left. \left( \frac{L_1}{2} + \frac{\beta}{2\theta_{1,s}}\|\mathbf{A}_1\|_2^2 \right)\left\|\mathbf{x}_1 - \mathbf{y}_{s,1}^k\right\|^2 \right]
\end{aligned}
\tag{5.46}
$$

and

$$
\begin{aligned}
\mathbf{x}_{s,2}^{k+1} = \operatorname*{argmin}_{\mathbf{x}_2} &\left\{ h_2(\mathbf{x}_2) + \left\langle \tilde{\nabla} f_2\left(\mathbf{y}_{s,2}^k\right), \mathbf{x}_2 \right\rangle \right. \\
&+ \left\langle \frac{\beta}{\theta_{1,s}}\left(\mathbf{A}_1\mathbf{x}_{s,1}^{k+1} + \mathbf{A}_2\mathbf{y}_{s,2}^k - \mathbf{b}\right) + \lambda_s^k, \mathbf{A}_2\mathbf{x}_2 \right\rangle \\
&+ \left. \left[ \frac{1}{2}\left(1 + \frac{1}{b\theta_2}\right)L_2 + \frac{\beta}{2\theta_{1,s}}\|\mathbf{A}_2\|_2^2 \right]\left\|\mathbf{x}_2 - \mathbf{y}_{s,2}^k\right\|^2 \right\},
\end{aligned}
\tag{5.47}
$$

where

$$
\tilde{\nabla} f_2\left(\mathbf{y}_{s,2}^k\right) = \frac{1}{b}\sum_{i_{k,s}\in\mathcal{I}_{k,s}}\left(\nabla F_{2,i_{k,s}}\left(\mathbf{y}_{s,2}^k\right) - \nabla F_{2,i_{k,s}}(\tilde{\mathbf{x}}_{s,2}) + \nabla f_2(\tilde{\mathbf{x}}_{s,2})\right),
$$

in which $\mathcal{I}_{k,s}$ is a mini-batch of indices randomly drawn from $[n]$ with a size of $b$.

We first bound the variance of the stochastic gradient using the technique proposed in Katyusha [1]:

**Algorithm 5.4** Accelerated Stochastic Alternating Direction Method of Multiplier (Acc-SADMM)

---

**Input:** epoch length $m > 2$, $\beta$, $\tau = 2$, $c = 2$, $\mathbf{x}_0^0 = \mathbf{0}$, $\tilde{\mathbf{b}}_0 = \mathbf{0}$, $\tilde{\boldsymbol{\lambda}}_0^0 = \mathbf{0}$, $\tilde{\mathbf{x}}_0 = \mathbf{x}_0^0$, $\mathbf{y}_0^0 = \mathbf{x}_0^0$, $\theta_{1,s} = \frac{1}{c+\tau s}$, and $\theta_2 = \frac{m-\tau}{\tau(m-1)}$.

**for** $s = 0$ to $S - 1$ **do**

  Do inner loop, as stated in Algorithm 5.3.

  Set primal variables: $\mathbf{x}_{s+1}^0 = \mathbf{x}_s^m$.

  Update $\tilde{\mathbf{x}}_{s+1}$ by $\tilde{\mathbf{x}}_{s+1} = \frac{1}{m}\left(\left[1 - \frac{(\tau-1)\theta_{1,s+1}}{\theta_2}\right]\mathbf{x}_s^m + \left[1 + \frac{(\tau-1)\theta_{1,s+1}}{(m-1)\theta_2}\right]\sum_{k=1}^{m-1}\mathbf{x}_s^k\right)$.

  Update dual variable:   $\tilde{\boldsymbol{\lambda}}_{s+1}^0 = \boldsymbol{\lambda}_s^{m-1} + \beta(1-\tau)\left(\mathbf{A}_1\mathbf{x}_{s,1}^m + \mathbf{A}_2\mathbf{x}_{s,2}^m - \mathbf{b}\right)$.

  Update dual snapshot variable:   $\tilde{\mathbf{b}}_{s+1} = \mathbf{A}_1\tilde{\mathbf{x}}_{s+1,1} + \mathbf{A}_2\tilde{\mathbf{x}}_{s+1,2}$.

  Update extrapolation terms $\mathbf{y}_{s+1}^0$ through

$$\mathbf{y}_{s+1}^0 = (1-\theta_2)\mathbf{x}_s^m + \theta_2\tilde{\mathbf{x}}_{s+1} + \frac{\theta_{1,s+1}}{\theta_{1,s}}\left[(1-\theta_{1,s})\mathbf{x}_s^m - (1-\theta_{1,s}-\theta_2)\mathbf{x}_s^{m-1} - \theta_2\tilde{\mathbf{x}}_s\right].$$

**end for** $s$

**Output:**

$$\hat{\mathbf{x}}_S = \frac{1}{(m-1)(\theta_{1,S}+\theta_2)+1}\mathbf{x}_S^m + \frac{\theta_{1,S}+\theta_2}{(m-1)(\theta_{1,S}+\theta_2)+1}\sum_{k=1}^{m-1}\mathbf{x}_S^k.$$

---

**Lemma 5.4** *For $f(\mathbf{x}) = \frac{1}{n}\sum_{i=1}^n F_i(\mathbf{x})$, with each $F_i$ being convex and $L$-smooth, $i \in [n]$. For any $\mathbf{u}$ and $\tilde{\mathbf{x}}$, defining*

$$\tilde{\nabla}f(\mathbf{u}) = \nabla F_k(\mathbf{u}) - \nabla F_k(\tilde{\mathbf{x}}) + \frac{1}{n}\sum_{i=1}^n \nabla F_i(\tilde{\mathbf{x}}),$$

*we have*

$$\mathbb{E}\left\|\tilde{\nabla}f(\mathbf{u}) - \nabla f(\mathbf{u})\right\|^2 \le 2L\left(f(\tilde{\mathbf{x}}) - f(\mathbf{u}) + \langle\nabla f(\mathbf{u}), \mathbf{u} - \tilde{\mathbf{x}}\rangle\right), \quad (5.48)$$

*where the expectation is taken on the random number $k$ under the condition that $\mathbf{u}$ and $\tilde{\mathbf{x}}$ are known.*

**Proof** We have

$$\mathbb{E}\left\|\tilde{\nabla}f(\mathbf{u}) - \nabla f(\mathbf{u})\right\|^2$$

$$= \mathbb{E}\left(\left\|\nabla F_k(\mathbf{u}) - \nabla F_k(\tilde{\mathbf{x}}) - \left(\nabla f(\mathbf{u}) - \nabla f(\tilde{\mathbf{x}})\right)\right\|^2\right)$$

$$\overset{a}{\le} \mathbb{E}\left\|\nabla F_k(\mathbf{u}) - \nabla F_k(\tilde{\mathbf{x}})\right\|^2,$$

where in $\overset{a}{\leq}$ we use

$$\mathbb{E}\left(\nabla F_k(\mathbf{u}) - \nabla F_k(\tilde{\mathbf{x}})\right) = \nabla f(\mathbf{u}) - \nabla f(\tilde{\mathbf{x}})$$

and $\mathbb{E}\|\boldsymbol{\xi} - \mathbb{E}\boldsymbol{\xi}\|^2 \leq \mathbb{E}\|\boldsymbol{\xi}\|^2$ for random vector $\boldsymbol{\xi}$ (Proposition A.3). Then by directly applying (A.5) to $F_k$ we obtain (5.48). □

Now, we give the convergence result. The analysis is much more complex than that in SVRG-ADMM (Algorithm 5.2). The main property of Acc-SADMM (Algorithm 5.4) in the inner loop is shown below.

**Lemma 5.5** *For Algorithm 5.3, in any epoch with fixed s (for simplicity we drop the subscript s throughout the proof unless necessary), we have*

$$\mathbb{E}_{i_k}\tilde{L}\left(\mathbf{x}_1^{k+1}, \mathbf{x}_2^{k+1}, \boldsymbol{\lambda}^*\right) - \theta_2\tilde{L}(\tilde{\mathbf{x}}_1, \tilde{\mathbf{x}}_2, \boldsymbol{\lambda}^*) - (1-\theta_1-\theta_2)\tilde{L}\left(\mathbf{x}_1^k, \mathbf{x}_2^k, \boldsymbol{\lambda}^*\right)$$

$$\leq \frac{\theta_1}{2\beta}\left(\left\|\hat{\boldsymbol{\lambda}}^k - \boldsymbol{\lambda}^*\right\|^2 - \mathbb{E}_{i_k}\left\|\hat{\boldsymbol{\lambda}}^{k+1} - \boldsymbol{\lambda}^*\right\|^2\right)$$

$$+ \frac{1}{2}\left\|\mathbf{y}_1^k - (1-\theta_1-\theta_2)\mathbf{x}_1^k - \theta_2\tilde{\mathbf{x}}_1 - \theta_1\mathbf{x}_1^*\right\|_{\mathbf{G}_1}^2$$

$$- \frac{1}{2}\mathbb{E}_{i_k}\left\|\mathbf{x}_1^{k+1} - (1-\theta_1-\theta_2)\mathbf{x}_1^k - \theta_2\tilde{\mathbf{x}}_1 - \theta_1\mathbf{x}_1^*\right\|_{\mathbf{G}_1}^2$$

$$+ \frac{1}{2}\left\|\mathbf{y}_2^k - (1-\theta_1-\theta_2)\mathbf{x}_2^k - \theta_2\tilde{\mathbf{x}}_2 - \theta_1\mathbf{x}_2^*\right\|_{\mathbf{G}_2}^2$$

$$- \frac{1}{2}\mathbb{E}_{i_k}\left\|\mathbf{x}_2^{k+1} - (1-\theta_1-\theta_2)\mathbf{x}_2^k - \theta_2\tilde{\mathbf{x}}_2 - \theta_1\mathbf{x}_2^*\right\|_{\mathbf{G}_2}^2, \tag{5.49}$$

*where $\mathbb{E}_{i_k}$ denotes that the expectation is taken over the random samples in the mini-batch $\mathcal{I}_{k,s}$,*

$$\tilde{L}(\mathbf{x}_1, \mathbf{x}_2, \boldsymbol{\lambda}) = L(\mathbf{x}_1, \mathbf{x}_2, \boldsymbol{\lambda}) - L\left(\mathbf{x}_1^*, \mathbf{x}_2^*, \boldsymbol{\lambda}^*\right)$$

*is the shifted Lagrangian function in which*

$$L(\mathbf{x}_1, \mathbf{x}_2, \boldsymbol{\lambda}) = J_1(\mathbf{x}_1) + J_2(\mathbf{x}_2) + \langle\boldsymbol{\lambda}, \mathbf{A}_1\mathbf{x}_1 + \mathbf{A}_2\mathbf{x}_2 - \mathbf{b}\rangle$$

*is the Lagrangian function,*

$$\hat{\boldsymbol{\lambda}}^k = \tilde{\boldsymbol{\lambda}}^k + \frac{\beta(1-\theta_1)}{\theta_1}(\mathbf{A}\mathbf{x}^k - \mathbf{b}),$$

$$\mathbf{G}_1 = \left(L_1 + \frac{\beta}{\theta_1}\|\mathbf{A}_1\|_2^2\right)\mathbf{I} - \frac{\beta}{\theta_1}\mathbf{A}_1^T\mathbf{A}_1, \text{ and}$$

$$\mathbf{G}_2 = \left[\left(1 + \frac{1}{b\theta_2}\right)L_2 + \frac{\beta}{\theta_1}\|\mathbf{A}_2\|_2^2\right]\mathbf{I}.$$

*Other notations can be found in Table 5.1 and Algorithms 5.3 and 5.4.*

**Proof** Step 1: We first analyze $\mathbf{x}_1$. By the optimality of $\mathbf{x}_1^{k+1}$ in (5.46) and the convexity of $J_1(\cdot)$, we can obtain

$$J_1\left(\mathbf{x}_1^{k+1}\right) \leq (1 - \theta_1 - \theta_2) J_1\left(\mathbf{x}_1^k\right) + \theta_2 J_1(\tilde{\mathbf{x}}_1) + \theta_1 J_1\left(\mathbf{x}_1^*\right)$$

$$-\left\langle \mathbf{A}_1^T \bar{\lambda}\left(\mathbf{x}_1^{k+1}, \mathbf{y}_2^k\right), \mathbf{x}_1^{k+1} - (1 - \theta_1 - \theta_2)\mathbf{x}_1^k - \theta_2 \tilde{\mathbf{x}}_1 - \theta_1 \mathbf{x}_1^* \right\rangle$$

$$+ \frac{L_1}{2} \left\| \mathbf{x}_1^{k+1} - \mathbf{y}_1^k \right\|^2$$

$$-\left\langle \mathbf{x}_1^{k+1} - \mathbf{y}_1^k, \mathbf{x}_1^{k+1} - (1 - \theta_1 - \theta_2)\mathbf{x}_1^k - \theta_2 \tilde{\mathbf{x}}_1 - \theta_1 \mathbf{x}^* \right\rangle_{\mathbf{G}_1}. \qquad (5.50)$$

We prove (5.50) below.
For brevity, we define

$$\bar{\lambda}(\mathbf{x}_1, \mathbf{x}_2) = \lambda^k + \frac{\beta}{\theta_1} (\mathbf{A}_1 \mathbf{x}_1 + \mathbf{A}_2 \mathbf{x}_2 - \mathbf{b}).$$

From the optimality solution of $\mathbf{x}_1^{k+1}$ in (5.46), we have

$$-\left[ \left( L_1 + \frac{\beta}{\theta_1} \|\mathbf{A}_1\|_2^2 \right) \left(\mathbf{x}_1^{k+1} - \mathbf{y}_1^k\right) + \nabla f_1\left(\mathbf{y}_1^k\right) + \mathbf{A}_1^T \bar{\lambda}\left(\mathbf{y}_1^k, \mathbf{y}_2^k\right) \right]$$

$$\in \partial h_1\left(\mathbf{x}_1^{k+1}\right). \qquad (5.51)$$

Since $f_1$ is $L_1$-smooth, we have

$$f_1\left(\mathbf{x}_1^{k+1}\right) \leq f_1\left(\mathbf{y}_1^k\right) + \left\langle \nabla f_1\left(\mathbf{y}_1^k\right), \mathbf{x}_1^{k+1} - \mathbf{y}_1^k \right\rangle + \frac{L_1}{2} \left\| \mathbf{x}_1^{k+1} - \mathbf{y}_1^k \right\|^2$$

$$\overset{a}{\leq} f_1(\mathbf{u}_1) + \left\langle \nabla f_1\left(\mathbf{y}_1^k\right), \mathbf{x}_1^{k+1} - \mathbf{u}_1 \right\rangle + \frac{L_1}{2} \left\| \mathbf{x}_1^{k+1} - \mathbf{y}_1^k \right\|^2$$

$$\overset{b}{\leq} f_1(\mathbf{u}_1) - \left\langle \tilde{\nabla} h_1\left(\mathbf{x}_1^{k+1}\right), \mathbf{x}_1^{k+1} - \mathbf{u}_1 \right\rangle - \left\langle \mathbf{A}_1^T \bar{\lambda}\left(\mathbf{y}_1^k, \mathbf{y}_2^k\right), \mathbf{x}_1^{k+1} - \mathbf{u}_1 \right\rangle$$

$$- \left( L_1 + \frac{\beta}{\theta_1} \|\mathbf{A}_1\|_2^2 \right) \left\langle \mathbf{x}_1^{k+1} - \mathbf{y}_1^k, \mathbf{x}_1^{k+1} - \mathbf{u}_1 \right\rangle + \frac{L_1}{2} \left\| \mathbf{x}_1^{k+1} - \mathbf{y}_1^k \right\|^2,$$

where $\mathbf{u}_1$ is an arbitrary vector, $\tilde{\nabla} h_1(\mathbf{x}_1^{k+1}) \in \partial h_1(\mathbf{x}_1^{k+1})$, in $\overset{a}{\leq}$ we use the fact that $f_1(\cdot)$ is convex and so

$$f_1\left(\mathbf{y}_1^k\right) \leq f_1(\mathbf{u}_1) + \left\langle \nabla f_1\left(\mathbf{y}_1^k\right), \mathbf{y}_1^k - \mathbf{u}_1 \right\rangle,$$

and $\overset{b}{\leq}$ uses (5.51). On the other hand, the convexity of $h_1(\cdot)$ gives

$$h_1\left(\mathbf{x}_1^{k+1}\right) \leq h_1(\mathbf{u}_1) + \left\langle \tilde{\nabla} h_1\left(\mathbf{x}_1^{k+1}\right), \mathbf{x}_1^{k+1} - \mathbf{u}_1\right\rangle.$$

So we have

$$J_1\left(\mathbf{x}_1^{k+1}\right) \leq J_1(\mathbf{u}_1) - \left\langle \mathbf{A}_1^T \bar{\lambda}\left(\mathbf{y}_1^k, \mathbf{y}_2^k\right), \mathbf{x}_1^{k+1} - \mathbf{u}_1\right\rangle + \frac{L_1}{2}\left\|\mathbf{x}_1^{k+1} - \mathbf{y}_1^k\right\|^2$$
$$- \left(L_1 + \frac{\beta}{\theta_1}\|\mathbf{A}_1\|_2^2\right)\left\langle \mathbf{x}_1^{k+1} - \mathbf{y}_1^k, \mathbf{x}_1^{k+1} - \mathbf{u}_1\right\rangle.$$

Setting $\mathbf{u}_1$ be $\mathbf{x}_1^k$, $\tilde{\mathbf{x}}_1$, and $\mathbf{x}_1^*$, respectively, then multiplying the three inequalities by $1 - \theta_1 - \theta_2$, $\theta_2$, and $\theta_1$, respectively, and adding them together, we have

$$J_1\left(\mathbf{x}_1^{k+1}\right)$$

$$\leq (1 - \theta_1 - \theta_2) J_1\left(\mathbf{x}_1^k\right) + \theta_2 J_1(\tilde{\mathbf{x}}_1) + \theta_1 J_1\left(\mathbf{x}_1^*\right) + \frac{L_1}{2}\left\|\mathbf{x}_1^{k+1} - \mathbf{y}_1^k\right\|^2$$
$$- \left\langle \mathbf{A}_1^T \bar{\lambda}\left(\mathbf{y}_1^k, \mathbf{y}_2^k\right), \mathbf{x}_1^{k+1} - (1 - \theta_1 - \theta_2)\mathbf{x}_1^k - \theta_2\tilde{\mathbf{x}}_1 - \theta_1\mathbf{x}_1^*\right\rangle$$
$$- \left(L_1 + \frac{\beta}{\theta_1}\|\mathbf{A}_1\|_2^2\right)\left\langle \mathbf{x}_1^{k+1} - \mathbf{y}_1^k, \mathbf{x}_1^{k+1} - (1 - \theta_1 - \theta_2)\mathbf{x}_1^k - \theta_2\tilde{\mathbf{x}}_1 - \theta_1\mathbf{x}_1^*\right\rangle$$
$$\overset{a}{=} (1 - \theta_1 - \theta_2) J_1\left(\mathbf{x}_1^k\right) + \theta_2 J_1(\tilde{\mathbf{x}}_1) + \theta_1 J_1\left(\mathbf{x}_1^*\right) + \frac{L_1}{2}\left\|\mathbf{x}_1^{k+1} - \mathbf{y}_1^k\right\|^2$$
$$- \left\langle \mathbf{A}_1^T \bar{\lambda}\left(\mathbf{x}_1^{k+1}, \mathbf{y}_2^k\right), \mathbf{x}_1^{k+1} - (1 - \theta_1 - \theta_2)\mathbf{x}_1^k - \theta_2\tilde{\mathbf{x}}_1 - \theta_1\mathbf{x}_1^*\right\rangle$$
$$- \left\langle \mathbf{x}_1^{k+1} - \mathbf{y}_1^k, \mathbf{x}_1^{k+1} - (1 - \theta_1 - \theta_2)\mathbf{x}_1^k - \theta_2\tilde{\mathbf{x}}_1 - \theta_1\mathbf{x}_1^*\right\rangle_{\mathbf{G}_1},$$

where in $\overset{a}{=}$ we replace $\mathbf{A}_1^T \bar{\lambda}(\mathbf{y}_1^k, \mathbf{y}_2^k)$ with $\mathbf{A}_1^T \bar{\lambda}(\mathbf{x}_1^{k+1}, \mathbf{y}_2^k) - \frac{\beta}{\theta_1}\mathbf{A}_1^T \mathbf{A}_1(\mathbf{x}_1^{k+1} - \mathbf{y}_1^k)$.

Step 2: We next analyze $\mathbf{x}_2$. By the optimality of $\mathbf{x}_2^{k+1}$ in (5.47) and the convexity of $J_2(\cdot)$, we can obtain

$$\mathbb{E}_{i_k} J_2\left(\mathbf{x}_2^{k+1}\right)$$

$$\leq -\mathbb{E}_{i_k}\left\langle \mathbf{A}_2^T \bar{\lambda}\left(\mathbf{x}_1^{k+1}, \mathbf{y}_2^k\right) + \left(\alpha L_2 + \frac{\beta}{\theta_1}\|\mathbf{A}_2\|_2^2\right)\left(\mathbf{x}_2^{k+1} - \mathbf{y}_2^k\right), \mathbf{x}_2^{k+1} - \theta_2\tilde{\mathbf{x}}_2\right\rangle$$
$$- \mathbb{E}_{i_k}\left\langle \mathbf{A}_2^T \bar{\lambda}\left(\mathbf{x}_1^{k+1}, \mathbf{y}_2^k\right) + \left(\alpha L_2 + \frac{\beta}{\theta_1}\|\mathbf{A}_2\|_2^2\right)\left(\mathbf{x}_2^{k+1} - \mathbf{y}_2^k\right),\right.$$
$$\left. - (1 - \theta_1 - \theta_2)\mathbf{x}_2^k - \theta_1\mathbf{x}_2^*\right\rangle$$

$$+ (1 - \theta_1 - \theta_2) J_2 \left(\mathbf{x}_2^k\right) + \theta_1 J_2 \left(\mathbf{x}_2^*\right) + \theta_2 J_2(\tilde{\mathbf{x}}_2)$$

$$+ \mathbb{E}_{i_k} \left[\frac{1}{2} \left(1 + \frac{1}{b\theta_2}\right) L_2 \left\|\mathbf{x}_2^{k+1} - \mathbf{y}_2^k\right\|^2\right], \tag{5.52}$$

where $\alpha = 1 + \frac{1}{b\theta_2}$. We prove (5.52) below.

From the optimality of $\mathbf{x}_2^{k+1}$ in (5.47), we have

$$- \left[\left(\alpha L_2 + \frac{\beta}{\theta_1} \|\mathbf{A}_2\|_2^2\right) \left(\mathbf{x}_2^{k+1} - \mathbf{y}_2^k\right) + \tilde{\nabla} f_2 \left(\mathbf{y}_2^k\right) + \mathbf{A}_2^T \bar{\lambda} \left(\mathbf{x}_1^{k+1}, \mathbf{y}_2^k\right)\right]$$

$$\in \partial h_2 \left(\mathbf{x}_2^{k+1}\right). \tag{5.53}$$

Since $f_2$ is $L_2$-smooth, we have

$$f_2 \left(\mathbf{x}_2^{k+1}\right) \le f_2 \left(\mathbf{y}_2^k\right) + \left\langle \nabla f_2 \left(\mathbf{y}_2^k\right), \mathbf{x}_2^{k+1} - \mathbf{y}_2^k\right\rangle + \frac{L_2}{2} \left\|\mathbf{x}_2^{k+1} - \mathbf{y}_2^k\right\|^2. \tag{5.54}$$

We first consider $\langle \nabla f_2(\mathbf{y}_2^k), \mathbf{x}_2^{k+1} - \mathbf{y}_2^k\rangle$ and have

$$\left\langle \nabla f_2 \left(\mathbf{y}_2^k\right), \mathbf{x}_2^{k+1} - \mathbf{y}_2^k\right\rangle$$

$$\overset{a}{=} \left\langle \nabla f_2 \left(\mathbf{y}_2^k\right), \left(\mathbf{u}_2 - \mathbf{y}_2^k\right) + \left(\mathbf{x}_2^{k+1} - \mathbf{u}_2\right)\right\rangle$$

$$\overset{b}{=} \left\langle \nabla f_2 \left(\mathbf{y}_2^k\right), \mathbf{u}_2 - \mathbf{y}_2^k\right\rangle - \theta_3 \left\langle \nabla f_2 \left(\mathbf{y}_2^k\right), \mathbf{y}_2^k - \tilde{\mathbf{x}}_2\right\rangle + \left\langle \nabla f_2 \left(\mathbf{y}_2^k\right), \mathbf{z}^{k+1} - \mathbf{u}_2\right\rangle$$

$$= \left\langle \nabla f_2 \left(\mathbf{y}_2^k\right), \mathbf{u}_2 - \mathbf{y}_2^k\right\rangle - \theta_3 \left\langle \nabla f_2 \left(\mathbf{y}_2^k\right), \mathbf{y}_2^k - \tilde{\mathbf{x}}_2\right\rangle$$

$$+ \left\langle \tilde{\nabla} f_2 \left(\mathbf{y}_2^k\right), \mathbf{z}^{k+1} - \mathbf{u}_2\right\rangle + \left\langle \nabla f_2 \left(\mathbf{y}_2^k\right) - \tilde{\nabla} f_2 \left(\mathbf{y}_2^k\right), \mathbf{z}^{k+1} - \mathbf{u}_2\right\rangle, \tag{5.55}$$

where in $\overset{a}{=}$ we introduce an arbitrary vector $\mathbf{u}_2$ (we will set it to be $\mathbf{x}_2^k$, $\tilde{\mathbf{x}}_2$, and $\mathbf{x}_2^*$, respectively) and in $\overset{b}{=}$ we set

$$\mathbf{z}^{k+1} = \mathbf{x}_2^{k+1} + \theta_3 \left(\mathbf{y}_2^k - \tilde{\mathbf{x}}_2\right), \tag{5.56}$$

in which $\theta_3$ is an absolute constant determined later.

For $\langle \tilde{\nabla} f_2 \left(\mathbf{y}_2^k\right), \mathbf{z}^{k+1} - \mathbf{u}_2\rangle$, we have

$$\left\langle \tilde{\nabla} f_2 \left(\mathbf{y}_2^k\right), \mathbf{z}^{k+1} - \mathbf{u}_2\right\rangle$$

$$\overset{a}{\le} -\left\langle \tilde{\nabla} h_2 \left(\mathbf{x}_2^{k+1}\right) + \mathbf{A}_2^T \bar{\lambda} \left(\mathbf{x}_1^{k+1}, \mathbf{y}_2^k\right) + \left(\alpha L_2 + \frac{\beta}{\theta_1} \|\mathbf{A}_2\|_2^2\right) \left(\mathbf{x}_2^{k+1} - \mathbf{y}_2^k\right),$$

$$
\left. \mathbf{z}^{k+1} - \mathbf{u}_2 \right\rangle
$$

$$
\overset{b}{=} -\left\langle \tilde{\nabla} h_2\left(\mathbf{x}_2^{k+1}\right), \mathbf{x}_2^{k+1} + \theta_3\left(\mathbf{y}_2^k - \tilde{\mathbf{x}}_2\right) - \mathbf{u}_2 \right\rangle
$$

$$
- \left\langle \mathbf{A}_2^T \bar{\boldsymbol{\lambda}}\left(\mathbf{x}_1^{k+1}, \mathbf{y}_2^k\right) + \left(\alpha L_2 + \frac{\beta}{\theta_1} \|\mathbf{A}_2\|_2^2\right)\left(\mathbf{x}_2^{k+1} - \mathbf{y}_2^k\right), \mathbf{z}^{k+1} - \mathbf{u}_2 \right\rangle
$$

$$
= -\left\langle \tilde{\nabla} h_2\left(\mathbf{x}_2^{k+1}\right), \mathbf{x}_2^{k+1} + \theta_3\left(\mathbf{y}_2^k - \mathbf{x}_2^{k+1} + \mathbf{x}_2^{k+1} - \tilde{\mathbf{x}}_2\right) - \mathbf{u}_2 \right\rangle
$$

$$
- \left\langle \mathbf{A}_2^T \bar{\boldsymbol{\lambda}}\left(\mathbf{x}_1^{k+1}, \mathbf{y}_2^k\right) + \left(\alpha L_2 + \frac{\beta}{\theta_1} \|\mathbf{A}_2\|_2^2\right)\left(\mathbf{x}_2^{k+1} - \mathbf{y}_2^k\right), \mathbf{z}^{k+1} - \mathbf{u}_2 \right\rangle
$$

$$
\overset{c}{\le} h_2(\mathbf{u}_2) - h_2\left(\mathbf{x}_2^{k+1}\right) + \theta_3 h_2(\tilde{\mathbf{x}}_2) - \theta_3 h_2\left(\mathbf{x}_2^{k+1}\right) - \theta_3 \left\langle \tilde{\nabla} h_2\left(\mathbf{x}_2^{k+1}\right), \mathbf{y}_2^k - \mathbf{x}_2^{k+1} \right\rangle
$$

$$
- \left\langle \mathbf{A}_2^T \bar{\boldsymbol{\lambda}}\left(\mathbf{x}_1^{k+1}, \mathbf{y}_2^k\right) + \left(\alpha L_2 + \frac{\beta}{\theta_1} \|\mathbf{A}_2\|_2^2\right)\left(\mathbf{x}_2^{k+1} - \mathbf{y}_2^k\right), \mathbf{z}^{k+1} - \mathbf{u}_2 \right\rangle
$$

$$
\overset{d}{=} h_2(\mathbf{u}_2) - h_2\left(\mathbf{x}_2^{k+1}\right) + \theta_3 h_2(\tilde{\mathbf{x}}_2) - \theta_3 h_2\left(\mathbf{x}_2^{k+1}\right)
$$

$$
- \left\langle \mathbf{A}_2^T \bar{\boldsymbol{\lambda}}\left(\mathbf{x}_1^{k+1}, \mathbf{y}_2^k\right) + \left(\alpha L_2 + \frac{\beta}{\theta_1} \|\mathbf{A}_2\|_2^2\right)\left(\mathbf{x}_2^{k+1} - \mathbf{y}_2^k\right), \mathbf{z}^{k+1} - \mathbf{u}_2 \right\rangle
$$

$$
- \theta_3 \left\langle \mathbf{A}_2^T \bar{\boldsymbol{\lambda}}\left(\mathbf{x}_1^{k+1}, \mathbf{y}_2^k\right) + \left(\alpha L_2 + \frac{\beta}{\theta_1} \|\mathbf{A}_2\|_2^2\right)\left(\mathbf{x}_2^{k+1} - \mathbf{y}_2^k\right) + \tilde{\nabla} f_2\left(\mathbf{y}_2^k\right), \right.
$$

$$
\left. \mathbf{x}_2^{k+1} - \mathbf{y}_2^k \right\rangle, \tag{5.57}
$$

where $\tilde{\nabla} h_2(\mathbf{x}_2^{k+1}) \in \partial h_2(\mathbf{x}_2^{k+1})$. In $\overset{a}{\le}$ and $\overset{b}{=}$ we use (5.53) and (5.56), respectively. The inequality $\overset{d}{=}$ uses (5.53) again. The inequality $\overset{c}{\le}$ uses the convexity of $h_2$:

$$
\left\langle \tilde{\nabla} h_2\left(\mathbf{x}_2^{k+1}\right), \mathbf{w} - \mathbf{x}_2^{k+1} \right\rangle \le h_2(\mathbf{w}) - h_2\left(\mathbf{x}_2^{k+1}\right), \quad \mathbf{w} = \mathbf{u}_2, \tilde{\mathbf{x}}_2.
$$

Rearranging terms in (5.57) and using

$$
\tilde{\nabla} f_2\left(\mathbf{y}_2^k\right) = \nabla f_2\left(\mathbf{y}_2^k\right) + \left(\tilde{\nabla} f_2\left(\mathbf{y}_2^k\right) - \nabla f_2\left(\mathbf{y}_2^k\right)\right),
$$

we have

$$
\left\langle \tilde{\nabla} f_2\left(\mathbf{y}_2^k\right), \mathbf{z}^{k+1} - \mathbf{u}_2 \right\rangle
$$

$$
\le h_2(\mathbf{u}_2) - h_2\left(\mathbf{x}_2^{k+1}\right) + \theta_3 h_2(\tilde{\mathbf{x}}_2) - \theta_3 h_2\left(\mathbf{x}_2^{k+1}\right)
$$

$$
- \left\langle \mathbf{A}_2^T \bar{\boldsymbol{\lambda}}\left(\mathbf{x}_1^{k+1}, \mathbf{y}_2^k\right) + \left(\alpha L_2 + \frac{\beta}{\theta_1} \|\mathbf{A}_2\|_2^2\right)\left(\mathbf{x}_2^{k+1} - \mathbf{y}_2^k\right), \right.
$$

$$\theta_3 \left( \mathbf{x}_2^{k+1} - \mathbf{y}_2^k \right) + \mathbf{z}^{k+1} - \mathbf{u}_2 \Big\rangle$$

$$- \theta_3 \left\langle \nabla f_2 \left( \mathbf{y}_2^k \right) + \left( \tilde{\nabla} f_2 \left( \mathbf{y}_2^k \right) - \nabla f_2 \left( \mathbf{y}_2^k \right) \right), \mathbf{x}_2^{k+1} - \mathbf{y}_2^k \right\rangle. \tag{5.58}$$

Substituting (5.58) in (5.55), we obtain

$$(1 + \theta_3) \left\langle \nabla f_2 \left( \mathbf{y}_2^k \right), \mathbf{x}_2^{k+1} - \mathbf{y}_2^k \right\rangle$$

$$\leq \left\langle \nabla f_2 \left( \mathbf{y}_2^k \right), \mathbf{u}_2 - \mathbf{y}_2^k \right\rangle - \theta_3 \left\langle \nabla f_2 \left( \mathbf{y}_2^k \right), \mathbf{y}_2^k - \tilde{\mathbf{x}}_2 \right\rangle + h_2(\mathbf{u}_2) - h_2 \left( \mathbf{x}_2^{k+1} \right)$$

$$+ \theta_3 h_2(\tilde{\mathbf{x}}_2) - \theta_3 h_2 \left( \mathbf{x}_2^{k+1} \right)$$

$$- \left\langle \mathbf{A}_2^T \bar{\lambda} \left( \mathbf{x}_1^{k+1}, \mathbf{y}_2^k \right) + \left( \alpha L_2 + \frac{\beta}{\theta_1} \|\mathbf{A}_2\|_2^2 \right) \left( \mathbf{x}_2^{k+1} - \mathbf{y}_2^k \right), \right.$$

$$\left. \mathbf{z}^{k+1} - \mathbf{u}_2 + \theta_3 \left( \mathbf{x}_2^{k+1} - \mathbf{y}_2^k \right) \right\rangle$$

$$+ \left\langle \nabla f_2 \left( \mathbf{y}_2^k \right) - \tilde{\nabla} f_2 \left( \mathbf{y}_2^k \right), \theta_3 \left( \mathbf{x}_2^{k+1} - \mathbf{y}_2^k \right) + \mathbf{z}^{k+1} - \mathbf{u}_2 \right\rangle. \tag{5.59}$$

Multiplying (5.54) by $(1 + \theta_3)$ and then adding (5.59), we can eliminate the term $\langle \nabla f_2 \left( \mathbf{y}_2^k \right), \mathbf{x}_2^{k+1} - \mathbf{y}_2^k \rangle$ and obtain

$$(1 + \theta_3) J_2 \left( \mathbf{x}_2^{k+1} \right)$$

$$\leq (1 + \theta_3) f_2 \left( \mathbf{y}_2^k \right) + \left\langle \nabla f_2 \left( \mathbf{y}_2^k \right), \mathbf{u}_2 - \mathbf{y}_2^k \right\rangle - \theta_3 \left\langle \nabla f_2 \left( \mathbf{y}_2^k \right), \mathbf{y}_2^k - \tilde{\mathbf{x}}_2 \right\rangle + h_2(\mathbf{u}_2)$$

$$+ \theta_3 h_2(\tilde{\mathbf{x}}_2) - \left\langle \mathbf{A}_2^T \bar{\lambda} \left( \mathbf{x}_1^{k+1}, \mathbf{y}_2^k \right) + \left( \alpha L_2 + \frac{\beta}{\theta_1} \|\mathbf{A}_2\|_2^2 \right) \left( \mathbf{x}_2^{k+1} - \mathbf{y}_2^k \right), \right.$$

$$\left. \mathbf{z}^{k+1} - \mathbf{u}_2 + \theta_3 \left( \mathbf{x}_2^{k+1} - \mathbf{y}_2^k \right) \right\rangle$$

$$+ \left\langle \nabla f_2 \left( \mathbf{y}_2^k \right) - \tilde{\nabla} f_2 \left( \mathbf{y}_2^k \right), \theta_3 \left( \mathbf{x}_2^{k+1} - \mathbf{y}_2^k \right) + \mathbf{z}^{k+1} - \mathbf{u}_2 \right\rangle$$

$$+ \frac{(1 + \theta_3) L_2}{2} \left\| \mathbf{x}_2^{k+1} - \mathbf{y}_2^k \right\|^2$$

$$\overset{a}{\leq} J_2(\mathbf{u}_2) - \theta_3 \left\langle \nabla f_2 \left( \mathbf{y}_2^k \right), \mathbf{y}_2^k - \tilde{\mathbf{x}}_2 \right\rangle + \theta_3 f_2 \left( \mathbf{y}_2^k \right) + \theta_3 h_2(\tilde{\mathbf{x}}_2)$$

$$- \left\langle \mathbf{A}_2^T \bar{\lambda} \left( \mathbf{x}_1^{k+1}, \mathbf{y}_2^k \right) + \left( \alpha L_2 + \frac{\beta}{\theta_1} \|\mathbf{A}_2\|_2^2 \right) \left( \mathbf{x}_2^{k+1} - \mathbf{y}_2^k \right), \right.$$

$$\left. \mathbf{z}^{k+1} - \mathbf{u}_2 + \theta_3 \left( \mathbf{x}_2^{k+1} - \mathbf{y}_2^k \right) \right\rangle$$

$$+ \left\langle \nabla f_2 \left( \mathbf{y}_2^k \right) - \tilde{\nabla} f_2 \left( \mathbf{y}_2^k \right), \theta_3 \left( \mathbf{x}_2^{k+1} - \mathbf{y}_2^k \right) + \mathbf{z}^{k+1} - \mathbf{u}_2 \right\rangle$$
$$+ \frac{(1 + \theta_3) L_2}{2} \left\| \mathbf{x}_2^{k+1} - \mathbf{y}_2^k \right\|^2, \tag{5.60}$$

where $\overset{a}{\leq}$ uses the convexity of $f_2$:

$$\left\langle \nabla f_2 \left( \mathbf{y}_2^k \right), \mathbf{u}_2 - \mathbf{y}_2^k \right\rangle \leq f_2(\mathbf{u}_2) - f_2 \left( \mathbf{y}_2^k \right).$$

We now consider the term

$$\left\langle \nabla f_2 \left( \mathbf{y}_2^k \right) - \tilde{\nabla} f_2 \left( \mathbf{y}_2^k \right), \theta_3 \left( \mathbf{x}_2^{k+1} - \mathbf{y}_2^k \right) + \mathbf{z}^{k+1} - \mathbf{u}_2 \right\rangle.$$

We will set $\mathbf{u}_2$ to be $\mathbf{x}_2^k$ and $\mathbf{x}_2^*$, which do not depend on $\mathcal{I}_{k,s}$. So we have

$$\mathbb{E}_{i_k} \left\langle \nabla f_2 \left( \mathbf{y}_2^k \right) - \tilde{\nabla} f_2(\mathbf{y}^k), \theta_3 \left( \mathbf{x}_2^{k+1} - \mathbf{y}_2^k \right) + \mathbf{z}^{k+1} - \mathbf{u}_2 \right\rangle$$

$$= \mathbb{E}_{i_k} \left\langle \nabla f_2 \left( \mathbf{y}_2^k \right) - \tilde{\nabla} f_2 \left( \mathbf{y}_2^k \right), \theta_3 \mathbf{z}^{k+1} + \mathbf{z}^{k+1} \right\rangle$$

$$- \mathbb{E}_{i_k} \left\langle \nabla f_2 \left( \mathbf{y}_2^k \right) - \tilde{\nabla} f_2 \left( \mathbf{y}_2^k \right), \theta_3^2 \left( \mathbf{y}_2^k - \tilde{\mathbf{x}}_2 \right) + \theta_3 \mathbf{y}_2^k + \mathbf{u}_2 \right\rangle$$

$$\overset{a}{=} (1 + \theta_3) \mathbb{E}_{i_k} \left\langle \nabla f_2 \left( \mathbf{y}_2^k \right) - \tilde{\nabla} f_2 \left( \mathbf{y}_2^k \right), \mathbf{z}^{k+1} \right\rangle$$

$$\overset{b}{=} (1 + \theta_3) \mathbb{E}_{i_k} \left\langle \nabla f_2 \left( \mathbf{y}_2^k \right) - \tilde{\nabla} f_2 \left( \mathbf{y}_2^k \right), \mathbf{x}_2^{k+1} \right\rangle$$

$$\overset{c}{=} (1 + \theta_3) \mathbb{E}_{i_k} \left\langle \nabla f_2 \left( \mathbf{y}_2^k \right) - \tilde{\nabla} f_2 \left( \mathbf{y}_2^k \right), \mathbf{x}_2^{k+1} - \mathbf{y}_2^k \right\rangle$$

$$\overset{d}{\leq} \mathbb{E}_{i_k} \left( \frac{\theta_3 b}{2 L_2} \left\| \nabla f_2 \left( \mathbf{y}_2^k \right) - \tilde{\nabla} f_2 \left( \mathbf{y}_2^k \right) \right\|^2 \right) + \mathbb{E}_{i_k} \left( \frac{(1 + \theta_3)^2 L_2}{2 \theta_3 b} \left\| \mathbf{x}_2^{k+1} - \mathbf{y}_2^k \right\|^2 \right)$$

$$\overset{e}{\leq} \theta_3 \left( f_2(\tilde{\mathbf{x}}_2) - f_2 \left( \mathbf{y}_2^k \right) - \left\langle \nabla f_2 \left( \mathbf{y}_2^k \right), \tilde{\mathbf{x}}_2 - \mathbf{y}_2^k \right\rangle \right)$$

$$+ \mathbb{E}_{i_k} \left( \frac{(1 + \theta_3)^2 L_2}{2 \theta_3 b} \left\| \mathbf{x}_2^{k+1} - \mathbf{y}_2^k \right\|^2 \right), \tag{5.61}$$

where in the equality $\overset{a}{=}$ we use the fact that

$$\mathbb{E}_{i_k} \left( \nabla f_2 \left( \mathbf{y}_2^k \right) - \tilde{\nabla} f_2 \left( \mathbf{y}_2^k \right) \right) = \mathbf{0},$$

and $\mathbf{x}_2^k$, $\mathbf{y}_2^k$, $\tilde{\mathbf{x}}_2$, and $\mathbf{u}_2$ are independent of $i_{k,s}$ (they are known), so

$$\mathbb{E}_{i_k}\left\langle \nabla f_2\left(\mathbf{y}_2^k\right) - \tilde{\nabla} f_2\left(\mathbf{y}_2^k\right), \mathbf{y}_2^k\right\rangle = 0,$$

$$\mathbb{E}_{i_k}\left\langle \nabla f_2\left(\mathbf{y}_2^k\right) - \tilde{\nabla} f_2\left(\mathbf{y}_2^k\right), \tilde{\mathbf{x}}_2\right\rangle = 0,$$

$$\mathbb{E}_{i_k}\left\langle \nabla f_2\left(\mathbf{y}_2^k\right) - \tilde{\nabla} f_2\left(\mathbf{y}_2^k\right), \mathbf{u}_2\right\rangle = 0;$$

the equalities $\overset{b}{=}$ and $\overset{c}{=}$ hold similarly; the inequality $\overset{d}{\leq}$ uses the Cauchy–Schwartz inequality; and $\overset{e}{\leq}$ uses

$$\mathbb{E}_{i_k}\left\| \nabla f_2\left(\mathbf{y}_2^k\right) - \tilde{\nabla} f_2\left(\mathbf{y}_2^k\right)\right\|^2$$

$$\overset{a}{=} \frac{1}{b}\mathbb{E}_i\left\| \nabla f_2\left(\mathbf{y}_2^k\right) - \left(\nabla f_{2,i}\left(\mathbf{y}_2^k\right) - \nabla f_{2,i}(\tilde{\mathbf{x}}_2) + \nabla f_2(\tilde{\mathbf{x}}_2)\right)\right\|^2$$

$$\overset{b}{\leq} \frac{2L_2}{b}\left( f_2(\tilde{\mathbf{x}}_2) - f_2\left(\mathbf{y}_2^k\right) - \left\langle \nabla f_2\left(\mathbf{y}_2^k\right), \tilde{\mathbf{x}}_2 - \mathbf{y}_2^k\right\rangle\right),$$

where $\overset{a}{=}$ is by the independence of $i_{k,s}$, $\mathbb{E}_i$ means taking expectation only on random variable $i$ that is uniformly sampled from $[n]$, and $\overset{b}{\leq}$ uses Lemma 5.4.

Taking expectation on (5.60) and adding (5.61), we obtain

$$(1+\theta_3)\mathbb{E}_{i_k} J_2\left(\mathbf{x}_2^{k+1}\right)$$

$$\leq -\mathbb{E}_{i_k}\Big\langle \mathbf{A}_2^T\tilde{\lambda}\left(\mathbf{x}_1^{k+1}, \mathbf{y}_2^k\right) + \left(\alpha L_2 + \frac{\beta}{\theta_1}\|\mathbf{A}_2\|_2^2\right)\left(\mathbf{x}_2^{k+1} - \mathbf{y}_2^k\right),$$

$$\mathbf{z}^{k+1} - \mathbf{u}_2 + \theta_3\left(\mathbf{x}_2^{k+1} - \mathbf{y}_2^k\right)\Big\rangle$$

$$+ J_2(\mathbf{u}_2) + \theta_3 J_2(\tilde{\mathbf{x}}_2) + \mathbb{E}_{i_k}\left[\frac{1}{2}(1+\theta_3)\left(1 + \frac{1+\theta_3}{b\theta_3}\right)L_2\left\|\mathbf{x}_2^{k+1} - \mathbf{y}_2^k\right\|^2\right]$$

$$\overset{a}{=} -\mathbb{E}_{i_k}\Big\langle \mathbf{A}_2^T\tilde{\lambda}\left(\mathbf{x}_1^{k+1}, \mathbf{y}_2^k\right) + \left(\alpha L_2 + \frac{\beta}{\theta_1}\|\mathbf{A}_2\|_2^2\right)\left(\mathbf{x}_2^{k+1} - \mathbf{y}_2^k\right),$$

$$(1+\theta_3)\mathbf{x}_2^{k+1} - \theta_3\tilde{\mathbf{x}}_2 - \mathbf{u}_2\Big\rangle$$

$$+ J_2(\mathbf{u}_2) + \theta_3 J_2(\tilde{\mathbf{x}}_2) + \mathbb{E}_{i_k}\left[\frac{1}{2}(1+\theta_3)\left(1 + \frac{1}{b\theta_2}\right)L_2\left\|\mathbf{x}_2^{k+1} - \mathbf{y}_2^k\right\|^2\right],$$

where in $\overset{a}{=}$ we use (5.56) and set $\theta_3$ satisfying $\theta_2 = \frac{\theta_3}{1+\theta_3}$. Setting $\mathbf{u}_2$ to be $\mathbf{x}_2^k$ and $\mathbf{x}_2^*$, respectively, then multiplying the two inequalities by $1 - \theta_1(1+\theta_3)$ and $\theta_1(1+\theta_3)$, respectively, and adding them, we obtain

$$(1 + \theta_3)\mathbb{E}_{i_k} J_2\left(\mathbf{x}_2^{k+1}\right)$$

$$\leq -\mathbb{E}_{i_k}\left\langle \mathbf{A}_2^T \bar{\lambda}\left(\mathbf{x}_1^{k+1}, \mathbf{y}_2^k\right) + \left(\alpha L_2 + \frac{\beta}{\theta_1}\|\mathbf{A}_2\|_2^2\right)\left(\mathbf{x}_2^{k+1} - \mathbf{y}_2^k\right),\right.$$

$$\left.(1 + \theta_3)\mathbf{x}_2^{k+1} - \theta_3\tilde{\mathbf{x}}_2\right\rangle$$

$$- \mathbb{E}_{i_k}\left\langle \mathbf{A}_2^T \bar{\lambda}\left(\mathbf{x}_1^{k+1}, \mathbf{y}_2^k\right) + \left(\alpha L_2 + \frac{\beta}{\theta_1}\|\mathbf{A}_2\|_2^2\right)\left(\mathbf{x}_2^{k+1} - \mathbf{y}_2^k\right),\right.$$

$$\left.- [1 - \theta_1(1 + \theta_3)]\,\mathbf{x}_2^k\right\rangle$$

$$- \mathbb{E}_{i_k}\left\langle \mathbf{A}_2^T \bar{\lambda}\left(\mathbf{x}_1^{k+1}, \mathbf{y}_2^k\right) + \left(\alpha L_2 + \frac{\beta}{\theta_1}\|\mathbf{A}_2\|_2^2\right)\left(\mathbf{x}_2^{k+1} - \mathbf{y}_2^k\right), -\theta_1(1 + \theta_3)\mathbf{x}_2^*\right\rangle$$

$$+ [1 - \theta_1(1 + \theta_3)] J_2\left(\mathbf{x}_2^k\right) + \theta_1(1 + \theta_3) J_2\left(\mathbf{x}_2^*\right) + \theta_3 J_2(\tilde{\mathbf{x}}_2)$$

$$+ \mathbb{E}_{i_k}\left[\frac{1}{2}(1 + \theta_3)\left(1 + \frac{1}{b\theta_2}\right) L_2\left\|\mathbf{x}_2^{k+1} - \mathbf{y}_2^k\right\|^2\right]. \tag{5.62}$$

Dividing (5.62) by $(1 + \theta_3)$, we obtain

$$\mathbb{E}_{i_k} J_2\left(\mathbf{x}_2^{k+1}\right)$$

$$\leq -\mathbb{E}_{i_k}\left\langle \mathbf{A}_2^T \bar{\lambda}\left(\mathbf{x}_1^{k+1}, \mathbf{y}_2^k\right) + \left(\alpha L_2 + \frac{\beta}{\theta_1}\|\mathbf{A}_2\|_2^2\right)\left(\mathbf{x}_2^{k+1} - \mathbf{y}_2^k\right), \mathbf{x}_2^{k+1} - \theta_2\tilde{\mathbf{x}}_2\right\rangle$$

$$- \mathbb{E}_{i_k}\left\langle \mathbf{A}_2^T \bar{\lambda}\left(\mathbf{x}_1^{k+1}, \mathbf{y}_2^k\right) + \left(\alpha L_2 + \frac{\beta}{\theta_1}\|\mathbf{A}_2\|_2^2\right)\left(\mathbf{x}_2^{k+1} - \mathbf{y}_2^k\right),\right.$$

$$\left.- (1 - \theta_1 - \theta_2)\mathbf{x}_2^k - \theta_1\mathbf{x}_2^*\right\rangle$$

$$+ (1 - \theta_1 - \theta_2) J_2\left(\mathbf{x}_2^k\right) + \theta_1 J_2\left(\mathbf{x}_2^*\right) + \theta_2 J_2(\tilde{\mathbf{x}}_2)$$

$$+ \mathbb{E}_{i_k}\left[\frac{1}{2}\left(1 + \frac{1}{b\theta_2}\right) L_2\left\|\mathbf{x}_2^{k+1} - \mathbf{y}_2^k\right\|^2\right],$$

where we use $\theta_2 = \frac{\theta_3}{1+\theta_3}$ and so $\frac{1-\theta_1(1+\theta_3)}{1+\theta_3} = 1 - \theta_1 - \theta_2$.
   Step 3: Setting

$$\hat{\lambda}^k = \tilde{\lambda}^k + \frac{\beta(1 - \theta_1)}{\theta_1}\left(\mathbf{A}_1\mathbf{x}_1^k + \mathbf{A}_2\mathbf{x}_2^k - \mathbf{b}\right), \tag{5.63}$$

we prove that it has the following properties:

$$\hat{\boldsymbol{\lambda}}^{k+1} = \bar{\boldsymbol{\lambda}}\left(\mathbf{x}_1^{k+1}, \mathbf{x}_2^{k+1}\right),\tag{5.64}$$

$$\hat{\boldsymbol{\lambda}}^{k+1} - \hat{\boldsymbol{\lambda}}^k = \frac{\beta}{\theta_1}\mathbf{A}_1\left[\mathbf{x}_1^{k+1} - (1 - \theta_1 - \theta_2)\mathbf{x}_1^k - \theta_2\tilde{\mathbf{x}}_1 - \theta_1\mathbf{x}_1^*\right]$$

$$+ \frac{\beta}{\theta_1}\mathbf{A}_2\left[\mathbf{x}_2^{k+1} - (1 - \theta_1 - \theta_2)\mathbf{x}_2^k - \theta_2\tilde{\mathbf{x}}_2 - \theta_1\mathbf{x}_2^*\right].\tag{5.65}$$

Indeed, for Algorithm 5.3 we have

$$\boldsymbol{\lambda}^k = \tilde{\boldsymbol{\lambda}}^k + \frac{\beta\theta_2}{\theta_1}\left(\mathbf{A}_1\mathbf{x}_1^k + \mathbf{A}_2\mathbf{x}_2^k - \tilde{\mathbf{b}}\right)\tag{5.66}$$

and

$$\tilde{\boldsymbol{\lambda}}^{k+1} = \boldsymbol{\lambda}^k + \beta\left(\mathbf{A}_1\mathbf{x}_1^{k+1} + \mathbf{A}_2\mathbf{x}_2^{k+1} - \mathbf{b}\right).\tag{5.67}$$

With (5.63) we have

$$\hat{\boldsymbol{\lambda}}^{k+1} = \tilde{\boldsymbol{\lambda}}^{k+1} + \beta\left(\frac{1}{\theta_1} - 1\right)\left(\mathbf{A}_1\mathbf{x}_1^{k+1} + \mathbf{A}_2\mathbf{x}_2^{k+1} - \mathbf{b}\right)$$

$$\overset{a}{=} \boldsymbol{\lambda}^k + \frac{\beta}{\theta_1}\left(\mathbf{A}_1\mathbf{x}_1^{k+1} + \mathbf{A}_2\mathbf{x}_2^{k+1} - \mathbf{b}\right)\tag{5.68}$$

$$\overset{b}{=} \tilde{\boldsymbol{\lambda}}^k + \frac{\beta}{\theta_1}\left\{\mathbf{A}_1\mathbf{x}_1^{k+1} + \mathbf{A}_2\mathbf{x}_2^{k+1} - \mathbf{b} + \theta_2\left[\mathbf{A}_1\left(\mathbf{x}_1^k - \tilde{\mathbf{x}}_1\right) + \mathbf{A}_2\left(\mathbf{x}_2^k - \tilde{\mathbf{x}}_2\right)\right]\right\},$$

where in $\overset{a}{=}$ we use (5.67) and $\overset{b}{=}$ is obtained by (5.66) and $\tilde{\mathbf{b}} = \mathbf{A}_1\tilde{\mathbf{x}}_1 + \mathbf{A}_2\tilde{\mathbf{x}}_2$ (see Algorithm 5.4). Together with (5.63) we obtain

$$\hat{\boldsymbol{\lambda}}^{k+1} - \hat{\boldsymbol{\lambda}}^k = \frac{\beta}{\theta_1}\mathbf{A}_1\left[\mathbf{x}_1^{k+1} - (1 - \theta_1)\mathbf{x}_1^k - \theta_1\mathbf{x}_1^* + \theta_2\left(\mathbf{x}_1^k - \tilde{\mathbf{x}}_1\right)\right]$$

$$+ \frac{\beta}{\theta_1}\mathbf{A}_2\left[\mathbf{x}_2^{k+1} - (1 - \theta_1)\mathbf{x}_2^k - \theta_1\mathbf{x}_2^* + \theta_2\left(\mathbf{x}_2^k - \tilde{\mathbf{x}}_2\right)\right],$$

where we use the fact that $\mathbf{A}_1\mathbf{x}_1^* + \mathbf{A}_2\mathbf{x}_2^* = \mathbf{b}$. So (5.65) is proven.

Since (5.68) equals $\bar{\boldsymbol{\lambda}}(\mathbf{x}_1^{k+1}, \mathbf{x}_2^{k+1})$, we obtain (5.64).

Step 4: We now are ready to prove (5.49). By the definition of $\tilde{L}(\mathbf{x}_1, \mathbf{x}_2, \boldsymbol{\lambda})$, we have

$$\tilde{L}\left(\mathbf{x}_1^{k+1}, \mathbf{x}_2^{k+1}, \boldsymbol{\lambda}^*\right) - \theta_2\tilde{L}\left(\tilde{\mathbf{x}}_1, \tilde{\mathbf{x}}_2, \boldsymbol{\lambda}^*\right) - (1 - \theta_1 - \theta_2)\tilde{L}\left(\mathbf{x}_1^k, \mathbf{x}_2^k, \boldsymbol{\lambda}^*\right)$$

$$= J_1\left(\mathbf{x}_1^{k+1}\right) - (1 - \theta_1 - \theta_2)J_1\left(\mathbf{x}_1^k\right) - \theta_1 J_1\left(\mathbf{x}_1^*\right) - \theta_2 J_1(\tilde{\mathbf{x}}_1)$$

$$+ J_2\left(\mathbf{x}_2^{k+1}\right) - (1-\theta_1-\theta_2)J_2\left(\mathbf{x}_2^k\right) - \theta_1 J_2\left(\mathbf{x}_2^*\right) - \theta_2 J_2(\tilde{\mathbf{x}}_2)$$

$$+ \left\langle \boldsymbol{\lambda}^*, \mathbf{A}_1\left[\mathbf{x}_1^{k+1} - (1-\theta_1-\theta_2)\mathbf{x}_1^k - \theta_2\tilde{\mathbf{x}}_1 - \theta_1\mathbf{x}_1^*\right]\right\rangle$$

$$+ \left\langle \boldsymbol{\lambda}^*, \mathbf{A}_2\left[\mathbf{x}_2^{k+1} - (1-\theta_1-\theta_2)\mathbf{x}_2^k - \theta_2\tilde{\mathbf{x}}_2 - \theta_1\mathbf{x}_2^*\right]\right\rangle.$$

Plugging (5.50) and (5.52) into the above, we have

$$\mathbb{E}_{i_k}\tilde{L}\left(\mathbf{x}_1^{k+1},\mathbf{x}_2^{k+1},\boldsymbol{\lambda}^*\right) - \theta_2\tilde{L}\left(\tilde{\mathbf{x}}_1,\tilde{\mathbf{x}}_2,\boldsymbol{\lambda}^*\right) - (1-\theta_1-\theta_2)\tilde{L}\left(\mathbf{x}_1^k,\mathbf{x}_2^k,\boldsymbol{\lambda}^*\right)$$

$$\leq \mathbb{E}_{i_k}\left\langle \boldsymbol{\lambda}^* - \bar{\boldsymbol{\lambda}}\left(\mathbf{x}_1^{k+1},\mathbf{y}_2^k\right), \mathbf{A}_1\left[\mathbf{x}_1^{k+1} - (1-\theta_1-\theta_2)\mathbf{x}_1^k - \theta_2\tilde{\mathbf{x}}_1 - \theta_1\mathbf{x}_1^*\right]\right\rangle$$

$$+ \mathbb{E}_{i_k}\left\langle \boldsymbol{\lambda}^* - \bar{\boldsymbol{\lambda}}\left(\mathbf{x}_1^{k+1},\mathbf{y}_2^k\right), \mathbf{A}_2\left[\mathbf{x}_2^{k+1} - (1-\theta_1-\theta_2)\mathbf{x}_2^k - \theta_2\tilde{\mathbf{x}}_2 - \theta_1\mathbf{x}_2^*\right]\right\rangle$$

$$- \mathbb{E}_{i_k}\left\langle \mathbf{x}_1^{k+1} - \mathbf{y}_1^k, \mathbf{x}_1^{k+1} - (1-\theta_1-\theta_2)\mathbf{x}_1^k - \theta_2\tilde{\mathbf{x}}_1 - \theta_1\mathbf{x}_1^*\right\rangle_{\mathbf{G}_1}$$

$$- \mathbb{E}_{i_k}\left\langle \mathbf{x}_2^{k+1} - \mathbf{y}_2^k, \mathbf{x}_2^{k+1} - (1-\theta_1-\theta_2)\mathbf{x}_2^k - \theta_2\tilde{\mathbf{x}}_2 - \theta_1\mathbf{x}_2^*\right\rangle_{\left(\alpha L_2 + \frac{\beta}{\theta_1}\|\mathbf{A}_2\|_2^2\right)\mathbf{I}}$$

$$+ \frac{L_1}{2}\mathbb{E}_{i_k}\left\|\mathbf{x}_1^{k+1} - \mathbf{y}_1^k\right\|^2 + \mathbb{E}_{i_k}\left[\frac{1}{2}\left(1+\frac{1}{b\theta_2}\right)L_2\left\|\mathbf{x}_2^{k+1} - \mathbf{y}_2^k\right\|^2\right]$$

$$\overset{a}{=} \mathbb{E}_{i_k}\left\langle \boldsymbol{\lambda}^* - \bar{\boldsymbol{\lambda}}\left(\mathbf{x}_1^{k+1},\mathbf{x}_2^{k+1}\right), \mathbf{A}_1\left[\mathbf{x}_1^{k+1} - (1-\theta_1-\theta_2)\mathbf{x}_1^k - \theta_2\tilde{\mathbf{x}}_1 - \theta_1\mathbf{x}_1^*\right]\right\rangle$$

$$+ \mathbb{E}_{i_k}\left\langle \boldsymbol{\lambda}^* - \bar{\boldsymbol{\lambda}}\left(\mathbf{x}_1^{k+1},\mathbf{x}_2^{k+1}\right), \mathbf{A}_2\left[\mathbf{x}_2^{k+1} - (1-\theta_1-\theta_2)\mathbf{x}_2^k - \theta_2\tilde{\mathbf{x}}_2 - \theta_1\mathbf{x}_2^*\right]\right\rangle$$

$$- \mathbb{E}_{i_k}\left\langle \mathbf{x}_1^{k+1} - \mathbf{y}_1^k, \mathbf{x}_1^{k+1} - (1-\theta_1-\theta_2)\mathbf{x}_1^k - \theta_2\tilde{\mathbf{x}}_1 - \theta_1\mathbf{x}_1^*\right\rangle_{\mathbf{G}_1}$$

$$- \mathbb{E}_{i_k}\left\langle \mathbf{x}_2^{k+1} - \mathbf{y}_2^k,\right.$$

$$\left.\mathbf{x}_2^{k+1} - (1-\theta_1-\theta_2)\mathbf{x}_2^k - \theta_2\tilde{\mathbf{x}}_2 - \theta_1\mathbf{x}_2^*\right\rangle_{\left(\alpha L_2 + \frac{\beta}{\theta_1}\|\mathbf{A}_2\|_2^2\right)\mathbf{I} - \frac{\beta}{\theta_1}\mathbf{A}_2^T\mathbf{A}_2}$$

$$+ \frac{L_1}{2}\mathbb{E}_{i_k}\left\|\mathbf{x}_1^{k+1} - \mathbf{y}_1^k\right\|^2 + \mathbb{E}_{i_k}\left[\frac{1}{2}\left(1+\frac{1}{b\theta_2}\right)L_2\left\|\mathbf{x}_2^{k+1} - \mathbf{y}_2^k\right\|^2\right]$$

$$+ \frac{\beta}{\theta_1}\mathbb{E}_{i_k}\left\langle \mathbf{A}_2\mathbf{x}_2^{k+1} - \mathbf{A}_2\mathbf{y}_2^k,\right.$$

$$\left.\mathbf{A}_1\left[\mathbf{x}_1^{k+1} - (1-\theta_1-\theta_2)\mathbf{x}_1^k - \theta_2\tilde{\mathbf{x}}_1 - \theta_1\mathbf{x}_1^*\right]\right\rangle, \tag{5.69}$$

where in the equality $\overset{a}{=}$ we change the term $\bar{\boldsymbol{\lambda}}(\mathbf{x}_1^{k+1},\mathbf{y}_2^k)$ to

$$\bar{\boldsymbol{\lambda}}\left(\mathbf{x}_1^{k+1},\mathbf{x}_2^{k+1}\right) - \frac{\beta}{\theta_1}\mathbf{A}_2\left(\mathbf{x}_2^{k+1} - \mathbf{y}_2^k\right).$$

For the first two terms in the right hand side of (5.69), we have

$$
\left\langle \boldsymbol{\lambda}^* - \bar{\boldsymbol{\lambda}}\left(\mathbf{x}_1^{k+1}, \mathbf{x}_2^{k+1}\right), \mathbf{A}_1 \left[\mathbf{x}_1^{k+1} - (1 - \theta_1 - \theta_2)\mathbf{x}_1^k - \theta_2\tilde{\mathbf{x}}_1 - \theta_1\mathbf{x}_1^*\right]\right\rangle
$$
$$
+ \left\langle \boldsymbol{\lambda}^* - \bar{\boldsymbol{\lambda}}\left(\mathbf{x}_1^{k+1}, \mathbf{x}_2^{k+1}\right), \mathbf{A}_2 \left[\mathbf{x}_2^{k+1} - (1 - \theta_1 - \theta_2)\mathbf{x}_2^k - \theta_2\tilde{\mathbf{x}}_2 - \theta_1\mathbf{x}_2^*\right]\right\rangle
$$
$$
\overset{a}{=} \frac{\theta_1}{\beta}\left\langle \boldsymbol{\lambda}^* - \hat{\boldsymbol{\lambda}}^{k+1}, \hat{\boldsymbol{\lambda}}^{k+1} - \hat{\boldsymbol{\lambda}}^k\right\rangle
$$
$$
\overset{b}{=} \frac{\theta_1}{2\beta}\left(\left\|\hat{\boldsymbol{\lambda}}^k - \boldsymbol{\lambda}^*\right\|^2 - \left\|\hat{\boldsymbol{\lambda}}^{k+1} - \boldsymbol{\lambda}^*\right\|^2 - \left\|\hat{\boldsymbol{\lambda}}^{k+1} - \hat{\boldsymbol{\lambda}}^k\right\|^2\right), \tag{5.70}
$$

where $\overset{a}{=}$ uses (5.64) and (5.65) and $\overset{b}{=}$ uses (A.2).

Substituting (5.70) into (5.69), we obtain

$$
\mathbb{E}_{i_k}\tilde{L}\left(\mathbf{x}_1^{k+1}, \mathbf{x}_2^{k+1}, \boldsymbol{\lambda}^*\right) - \theta_2\tilde{L}\left(\tilde{\mathbf{x}}_1, \tilde{\mathbf{x}}_2, \boldsymbol{\lambda}^*\right) - (1 - \theta_1 - \theta_2)\tilde{L}\left(\mathbf{x}_1^k, \mathbf{x}_2^k, \boldsymbol{\lambda}^*\right)
$$
$$
\leq \frac{\theta_1}{2\beta}\left(\left\|\hat{\boldsymbol{\lambda}}^k - \boldsymbol{\lambda}^*\right\|^2 - \mathbb{E}_{i_k}\left\|\hat{\boldsymbol{\lambda}}^{k+1} - \boldsymbol{\lambda}^*\right\|^2 - \mathbb{E}_{i_k}\left\|\hat{\boldsymbol{\lambda}}^{k+1} - \hat{\boldsymbol{\lambda}}^k\right\|^2\right)
$$
$$
- \mathbb{E}_{i_k}\left\langle \mathbf{x}_1^{k+1} - \mathbf{y}_1^k, \mathbf{x}_1^{k+1} - (1 - \theta_1 - \theta_2)\mathbf{x}_1^k - \theta_2\tilde{\mathbf{x}}_1 - \theta_1\mathbf{x}_1^*\right\rangle_{\mathbf{G}_1}
$$
$$
- \mathbb{E}_{i_k}\left\langle \mathbf{x}_2^{k+1} - \mathbf{y}_2^k, \right.
$$
$$
\left. \mathbf{x}_2^{k+1} - (1 - \theta_1 - \theta_2)\mathbf{x}_2^k - \theta_2\tilde{\mathbf{x}}_2 - \theta_1\mathbf{x}_2^*\right\rangle_{\left(\alpha L_2 + \frac{\beta}{\theta_1}\|\mathbf{A}_2\|_2^2\right)\mathbf{I} - \frac{\beta}{\theta_1}\mathbf{A}_2^T\mathbf{A}_2}
$$
$$
+ \frac{L_1}{2}\mathbb{E}_{i_k}\left\|\mathbf{x}_1^{k+1} - \mathbf{y}_1^k\right\|^2 + \mathbb{E}_{i_k}\left[\frac{1}{2}\left(1 + \frac{1}{b\theta_2}\right)L_2\left\|\mathbf{x}_2^{k+1} - \mathbf{y}_2^k\right\|^2\right]
$$
$$
+ \frac{\beta}{\theta_1}\mathbb{E}_{i_k}\left\langle \mathbf{A}_2\mathbf{x}_2^{k+1} - \mathbf{A}_2\mathbf{y}_2^k, \right.
$$
$$
\left. \mathbf{A}_1\left[\mathbf{x}_1^{k+1} - (1 - \theta_1 - \theta_2)\mathbf{x}_1^k - \theta_2\tilde{\mathbf{x}}_1 - \theta_1\mathbf{x}_1^*\right]\right\rangle. \tag{5.71}
$$

Then applying identity (A.1) to the second and the third terms in the right hand of (5.71) and rearranging terms, we have

$$
\mathbb{E}_{i_k}\tilde{L}\left(\mathbf{x}_1^{k+1}, \mathbf{x}_2^{k+1}, \boldsymbol{\lambda}^*\right) - \theta_2\tilde{L}\left(\tilde{\mathbf{x}}_1, \tilde{\mathbf{x}}_2, \boldsymbol{\lambda}^*\right) - (1 - \theta_1 - \theta_2)\tilde{L}\left(\mathbf{x}_1^k, \mathbf{x}_2^k, \boldsymbol{\lambda}^*\right)
$$
$$
\leq \frac{\theta_1}{2\beta}\left(\left\|\hat{\boldsymbol{\lambda}}^k - \boldsymbol{\lambda}^*\right\|^2 - \mathbb{E}_{i_k}\left\|\hat{\boldsymbol{\lambda}}^{k+1} - \boldsymbol{\lambda}^*\right\|^2 - \mathbb{E}_{i_k}\left\|\hat{\boldsymbol{\lambda}}^{k+1} - \hat{\boldsymbol{\lambda}}^k\right\|^2\right)
$$
$$
+ \frac{1}{2}\left\|\mathbf{y}_1^k - (1 - \theta_1 - \theta_2)\mathbf{x}_1^k - \theta_2\tilde{\mathbf{x}}_1 - \theta_1\mathbf{x}_1^*\right\|_{\mathbf{G}_1}^2
$$

$$-\frac{1}{2}\mathbb{E}_{i_k}\left\|\mathbf{x}_1^{k+1}-(1-\theta_1-\theta_2)\mathbf{x}_1^k-\theta_2\tilde{\mathbf{x}}_1-\theta_1\mathbf{x}_1^*\right\|_{\mathbf{G}_1}^2$$

$$+\frac{1}{2}\left\|\mathbf{y}_2^k-(1-\theta_1-\theta_2)\mathbf{x}_2^k-\theta_2\tilde{\mathbf{x}}_2-\theta_1\mathbf{x}_2^*\right\|_{\left(\alpha L_2+\frac{\beta}{\theta_1}\|\mathbf{A}_2\|_2^2\right)\mathbf{I}-\frac{\beta}{\theta_1}\mathbf{A}_2^T\mathbf{A}_2}^2$$

$$-\frac{1}{2}\mathbb{E}_{i_k}\left\|\mathbf{x}_2^{k+1}-(1-\theta_1-\theta_2)\mathbf{x}_2^k-\theta_2\tilde{\mathbf{x}}_2-\theta_1\mathbf{x}_2^*\right\|_{\left(\alpha L_2+\frac{\beta}{\theta_1}\|\mathbf{A}_2\|_2^2\right)\mathbf{I}-\frac{\beta}{\theta_1}\mathbf{A}_2^T\mathbf{A}_2}^2$$

$$-\frac{1}{2}\mathbb{E}_{i_k}\left\|\mathbf{x}_1^{k+1}-\mathbf{y}_1^k\right\|_{\frac{\beta}{\theta_1}\|\mathbf{A}_1\|_2^2\mathbf{I}-\frac{\beta}{\theta_1}\mathbf{A}_1^T\mathbf{A}_1}^2-\frac{1}{2}\mathbb{E}_{i_k}\left\|\mathbf{x}_2^{k+1}-\mathbf{y}_2^k\right\|_{\frac{\beta}{\theta_1}\|\mathbf{A}_2\|_2^2\mathbf{I}-\frac{\beta}{\theta_1}\mathbf{A}_2^T\mathbf{A}_2}^2$$

$$+\frac{\beta}{\theta_1}\mathbb{E}_{i_k}\left\langle\mathbf{A}_2\mathbf{x}_2^{k+1}-\mathbf{A}_2\mathbf{y}_2^k,\mathbf{A}_1\left[\mathbf{x}_1^{k+1}-(1-\theta_1-\theta_2)\mathbf{x}_1^k-\theta_2\tilde{\mathbf{x}}_1-\theta_1\mathbf{x}_1^*\right]\right\rangle. \tag{5.72}$$

For the last term in the right hand of (5.72), we have

$$\frac{\beta}{\theta_1}\left\langle\mathbf{A}_2\mathbf{x}_2^{k+1}-\mathbf{A}_2\mathbf{y}_2^k,\mathbf{A}_1\left[\mathbf{x}_1^{k+1}-(1-\theta_1-\theta_2)\mathbf{x}_1^k-\theta_2\tilde{\mathbf{x}}_1-\theta_1\mathbf{x}_1^*\right]\right\rangle$$

$$\stackrel{a}{=}\frac{\beta}{\theta_1}\left\langle\mathbf{A}_2\mathbf{x}_2^{k+1}-\mathbf{A}_2\mathbf{v}-\left(\mathbf{A}_2\mathbf{y}_2^k-\mathbf{A}_2\mathbf{v}\right),\right.$$

$$\left.\mathbf{A}_1\left[\mathbf{x}_1^{k+1}-(1-\theta_1-\theta_2)\mathbf{x}_1^k-\theta_2\tilde{\mathbf{x}}_1-\theta_1\mathbf{x}_1^*\right]-\mathbf{0}\right\rangle$$

$$\stackrel{b}{=}\frac{\beta}{2\theta_1}\left\|\mathbf{A}_2\mathbf{x}_2^{k+1}-\mathbf{A}_2\mathbf{v}+\mathbf{A}_1\left[\mathbf{x}_1^{k+1}-(1-\theta_1-\theta_2)\mathbf{x}_1^k-\theta_2\tilde{\mathbf{x}}_1-\theta_1\mathbf{x}_1^*\right]\right\|^2$$

$$-\frac{\beta}{2\theta_1}\left\|\mathbf{A}_2\mathbf{x}_2^{k+1}-\mathbf{A}_2\mathbf{v}\right\|^2+\frac{\beta}{2\theta_1}\left\|\mathbf{A}_2\mathbf{y}_2^k-\mathbf{A}_2\mathbf{v}\right\|^2$$

$$-\frac{\beta}{2\theta_1}\left\|\mathbf{A}_2\mathbf{y}_2^k-\mathbf{A}_2\mathbf{v}+\mathbf{A}_1\left[\mathbf{x}_1^{k+1}-(1-\theta_1-\theta_2)\mathbf{x}_1^k-\theta_2\tilde{\mathbf{x}}_1-\theta_1\mathbf{x}_1^*\right]\right\|^2$$

$$\stackrel{c}{=}\frac{\theta_1}{2\beta}\left\|\hat{\boldsymbol{\lambda}}^{k+1}-\hat{\boldsymbol{\lambda}}^k\right\|^2-\frac{\beta}{2\theta_1}\left\|\mathbf{A}_2\mathbf{x}_2^{k+1}-\mathbf{A}_2\mathbf{v}\right\|^2+\frac{\beta}{2\theta_1}\left\|\mathbf{A}_2\mathbf{y}_2^k-\mathbf{A}_2\mathbf{v}\right\|^2$$

$$-\frac{\beta}{2\theta_1}\left\|\mathbf{A}_2\mathbf{y}_2^k-\mathbf{A}_2\mathbf{v}+\mathbf{A}_1\left[\mathbf{x}_1^{k+1}-(1-\theta_1-\theta_2)\mathbf{x}_1^k-\theta_2\tilde{\mathbf{x}}_1-\theta_1\mathbf{x}_1^*\right]\right\|^2, \tag{5.73}$$

where in $\stackrel{a}{=}$ we set

$$\mathbf{v}=(1-\theta_1-\theta_2)\mathbf{x}_2^k+\theta_2\tilde{\mathbf{x}}_2+\theta_1\mathbf{x}_2^*,$$

$\overset{b}{=}$ uses (A.3), and $\overset{c}{=}$ uses (5.65). Substituting (5.73) into (5.72), we have

$$\mathbb{E}_{i_k} \tilde{L} \left( \mathbf{x}_1^{k+1}, \mathbf{x}_2^{k+1}, \boldsymbol{\lambda}^* \right) - \theta_2 \tilde{L} \left( \tilde{\mathbf{x}}_1, \tilde{\mathbf{x}}_2, \boldsymbol{\lambda}^* \right) - (1 - \theta_1 - \theta_2) \tilde{L} \left( \mathbf{x}_1^k, \mathbf{x}_2^k, \boldsymbol{\lambda}^* \right)$$

$$\leq \frac{\theta_1}{2\beta} \left( \left\| \hat{\boldsymbol{\lambda}}^k - \boldsymbol{\lambda}^* \right\|^2 - \mathbb{E}_{i_k} \left\| \hat{\boldsymbol{\lambda}}^{k+1} - \boldsymbol{\lambda}^* \right\|^2 \right)$$

$$+ \frac{1}{2} \left\| \mathbf{y}_1^k - (1 - \theta_1 - \theta_2) \mathbf{x}_1^k - \theta_2 \tilde{\mathbf{x}}_1 - \theta_1 \mathbf{x}_1^* \right\|_{\mathbf{G}_1}^2$$

$$- \frac{1}{2} \mathbb{E}_{i_k} \left\| \mathbf{x}_1^{k+1} - (1 - \theta_1 - \theta_2) \mathbf{x}_1^k - \theta_2 \tilde{\mathbf{x}}_1 - \theta_1 \mathbf{x}_1^* \right\|_{\mathbf{G}_1}^2$$

$$+ \frac{1}{2} \left\| \mathbf{y}_2^k - (1 - \theta_1 - \theta_2) \mathbf{x}_2^k - \theta_2 \tilde{\mathbf{x}}_2 - \theta_1 \mathbf{x}_2^* \right\|_{\left( \alpha L_2 + \frac{\beta}{\theta_1} \|\mathbf{A}_2\|_2^2 \right) \mathbf{I}}^2$$

$$- \frac{1}{2} \mathbb{E}_{i_k} \left\| \mathbf{x}_2^{k+1} - (1 - \theta_1 - \theta_2) \mathbf{x}_2^k - \theta_2 \tilde{\mathbf{x}}_2 - \theta_1 \mathbf{x}_2^* \right\|_{\left( \alpha L_2 + \frac{\beta}{\theta_1} \|\mathbf{A}_2\|_2^2 \right) \mathbf{I}}^2$$

$$- \frac{1}{2} \mathbb{E}_{i_k} \left\| \mathbf{x}_1^{k+1} - \mathbf{y}_1^k \right\|_{\frac{\beta}{\theta_1} \|\mathbf{A}_1\|_2^2 \mathbf{I} - \frac{\beta}{\theta_1} \mathbf{A}_1^T \mathbf{A}_1}^2$$

$$- \frac{1}{2} \mathbb{E}_{i_k} \left\| \mathbf{x}_2^{k+1} - \mathbf{y}_2^k \right\|_{\frac{\beta}{\theta_1} \|\mathbf{A}_2\|_2^2 \mathbf{I} - \frac{\beta}{\theta_1} \mathbf{A}_2^T \mathbf{A}_2}^2$$

$$- \frac{\beta}{2\theta_1} \mathbb{E}_{i_k} \left\| \mathbf{A}_2 \mathbf{y}_2^k - \mathbf{A}_2 \mathbf{v} \right\|$$

$$+ \mathbf{A}_1 \left[ \mathbf{x}_1^{k+1} - (1 - \theta_1 - \theta_2) \mathbf{x}_1^k - \theta_2 \tilde{\mathbf{x}}_1 - \theta_1 \mathbf{x}_1^* \right] \Big\|^2. \tag{5.74}$$

Since the last three terms in the right hand of (5.74) are non-positive, we obtain (5.49). □

For the $\hat{\boldsymbol{\lambda}}^k$ defined in (5.63), besides properties (5.64) and (5.65), we further prove that it has the following property.

**Lemma 5.6**

$$\hat{\boldsymbol{\lambda}}_s^0 = \hat{\boldsymbol{\lambda}}_{s-1}^m, \quad s \geq 1. \tag{5.75}$$

*Proof*

$$\hat{\boldsymbol{\lambda}}_s^0 \overset{a}{=} \tilde{\boldsymbol{\lambda}}_s^0 + \frac{\beta(1 - \theta_{1,s})}{\theta_{1,s}} \left( \mathbf{A}_1 \mathbf{x}_{s-1,1}^m + \mathbf{A}_2 \mathbf{x}_{s-1,2}^m - \mathbf{b} \right)$$

$$\overset{b}{=} \tilde{\boldsymbol{\lambda}}_s^0 + \beta \left( \frac{1}{\theta_{1,s-1}} + \tau - 1 \right) \left( \mathbf{A}_1 \mathbf{x}_{s-1,1}^m + \mathbf{A}_2 \mathbf{x}_{s-1,2}^m - \mathbf{b} \right)$$

$$\overset{c}{=} \boldsymbol{\lambda}_{s-1}^{m-1} - \beta(\tau - 1)\left(\mathbf{A}_1\mathbf{x}_{s-1,1}^m + \mathbf{A}_2\mathbf{x}_{s-1,2}^m - \mathbf{b}\right)$$

$$+ \beta\left(\frac{1}{\theta_{1,s-1}} + \tau - 1\right)\left(\mathbf{A}_1\mathbf{x}_{s-1,1}^m + \mathbf{A}_2\mathbf{x}_{s-1,2}^m - \mathbf{b}\right)$$

$$= \boldsymbol{\lambda}_{s-1}^{m-1} + \frac{\beta}{\theta_{1,s-1}}\left(\mathbf{A}_1\mathbf{x}_{s-1,1}^m + \mathbf{A}_2\mathbf{x}_{s-1,2}^m - \mathbf{b}\right)$$

$$\overset{d}{=} \tilde{\boldsymbol{\lambda}}_{s-1}^m - \left(\beta - \frac{\beta}{\theta_{1,s-1}}\right)\left(\mathbf{A}_1\mathbf{x}_{s-1,1}^m + \mathbf{A}_2\mathbf{x}_{s-1,2}^m - \mathbf{b}\right)$$

$$= \hat{\boldsymbol{\lambda}}_{s-1}^m,$$

where $\overset{a}{=}$ uses (5.63) and $\mathbf{x}_s^0 = \mathbf{x}_{s-1}^m$, $\overset{b}{=}$ uses the fact that $\frac{1}{\theta_{1,s}} = \frac{1}{\theta_{1,s-1}} + \tau$, $\overset{c}{=}$ uses

$$\tilde{\boldsymbol{\lambda}}_{s+1}^0 = \boldsymbol{\lambda}_s^{m-1} + \beta(1 - \tau)\left(\mathbf{A}_1\mathbf{x}_{s,1}^m + \mathbf{A}_2\mathbf{x}_{s,2}^m - \mathbf{b}\right)$$

in Algorithm 5.4, and $\overset{d}{=}$ uses (5.67).                                                                    □

Now we can prove the following result.

**Theorem 5.3** *For Algorithm 5.4, we have*

$$\mathbb{E}\left[\frac{1}{2\beta}\left\|\frac{\beta(m-1)(\theta_2 + \theta_{1,S}) + \beta}{\theta_{1,S}}\left(\mathbf{A}\hat{\mathbf{x}}_S - \mathbf{b}\right)\right.\right.$$

$$\left.\left. -\frac{\beta(m-1)\theta_2}{\theta_{1,0}}\left(\mathbf{A}\mathbf{x}_0^0 - \mathbf{b}\right) + \tilde{\boldsymbol{\lambda}}_0^0 - \boldsymbol{\lambda}^*\right\|^2\right]$$

$$+ \mathbb{E}\left[\frac{(m-1)(\theta_2 + \theta_{1,S}) + 1}{\theta_{1,S}}\left(J(\hat{\mathbf{x}}_S) - J(\mathbf{x}^*) + \langle\boldsymbol{\lambda}^*, \mathbf{A}\hat{\mathbf{x}}_S - \mathbf{b}\rangle\right)\right]$$

$$\leq C_3\left(J\left(\mathbf{x}_0^0\right) - J\left(\mathbf{x}^*\right) + \langle\boldsymbol{\lambda}^*, \mathbf{A}\mathbf{x}_0^0 - \mathbf{b}\rangle\right)$$

$$+ \frac{1}{2\beta}\left\|\tilde{\boldsymbol{\lambda}}_0^0 + \frac{\beta(1 - \theta_{1,0})}{\theta_{1,0}}\left(\mathbf{A}\mathbf{x}_0^0 - \mathbf{b}\right) - \boldsymbol{\lambda}^*\right\|^2$$

$$+ \frac{1}{2}\left\|\mathbf{x}_{0,1}^0 - \mathbf{x}_1^*\right\|_{(\theta_{1,0}L_1 + \beta\|\mathbf{A}_1\|_2^2)\mathbf{I} - \beta\mathbf{A}_1^T\mathbf{A}_1}^2$$

$$+ \frac{1}{2}\left\|\mathbf{x}_{0,2}^0 - \mathbf{x}_2^*\right\|_{\left[\left(1 + \frac{1}{b\theta_2}\right)\theta_{1,0}L_2 + \beta\|\mathbf{A}_2\|_2^2\right]\mathbf{I}}^2, \qquad (5.76)$$

*where* $C_3 = \frac{1 - \theta_{1,0} + (m-1)\theta_2}{\theta_{1,0}}$.

**Proof** Taking full expectation over the first $k + 1$ iterations for (5.49) and dividing $\theta_1$ on both sides of it, we obtain

$$
\frac{1}{\theta_1} \mathbb{E}_{,s} \tilde{L}\left(\mathbf{x}_1^{k+1}, \mathbf{x}_2^{k+1}, \boldsymbol{\lambda}^*\right) - \frac{\theta_2}{\theta_1} \tilde{L}\left(\tilde{\mathbf{x}}_1, \tilde{\mathbf{x}}_2, \boldsymbol{\lambda}^*\right) - \frac{1 - \theta_1 - \theta_2}{\theta_1} \mathbb{E}_{,s} \tilde{L}\left(\mathbf{x}_1^k, \mathbf{x}_2^k, \boldsymbol{\lambda}^*\right)
$$

$$
\leq \frac{1}{2\beta}\left(\mathbb{E}_{,s}\left\|\hat{\boldsymbol{\lambda}}^k - \boldsymbol{\lambda}^*\right\|^2 - \mathbb{E}_{,s}\left\|\hat{\boldsymbol{\lambda}}^{k+1} - \boldsymbol{\lambda}^*\right\|^2\right)
$$

$$
+ \frac{\theta_1}{2} \mathbb{E}_{,s}\left\|\frac{1}{\theta_1}\left[\mathbf{y}_1^k - (1 - \theta_1 - \theta_2)\mathbf{x}_1^k - \theta_2\tilde{\mathbf{x}}_1\right] - \mathbf{x}_1^*\right\|^2_{\left(L_1 + \frac{\beta}{\theta_1}\|\mathbf{A}_1\|_2^2\right)\mathbf{I} - \frac{\beta}{\theta_1}\mathbf{A}_1^T\mathbf{A}_1}
$$

$$
- \frac{\theta_1}{2} \mathbb{E}_{,s}\left\|\frac{1}{\theta_1}\left[\mathbf{x}_1^{k+1} - (1 - \theta_1 - \theta_2)\mathbf{x}_1^k - \theta_2\tilde{\mathbf{x}}_1\right] - \mathbf{x}_1^*\right\|^2_{\left(L_1 + \frac{\beta}{\theta_1}\|\mathbf{A}_1\|_2^2\right)\mathbf{I} - \frac{\beta}{\theta_1}\mathbf{A}_1^T\mathbf{A}_1}
$$

$$
+ \frac{\theta_1}{2} \mathbb{E}_{,s}\left\|\frac{1}{\theta_1}\left[\mathbf{y}_2^k - (1 - \theta_1 - \theta_2)\mathbf{x}_2^k - \theta_2\tilde{\mathbf{x}}_2\right] - \mathbf{x}_2^*\right\|^2_{\left(\alpha L_2 + \frac{\beta}{\theta_1}\|\mathbf{A}_2\|_2^2\right)\mathbf{I}}
$$

$$
- \frac{\theta_1}{2} \mathbb{E}_{,s}\left\|\frac{1}{\theta_1}\left[\mathbf{x}_2^{k+1} - (1 - \theta_1 - \theta_2)\mathbf{x}_2^k - \theta_2\tilde{\mathbf{x}}_2\right] - \mathbf{x}_2^*\right\|^2_{\left(\alpha L_2 + \frac{\beta}{\theta_1}\|\mathbf{A}_2\|_2^2\right)\mathbf{I}},
$$

$$
\tag{5.77}
$$

where $\mathbb{E}_{,s}$ denotes taking full expectation on $k + 1$ inner iterations when fixing the first $s - 1$ epochs. Since

$$
\mathbf{y}^k = \mathbf{x}^k + (1 - \theta_1 - \theta_2)\left(\mathbf{x}^k - \mathbf{x}^{k-1}\right), \quad k \geq 1,
$$

we obtain

$$
\frac{1}{\theta_1} \mathbb{E}_{,s} \tilde{L}\left(\mathbf{x}_1^{k+1}, \mathbf{x}_2^{k+1}, \boldsymbol{\lambda}^*\right) - \frac{\theta_2}{\theta_1} \tilde{L}\left(\tilde{\mathbf{x}}_1, \tilde{\mathbf{x}}_2, \boldsymbol{\lambda}^*\right) - \frac{1 - \theta_1 - \theta_2}{\theta_1} \mathbb{E}_{,s} \tilde{L}\left(\mathbf{x}_1^k, \mathbf{x}_2^k, \boldsymbol{\lambda}^*\right)
$$

$$
\leq \frac{1}{2\beta}\left(\mathbb{E}_{,s}\left\|\hat{\boldsymbol{\lambda}}^k - \boldsymbol{\lambda}^*\right\|^2 - \mathbb{E}_{,s}\left\|\hat{\boldsymbol{\lambda}}^{k+1} - \boldsymbol{\lambda}^*\right\|^2\right)
$$

$$
+ \frac{\theta_1}{2} \mathbb{E}_{,s}\left\|\frac{1}{\theta_1}\left[\mathbf{x}_1^k - (1 - \theta_1 - \theta_2)\mathbf{x}_1^{k-1} - \theta_2\tilde{\mathbf{x}}_1\right] - \mathbf{x}_1^*\right\|^2_{\left(L_1 + \frac{\beta}{\theta_1}\|\mathbf{A}_1\|_2^2\right)\mathbf{I} - \frac{\beta}{\theta_1}\mathbf{A}_1^T\mathbf{A}_1}
$$

$$
- \frac{\theta_1}{2} \mathbb{E}_{,s}\left\|\frac{1}{\theta_1}\left[\mathbf{x}_1^{k+1} - (1 - \theta_1 - \theta_2)\mathbf{x}_1^k - \theta_2\tilde{\mathbf{x}}_1\right] - \mathbf{x}_1^*\right\|^2_{\left(L_1 + \frac{\beta}{\theta_1}\|\mathbf{A}_1\|_2^2\right)\mathbf{I} - \frac{\beta}{\theta_1}\mathbf{A}_1^T\mathbf{A}_1}
$$

$$
+ \frac{\theta_1}{2} \mathbb{E}_{,s}\left\|\frac{1}{\theta_1}\left[\mathbf{x}_2^k - (1 - \theta_1 - \theta_2)\mathbf{x}_2^{k-1} - \theta_2\tilde{\mathbf{x}}_2\right] - \mathbf{x}_2^*\right\|^2_{\left(\alpha L_2 + \frac{\beta}{\theta_1}\|\mathbf{A}_2\|_2^2\right)\mathbf{I}}
$$

$$-\frac{\theta_1}{2}\mathbb{E}_{,s}\left\|\frac{1}{\theta_1}\left[\mathbf{x}_2^{k+1}-(1-\theta_1-\theta_2)\mathbf{x}_2^k-\theta_2\tilde{\mathbf{x}}_2\right]-\mathbf{x}_2^*\right\|^2_{\left(\alpha L_2+\frac{\beta}{\theta_1}\|\mathbf{A}_2\|_2^2\right)\mathbf{I}},$$

$$k\geq 1. \qquad (5.78)$$

Adding back the subscript $s$, taking full expectation on the first $s$ epochs, and then summing (5.77) with $k$ from 0 to $m-1$ (for $k\geq 1$, using (5.78)), we have

$$\frac{1}{\theta_{1,s}}\mathbb{E}\left(L(\mathbf{x}_s^m,\boldsymbol{\lambda}^*)-L(\mathbf{x}^*,\boldsymbol{\lambda}^*)\right)+\frac{\theta_2+\theta_{1,s}}{\theta_{1,s}}\sum_{k=1}^{m-1}\mathbb{E}\left(L\left(\mathbf{x}_s^k,\boldsymbol{\lambda}^*\right)-L\left(\mathbf{x}^*,\boldsymbol{\lambda}^*\right)\right)$$

$$\leq\frac{1-\theta_{1,s}-\theta_2}{\theta_{1,s}}\mathbb{E}\left(L\left(\mathbf{x}_s^0,\boldsymbol{\lambda}^*\right)-L\left(\mathbf{x}^*,\boldsymbol{\lambda}^*\right)\right)+\frac{m\theta_2}{\theta_{1,s}}\mathbb{E}\left(L\left(\tilde{\mathbf{x}}_s,\boldsymbol{\lambda}^*\right)-L\left(\mathbf{x}^*,\boldsymbol{\lambda}^*\right)\right)$$

$$+\frac{1}{2}\mathbb{E}\left\|\frac{1}{\theta_{1,s}}\left[\mathbf{y}_{s,1}^0-\theta_2\tilde{\mathbf{x}}_{s,1}-(1-\theta_{1,s}-\theta_2)\mathbf{x}_{s,1}^0\right]\right.$$

$$\left.-\mathbf{x}_1^*\right\|^2_{\left(\theta_{1,s}L_1+\beta\|\mathbf{A}_1\|_2^2\right)\mathbf{I}-\beta\mathbf{A}_1^T\mathbf{A}_1}$$

$$-\frac{1}{2}\mathbb{E}\left\|\frac{1}{\theta_{1,s}}\left[\mathbf{x}_{s,1}^m-\theta_2\tilde{\mathbf{x}}_{s,1}-(1-\theta_{1,s}-\theta_2)\mathbf{x}_{s,1}^{m-1}\right]\right.$$

$$\left.-\mathbf{x}_1^*\right\|^2_{\left(\theta_{1,s}L_1+\beta\|\mathbf{A}_1\|_2^2\right)\mathbf{I}-\beta\mathbf{A}_1^T\mathbf{A}_1}$$

$$+\frac{1}{2}\mathbb{E}\left\|\frac{1}{\theta_{1,s}}\left[\mathbf{y}_{s,2}^0-\theta_2\tilde{\mathbf{x}}_{s,2}-(1-\theta_{1,s}-\theta_2)\mathbf{x}_{s,2}^0\right]-\mathbf{x}_2^*\right\|^2_{\left(\alpha\theta_{1,s}L_2+\beta\|\mathbf{A}_2\|_2^2\right)\mathbf{I}}$$

$$-\frac{1}{2}\mathbb{E}\left\|\frac{1}{\theta_{1,s}}\left[\mathbf{x}_{s,2}^m-\theta_2\tilde{\mathbf{x}}_{s,2}-(1-\theta_{1,s}-\theta_2)\mathbf{x}_{s,2}^{m-1}\right]-\mathbf{x}_2^*\right\|^2_{\left(\alpha\theta_{1,s}L_2+\beta\|\mathbf{A}_2\|_2^2\right)\mathbf{I}}$$

$$+\frac{1}{2\beta}\left(\mathbb{E}\left\|\hat{\boldsymbol{\lambda}}_s^0-\boldsymbol{\lambda}^*\right\|^2-\mathbb{E}\left\|\hat{\boldsymbol{\lambda}}_s^m-\boldsymbol{\lambda}^*\right\|^2\right),\quad s\geq 0, \qquad (5.79)$$

where we use $L(\mathbf{x}_s^k,\boldsymbol{\lambda}^*)$ and $L(\tilde{\mathbf{x}}_s,\boldsymbol{\lambda}^*)$ to denote $L(\mathbf{x}_{s,1}^k,\mathbf{x}_{s,2}^k,\boldsymbol{\lambda}^*)$ and $L(\tilde{\mathbf{x}}_{s,1},\tilde{\mathbf{x}}_{s,2},\boldsymbol{\lambda}^*)$, respectively. Since $L(\mathbf{x},\boldsymbol{\lambda}^*)$ is convex for $\mathbf{x}$, we have

$$mL(\tilde{\mathbf{x}}_s,\boldsymbol{\lambda}^*)$$

$$=mL\left(\frac{1}{m}\left[\left(1-\frac{(\tau-1)\theta_{1,s}}{\theta_2}\right)\mathbf{x}_{s-1}^m+\left(1+\frac{(\tau-1)\theta_{1,s}}{(m-1)\theta_2}\right)\sum_{k=1}^{m-1}\mathbf{x}_{s-1}^k\right],\boldsymbol{\lambda}^*\right)$$

$$\leq\left[1-\frac{(\tau-1)\theta_{1,s}}{\theta_2}\right]L\left(\mathbf{x}_{s-1}^m,\boldsymbol{\lambda}^*\right)+\left[1+\frac{(\tau-1)\theta_{1,s}}{(m-1)\theta_2}\right]\sum_{k=1}^{m-1}L\left(\mathbf{x}_{s-1}^k,\boldsymbol{\lambda}^*\right).$$

$$(5.80)$$

Substituting (5.80) into (5.79), and using $\mathbf{x}_{s-1}^m = \mathbf{x}_s^0$, we have

$$
\frac{1}{\theta_{1,s}} \mathbb{E}\left(L\left(\mathbf{x}_s^m, \boldsymbol{\lambda}^*\right) - L\left(\mathbf{x}^*, \boldsymbol{\lambda}^*\right)\right) + \frac{\theta_2 + \theta_{1,s}}{\theta_{1,s}} \sum_{k=1}^{m-1} \mathbb{E}\left(L\left(\mathbf{x}_s^k, \boldsymbol{\lambda}^*\right) - L\left(\mathbf{x}^*, \boldsymbol{\lambda}^*\right)\right)
$$

$$
\leq \frac{1 - \tau\theta_{1,s}}{\theta_{1,s}} \mathbb{E}\left(L\left(\mathbf{x}_{s-1}^m, \boldsymbol{\lambda}^*\right) - L\left(\mathbf{x}^*, \boldsymbol{\lambda}^*\right)\right)
$$

$$
+ \frac{\theta_2 + \frac{\tau-1}{m-1}\theta_{1,s}}{\theta_{1,s}} \sum_{k=1}^{m-1} \mathbb{E}\left(L\left(\mathbf{x}_{s-1}^k, \boldsymbol{\lambda}^*\right) - L\left(\mathbf{x}^*, \boldsymbol{\lambda}^*\right)\right)
$$

$$
+ \frac{1}{2}\mathbb{E}\left\| \frac{1}{\theta_{1,s}}\left[\mathbf{y}_{s,1}^0 - \theta_2\tilde{\mathbf{x}}_{s,1} - (1 - \theta_{1,s} - \theta_2)\mathbf{x}_{s,1}^0\right] \right.
$$

$$
\left. - \mathbf{x}_1^* \right\|_{(\theta_{1,s}L_1 + \beta\|\mathbf{A}_1\|_2^2)\mathbf{I} - \beta\mathbf{A}_1^T\mathbf{A}_1}^2
$$

$$
- \frac{1}{2}\mathbb{E}\left\| \frac{1}{\theta_{1,s}}\left[\mathbf{x}_{s,1}^m - \theta_2\tilde{\mathbf{x}}_{s,1} - (1 - \theta_{1,s} - \theta_2)\mathbf{x}_{s,1}^{m-1}\right] \right.
$$

$$
\left. - \mathbf{x}_1^* \right\|_{(\theta_{1,s}L_1 + \beta\|\mathbf{A}_1\|_2^2)\mathbf{I} - \beta\mathbf{A}_1^T\mathbf{A}_1}^2
$$

$$
+ \frac{1}{2}\mathbb{E}\left\| \frac{1}{\theta_{1,s}}\left[\mathbf{y}_{s,2}^0 - \theta_2\tilde{\mathbf{x}}_{s,2} - (1 - \theta_{1,s} - \theta_2)\mathbf{x}_{s,2}^0\right] - \mathbf{x}_2^* \right\|_{(\alpha\theta_{1,s}L_2 + \beta\|\mathbf{A}_2\|_2^2)\mathbf{I}}^2
$$

$$
- \frac{1}{2}\mathbb{E}\left\| \frac{1}{\theta_{1,s}}\left[\mathbf{x}_{s,2}^m - \theta_2\tilde{\mathbf{x}}_{s,2} - (1 - \theta_{1,s} - \theta_2)\mathbf{x}_{s,2}^{m-1}\right] - \mathbf{x}_2^* \right\|_{(\alpha\theta_{1,s}L_2 + \beta\|\mathbf{A}_2\|_2^2)\mathbf{I}}^2
$$

$$
+ \frac{1}{2\beta}\left(\mathbb{E}\left\|\hat{\boldsymbol{\lambda}}_s^0 - \boldsymbol{\lambda}^*\right\|^2 - \mathbb{E}\left\|\hat{\boldsymbol{\lambda}}_s^m - \boldsymbol{\lambda}^*\right\|^2\right), \quad s \geq 1. \tag{5.81}
$$

Then from the setting of $\theta_{1,s} = \frac{1}{c+\tau s}$, where $c = 2$, and $\theta_2 = \frac{m-\tau}{\tau(m-1)}$, we have

$$
\frac{1}{\theta_{1,s}} = \frac{1 - \tau\theta_{1,s+1}}{\theta_{1,s+1}}, \quad s \geq 0, \tag{5.82}
$$

and

$$
\frac{\theta_2 + \theta_{1,s}}{\theta_{1,s}} = \frac{\theta_2}{\theta_{1,s+1}} - \tau\theta_2 + 1 = \frac{\theta_2 + \frac{\tau-1}{m-1}\theta_{1,s+1}}{\theta_{1,s+1}}, \quad s \geq 0. \tag{5.83}
$$

Substituting (5.82) into the first term and (5.83) into the second term in the right hand side of (5.81), we obtain

$$
\frac{1}{\theta_{1,s}} \mathbb{E}\left(L\left(\mathbf{x}_s^m, \boldsymbol{\lambda}^*\right) - L\left(\mathbf{x}^*, \boldsymbol{\lambda}^*\right)\right) + \frac{\theta_2 + \theta_{1,s}}{\theta_{1,s}} \sum_{k=1}^{m-1} \mathbb{E}\left(L\left(\mathbf{x}_s^k, \boldsymbol{\lambda}^*\right) - L\left(\mathbf{x}^*, \boldsymbol{\lambda}^*\right)\right)
$$

$$
\leq \frac{1}{\theta_{1,s-1}} \mathbb{E}\left(L\left(\mathbf{x}_{s-1}^m, \boldsymbol{\lambda}^*\right) - L\left(\mathbf{x}^*, \boldsymbol{\lambda}^*\right)\right)
$$

$$
+ \frac{\theta_2 + \theta_{1,s-1}}{\theta_{1,s-1}} \sum_{k=1}^{m-1} \mathbb{E}\left(L\left(\mathbf{x}_{s-1}^k, \boldsymbol{\lambda}^*\right) - L\left(\mathbf{x}^*, \boldsymbol{\lambda}^*\right)\right)
$$

$$
+ \frac{1}{2}\mathbb{E}\left\|\frac{1}{\theta_{1,s}}\left[\mathbf{y}_{s,1}^0 - \theta_2\tilde{\mathbf{x}}_{s,1} - (1 - \theta_{1,s} - \theta_2)\mathbf{x}_{s,1}^0\right]\right.
$$

$$
\left. - \mathbf{x}_1^* \right\|^2_{\left(\theta_{1,s}L_1 + \beta\|\mathbf{A}_1\|_2^2\right)\mathbf{I} - \beta\mathbf{A}_1^T\mathbf{A}_1}
$$

$$
- \frac{1}{2}\mathbb{E}\left\|\frac{1}{\theta_{1,s}}\left[\mathbf{x}_{s,1}^m - \theta_2\tilde{\mathbf{x}}_{s,1} - (1 - \theta_{1,s} - \theta_2)\mathbf{x}_{s,1}^{m-1}\right]\right.
$$

$$
\left. - \mathbf{x}_1^* \right\|^2_{\left(\theta_{1,s}L_1 + \beta\|\mathbf{A}_1\|_2^2\right)\mathbf{I} - \beta\mathbf{A}_1^T\mathbf{A}_1}
$$

$$
+ \frac{1}{2}\mathbb{E}\left\|\frac{1}{\theta_{1,s}}\left[\mathbf{y}_{s,2}^0 - \theta_2\tilde{\mathbf{x}}_{s,2} - (1 - \theta_{1,s} - \theta_2)\mathbf{x}_{s,2}^0\right] - \mathbf{x}_2^* \right\|^2_{\left(\alpha\theta_{1,s}L_2 + \beta\|\mathbf{A}_2\|_2^2\right)\mathbf{I}}
$$

$$
- \frac{1}{2}\mathbb{E}\left\|\frac{1}{\theta_{1,s}}\left[\mathbf{x}_{s,2}^m - \theta_2\tilde{\mathbf{x}}_{s,2} - (1 - \theta_{1,s} - \theta_2)\mathbf{x}_{s,2}^{m-1}\right] - \mathbf{x}_2^* \right\|^2_{\left(\alpha\theta_{1,s}L_2 + \beta\|\mathbf{A}_2\|_2^2\right)\mathbf{I}}
$$

$$
+ \frac{1}{2\beta}\left(\mathbb{E}\left\|\hat{\boldsymbol{\lambda}}_s^0 - \boldsymbol{\lambda}^*\right\|^2 - \mathbb{E}\left\|\hat{\boldsymbol{\lambda}}_s^m - \boldsymbol{\lambda}^*\right\|^2\right), \quad s \geq 1. \tag{5.84}
$$

When $k = 0$, for

$$
\mathbf{y}_{s+1}^0 = (1 - \theta_2)\mathbf{x}_s^m + \theta_2\tilde{\mathbf{x}}_{s+1}
$$

$$
+ \frac{\theta_{1,s+1}}{\theta_{1,s}}\left[(1 - \theta_{1,s})\mathbf{x}_s^m - (1 - \theta_{1,s} - \theta_2)\mathbf{x}_s^{m-1} - \theta_2\tilde{\mathbf{x}}_s\right],
$$

we obtain

$$
\frac{1}{\theta_{1,s}}\left[\mathbf{x}_s^m - \theta_2\tilde{\mathbf{x}}_s - (1 - \theta_{1,s} - \theta_2)\mathbf{x}_s^{m-1}\right]
$$

$$
= \frac{1}{\theta_{1,s+1}}\left[\mathbf{y}_{s+1}^0 - \theta_2\tilde{\mathbf{x}}_{s+1} - (1 - \theta_{1,s+1} - \theta_2)\mathbf{x}_{s+1}^0\right]. \tag{5.85}
$$

Substituting (5.85) into the third and the fifth terms in the right hand side of (5.84) and substituting (5.75) into the last term in the right hand of (5.84), we obtain

$$
\frac{1}{\theta_{1,s}} \mathbb{E}\left(L\left(\mathbf{x}_s^m, \boldsymbol{\lambda}^*\right) - L\left(\mathbf{x}^*, \boldsymbol{\lambda}^*\right)\right) + \frac{\theta_2 + \theta_{1,s}}{\theta_{1,s}} \sum_{k=1}^{m-1} \mathbb{E}\left(L\left(\mathbf{x}_s^k, \boldsymbol{\lambda}^*\right) - L\left(\mathbf{x}^*, \boldsymbol{\lambda}^*\right)\right)
$$

$$
\leq \frac{1}{\theta_{1,s-1}} \mathbb{E}\left(L\left(\mathbf{x}_{s-1}^m, \boldsymbol{\lambda}^*\right) - L\left(\mathbf{x}^*, \boldsymbol{\lambda}^*\right)\right)
$$

$$
+ \frac{\theta_2 + \theta_{1,s-1}}{\theta_{1,s-1}} \sum_{k=1}^{m-1} \mathbb{E}\left(L\left(\mathbf{x}_{s-1}^k, \boldsymbol{\lambda}^*\right) - L\left(\mathbf{x}^*, \boldsymbol{\lambda}^*\right)\right)
$$

$$
+ \frac{1}{2} \mathbb{E}\left\| \frac{1}{\theta_{1,s-1}}\left[\mathbf{x}_{s-1,1}^m - \theta_2\tilde{\mathbf{x}}_{s-1,1} - (1 - \theta_{1,s-1} - \theta_2)\mathbf{x}_{s-1,1}^{m-1}\right] \right.
$$

$$
\left. - \mathbf{x}_1^* \right\|_{\left(\theta_{1,s}L_1 + \beta\|\mathbf{A}_1\|_2^2\right)\mathbf{I} - \beta\mathbf{A}_1^T\mathbf{A}_1}^2
$$

$$
- \frac{1}{2} \mathbb{E}\left\| \frac{1}{\theta_{1,s}}\left[\mathbf{x}_{s,1}^m - \theta_2\tilde{\mathbf{x}}_{s,1} - (1 - \theta_{1,s} - \theta_2)\mathbf{x}_{s,1}^{m-1}\right] \right.
$$

$$
\left. - \mathbf{x}_1^* \right\|_{\left(\theta_{1,s}L_1 + \beta\|\mathbf{A}_1\|_2^2\right)\mathbf{I} - \beta\mathbf{A}_1^T\mathbf{A}_1}^2
$$

$$
+ \frac{1}{2} \mathbb{E}\left\| \frac{1}{\theta_{1,s-1}}\left[\mathbf{x}_{s-1,2}^m - \theta_2\tilde{\mathbf{x}}_{s-1,2} - (1 - \theta_{1,s-1} - \theta_2)\mathbf{x}_{s-1,2}^{m-1}\right] \right.
$$

$$
\left. - \mathbf{x}_2^* \right\|_{\left(\alpha\theta_{1,s}L_2 + \beta\|\mathbf{A}_2\|_2^2\right)\mathbf{I}}^2
$$

$$
- \frac{1}{2} \mathbb{E}\left\| \frac{1}{\theta_{1,s}}\left[\mathbf{x}_{s,2}^m - \theta_2\tilde{\mathbf{x}}_{s,2} - (1 - \theta_{1,s} - \theta_2)\mathbf{x}_{s,2}^{m-1}\right] \right.
$$

$$
\left. - \mathbf{x}_2^* \right\|_{\left(\alpha\theta_{1,s}L_2 + \beta\|\mathbf{A}_2\|_2^2\right)\mathbf{I}}^2
$$

$$
+ \frac{1}{2\beta}\left(\mathbb{E}\left\|\hat{\boldsymbol{\lambda}}_{s-1}^m - \boldsymbol{\lambda}^*\right\|^2 - \mathbb{E}\left\|\hat{\boldsymbol{\lambda}}_s^m - \boldsymbol{\lambda}^*\right\|^2\right), \quad s \geq 1.
$$

Since $\|\mathbf{x}\|_{\mathbf{M}_1}^2 \geq \|\mathbf{x}\|_{\mathbf{M}_2}^2$ if $\mathbf{M}_1 \succeq \mathbf{M}_2$, and $\theta_{1,s-1} \geq \theta_{1,s}$, we have

$$
\frac{1}{\theta_{1,s}} \mathbb{E}\left(L\left(\mathbf{x}_s^m, \boldsymbol{\lambda}^*\right) - L\left(\mathbf{x}^*, \boldsymbol{\lambda}^*\right)\right) + \frac{\theta_2 + \theta_{1,s}}{\theta_{1,s}} \sum_{k=1}^{m-1} \mathbb{E}\left(L\left(\mathbf{x}_s^k, \boldsymbol{\lambda}^*\right) - L\left(\mathbf{x}^*, \boldsymbol{\lambda}^*\right)\right)
$$

$$
\leq \frac{1}{\theta_{1,s-1}} \mathbb{E}\left(L\left(\mathbf{x}_{s-1}^m, \boldsymbol{\lambda}^*\right) - L\left(\mathbf{x}^*, \boldsymbol{\lambda}^*\right)\right)
$$

$$+ \frac{\theta_2 + \theta_{1,s-1}}{\theta_{1,s-1}} \sum_{k=1}^{m-1} \mathbb{E}\left( L\left(\mathbf{x}_{s-1}^k, \boldsymbol{\lambda}^*\right) - L\left(\mathbf{x}^*, \boldsymbol{\lambda}^*\right)\right)$$

$$+ \frac{1}{2}\mathbb{E}\left\| \frac{1}{\theta_{1,s-1}} \left[ \mathbf{x}_{s-1,1}^m - \theta_2\tilde{\mathbf{x}}_{s-1,1} - (1 - \theta_{1,s-1} - \theta_2)\mathbf{x}_{s-1,1}^{m-1}\right]\right.$$
$$\left. - \mathbf{x}_1^* \vphantom{\frac{1}{2}}\right\|^2_{\left(\theta_{1,s-1}L_1 + \beta\|\mathbf{A}_1\|_2^2\right)\mathbf{I} - \beta\mathbf{A}_1^T\mathbf{A}_1}$$

$$- \frac{1}{2}\mathbb{E}\left\| \frac{1}{\theta_{1,s}} \left[ \mathbf{x}_{s,1}^m - \theta_2\tilde{\mathbf{x}}_{s,1} - (1 - \theta_{1,s} - \theta_2)\mathbf{x}_{s,1}^{m-1}\right]\right.$$
$$\left. - \mathbf{x}_1^* \vphantom{\frac{1}{2}}\right\|^2_{\left(\theta_{1,s}L_1 + \beta\|\mathbf{A}_1\|_2^2\right)\mathbf{I} - \beta\mathbf{A}_1^T\mathbf{A}_1}$$

$$+ \frac{1}{2}\mathbb{E}\left\| \frac{1}{\theta_{1,s-1}} \left[ \mathbf{x}_{s-1,2}^m - \theta_2\tilde{\mathbf{x}}_{s-1,2} - (1 - \theta_{1,s-1} - \theta_2)\mathbf{x}_{s-1,2}^{m-1}\right]\right.$$
$$\left. - \mathbf{x}_2^* \vphantom{\frac{1}{2}}\right\|^2_{\left(\alpha\theta_{1,s-1}L_2 + \beta\|\mathbf{A}_2\|_2^2\right)\mathbf{I}}$$

$$- \frac{1}{2}\mathbb{E}\left\| \frac{1}{\theta_{1,s}} \left[ \mathbf{x}_{s,2}^m - \theta_2\tilde{\mathbf{x}}_{s,2} - (1 - \theta_{1,s} - \theta_2)\mathbf{x}_{s,2}^{m-1}\right]\right.$$
$$\left. - \mathbf{x}_2^* \vphantom{\frac{1}{2}}\right\|^2_{\left(\alpha\theta_{1,s}L_2 + \beta\|\mathbf{A}_2\|_2^2\right)\mathbf{I}}$$

$$+ \frac{1}{2\beta}\left( \mathbb{E}\left\|\hat{\boldsymbol{\lambda}}_{s-1}^m - \boldsymbol{\lambda}^*\right\|^2 - \mathbb{E}\left\|\hat{\boldsymbol{\lambda}}_s^m - \boldsymbol{\lambda}^*\right\|^2\right), \quad s \geq 1. \tag{5.86}$$

When $s = 0$, via (5.79) and using

$$\mathbf{y}_{0,1}^0 = \tilde{\mathbf{x}}_{0,1} = \mathbf{x}_{0,1}^0, \quad \mathbf{y}_{0,2}^0 = \tilde{\mathbf{x}}_{0,2} = \mathbf{x}_{0,2}^0, \quad \text{and} \quad \theta_{1,0} \geq \theta_{1,1},$$

we obtain

$$\frac{1}{\theta_{1,0}}\mathbb{E}\left( L\left(\mathbf{x}_0^m, \boldsymbol{\lambda}^*\right) - L\left(\mathbf{x}^*, \boldsymbol{\lambda}^*\right)\right) + \frac{\theta_2 + \theta_{1,0}}{\theta_{1,0}} \sum_{k=1}^{m-1} \mathbb{E}\left( L\left(\mathbf{x}_0^k, \boldsymbol{\lambda}^*\right) - L\left(\mathbf{x}^*, \boldsymbol{\lambda}^*\right)\right)$$

$$\leq \frac{1 - \theta_{1,0} + (m-1)\theta_2}{\theta_{1,0}}\left( L\left(\mathbf{x}_0^0, \boldsymbol{\lambda}^*\right) - L\left(\mathbf{x}^*, \boldsymbol{\lambda}^*\right)\right)$$

$$+ \frac{1}{2}\left\| \mathbf{x}_{0,1}^0 - \mathbf{x}_1^*\right\|^2_{\left(\theta_{1,0}L_1 + \beta\|\mathbf{A}_1\|_2^2\right)\mathbf{I} - \beta\mathbf{A}_1^T\mathbf{A}_1}$$

$$-\frac{1}{2}\mathbb{E}\left\|\frac{1}{\theta_{1,0}}\left[\mathbf{x}_{0,1}^m - \theta_2\tilde{\mathbf{x}}_{0,1} - (1-\theta_{1,0}-\theta_2)\mathbf{x}_{0,1}^{m-1}\right]\right.$$

$$\left. - \mathbf{x}_1^*\right\|_{\left(\theta_{1,1}L_1+\beta\|\mathbf{A}_1\|_2^2\right)\mathbf{I}-\beta\mathbf{A}_1^T\mathbf{A}_1}^2$$

$$+\frac{1}{2}\left\|\mathbf{x}_{0,2}^0 - \mathbf{x}_2^*\right\|_{\left(\alpha\theta_{1,0}L_2+\beta\|\mathbf{A}_2\|_2^2\right)\mathbf{I}}^2$$

$$-\frac{1}{2}\mathbb{E}\left\|\frac{1}{\theta_{1,0}}\left[\mathbf{x}_{0,2}^m - \theta_2\tilde{\mathbf{x}}_{0,2} - (1-\theta_{1,0}-\theta_2)\mathbf{x}_{0,2}^{m-1}\right]\right.$$

$$\left. - \mathbf{x}_2^*\right\|_{\left(\alpha\theta_{1,1}L_2+\beta\|\mathbf{A}_2\|_2^2\right)\mathbf{I}}^2$$

$$+\frac{1}{2\beta}\left(\left\|\hat{\boldsymbol{\lambda}}_0^0 - \boldsymbol{\lambda}^*\right\|^2 - \mathbb{E}\left\|\hat{\boldsymbol{\lambda}}_0^m - \boldsymbol{\lambda}^*\right\|^2\right), \tag{5.87}$$

where we use $\theta_{1,0} \geq \theta_{1,1}$ in the fifth and the eighth lines.

Summing (5.86) with $s$ from 1 to $S$ and adding (5.87), we have

$$\frac{1}{\theta_{1,S}}\mathbb{E}\left(L\left(\mathbf{x}_S^m, \boldsymbol{\lambda}^*\right) - L\left(\mathbf{x}^*, \boldsymbol{\lambda}^*\right)\right) + \frac{\theta_{1,S}+\theta_2}{\theta_{1,S}}\sum_{k=1}^{m-1}\mathbb{E}\left(L\left(\mathbf{x}_S^k, \boldsymbol{\lambda}^*\right) - L\left(\mathbf{x}^*, \boldsymbol{\lambda}^*\right)\right)$$

$$\leq \frac{1-\theta_{1,0}+(m-1)\theta_2}{\theta_{1,0}}\left(L\left(\mathbf{x}_0^0, \boldsymbol{\lambda}^*\right) - L\left(\mathbf{x}^*, \boldsymbol{\lambda}^*\right)\right)$$

$$+\frac{1}{2}\left\|\mathbf{x}_{0,1}^0 - \mathbf{x}_1^*\right\|_{\left(\theta_{1,0}L_1+\beta\|\mathbf{A}_1\|_2^2\right)\mathbf{I}-\beta\mathbf{A}_1^T\mathbf{A}_1}^2 + \frac{1}{2}\left\|\mathbf{x}_{0,2}^0 - \mathbf{x}_2^*\right\|_{\left(\alpha\theta_{1,0}L_2+\beta\|\mathbf{A}_2\|_2^2\right)\mathbf{I}}^2$$

$$+\frac{1}{2\beta}\left(\left\|\hat{\boldsymbol{\lambda}}_0^0 - \boldsymbol{\lambda}^*\right\|^2 - \mathbb{E}\left\|\hat{\boldsymbol{\lambda}}_S^m - \boldsymbol{\lambda}^*\right\|^2\right)$$

$$-\frac{1}{2}\mathbb{E}\left\|\frac{1}{\theta_{1,S}}\left[\mathbf{x}_{S,1}^m - \theta_2\tilde{\mathbf{x}}_{S,1} - (1-\theta_{1,S}-\theta_2)\mathbf{x}_{S,1}^{m-1}\right]\right.$$

$$\left. - \mathbf{x}_1^*\right\|_{\left(\theta_{1,S}L_1+\beta\|\mathbf{A}_1\|_2^2\right)\mathbf{I}-\beta\mathbf{A}_1^T\mathbf{A}_1}^2$$

$$-\frac{1}{2}\mathbb{E}\left\|\frac{1}{\theta_{1,S}}\left[\mathbf{x}_{S,2}^m - \theta_2\tilde{\mathbf{x}}_{S,2} - (1-\theta_{1,S}-\theta_2)\mathbf{x}_{S,2}^{m-1}\right]\right.$$

$$\left. - \mathbf{x}_2^*\right\|_{\left(\alpha\theta_{1,S}L_2+\beta\|\mathbf{A}_2\|_2^2\right)\mathbf{I}}^2$$

$$\leq \frac{1-\theta_{1,0}+(m-1)\theta_2}{\theta_{1,0}}\left(L\left(\mathbf{x}_0^0, \boldsymbol{\lambda}^*\right) - L\left(\mathbf{x}^*, \boldsymbol{\lambda}^*\right)\right)$$

$$+ \frac{1}{2} \left\| \mathbf{x}_{0,1}^0 - \mathbf{x}_1^* \right\|^2_{\left( \theta_{1,0} L_1 + \beta \|\mathbf{A}_1\|_2^2 \right) \mathbf{I} - \beta \mathbf{A}_1^T \mathbf{A}_1} + \frac{1}{2} \left\| \mathbf{x}_{0,2}^0 - \mathbf{x}_2^* \right\|^2_{\left( \alpha \theta_{1,0} L_2 + \beta \|\mathbf{A}_2\|_2^2 \right) \mathbf{I}}$$

$$+ \frac{1}{2\beta} \left( \left\| \hat{\boldsymbol{\lambda}}_0^0 - \boldsymbol{\lambda}^* \right\|^2 - \mathbb{E} \left\| \hat{\boldsymbol{\lambda}}_S^m - \boldsymbol{\lambda}^* \right\|^2 \right). \tag{5.88}$$

Now we analyze $\|\hat{\boldsymbol{\lambda}}_S^m - \boldsymbol{\lambda}^*\|^2$. From (5.75), for $s \geq 1$ we have

$$\hat{\boldsymbol{\lambda}}_s^m - \hat{\boldsymbol{\lambda}}_{s-1}^m$$

$$= \hat{\boldsymbol{\lambda}}_s^m - \hat{\boldsymbol{\lambda}}_s^0$$

$$= \sum_{k=1}^m \left( \hat{\boldsymbol{\lambda}}_s^k - \hat{\boldsymbol{\lambda}}_s^{k-1} \right)$$

$$\overset{a}{=} \beta \sum_{k=1}^m \left[ \frac{1}{\theta_{1,s}} \left( \mathbf{A}\mathbf{x}_s^k - \mathbf{b} \right) - \frac{1 - \theta_{1,s} - \theta_2}{\theta_{1,s}} \left( \mathbf{A}\mathbf{x}_s^{k-1} - \mathbf{b} \right) - \frac{\theta_2}{\theta_{1,s}} \left( \mathbf{A}\tilde{\mathbf{x}}_s - \mathbf{b} \right) \right]$$

$$= \frac{\beta}{\theta_{1,s}} \left( \mathbf{A}\mathbf{x}_s^m - \mathbf{b} \right) + \frac{\beta(\theta_2 + \theta_{1,s})}{\theta_{1,s}} \sum_{k=1}^{m-1} \left( \mathbf{A}\mathbf{x}_s^k - \mathbf{b} \right)$$

$$- \frac{\beta(1 - \theta_{1,s} - \theta_2)}{\theta_{1,s}} \left( \mathbf{A}\mathbf{x}_{s-1}^m - \mathbf{b} \right) - \frac{m\beta\theta_2}{\theta_{1,s}} \left( \mathbf{A}\tilde{\mathbf{x}}_s - \mathbf{b} \right)$$

$$\overset{b}{=} \frac{\beta}{\theta_{1,s}} \left( \mathbf{A}\mathbf{x}_s^m - \mathbf{b} \right) + \frac{\beta(\theta_2 + \theta_{1,s})}{\theta_{1,s}} \sum_{k=1}^{m-1} \left( \mathbf{A}\mathbf{x}_s^k - \mathbf{b} \right)$$

$$- \beta \left[ \frac{1 - \theta_{1,s} - (\tau - 1)\theta_{1,s}}{\theta_{1,s}} \left( \mathbf{A}\mathbf{x}_{s-1}^m - \mathbf{b} \right) \right.$$

$$\left. + \frac{\theta_2 + \frac{\tau-1}{m-1}\theta_{1,s}}{\theta_{1,s}} \sum_{k=1}^{m-1} \left( \mathbf{A}\mathbf{x}_{s-1}^k - \mathbf{b} \right) \right]$$

$$\overset{c}{=} \frac{\beta}{\theta_{1,s}} \left( \mathbf{A}\mathbf{x}_s^m - \mathbf{b} \right) + \frac{\beta(\theta_2 + \theta_{1,s})}{\theta_{1,s}} \sum_{k=1}^{m-1} \left( \mathbf{A}\mathbf{x}_s^k - \mathbf{b} \right)$$

$$- \frac{\beta}{\theta_{1,s-1}} \left( \mathbf{A}\mathbf{x}_{s-1}^m - \mathbf{b} \right) - \frac{\beta(\theta_2 + \theta_{1,s-1})}{\theta_{1,s-1}} \sum_{k=1}^{m-1} \left( \mathbf{A}\mathbf{x}_{s-1}^k - \mathbf{b} \right), \tag{5.89}$$

where $\overset{a}{=}$ uses (5.65), $\overset{b}{=}$ uses the definition of $\tilde{\mathbf{x}}_s$, and $\overset{c}{=}$ uses (5.82) and (5.83). When $s = 0$, we can obtain

$$\hat{\lambda}_0^m - \hat{\lambda}_0^0$$

$$= \sum_{k=1}^{m} \left( \hat{\lambda}_0^k - \hat{\lambda}_0^{k-1} \right)$$

$$= \sum_{k=1}^{m} \left[ \frac{\beta}{\theta_{1,0}} \left( \mathbf{A}\mathbf{x}_0^k - \mathbf{b} \right) - \frac{\beta(1 - \theta_{1,0} - \theta_2)}{\theta_{1,0}} \left( \mathbf{A}\mathbf{x}_0^{k-1} - \mathbf{b} \right) - \frac{\theta_2 \beta}{\theta_{1,0}} \left( \mathbf{A}\mathbf{x}_0^0 - \mathbf{b} \right) \right]$$

$$= \frac{\beta}{\theta_{1,0}} \left( \mathbf{A}\mathbf{x}_0^m - \mathbf{b} \right) + \frac{\beta(\theta_2 + \theta_{1,0})}{\theta_{1,0}} \sum_{k=1}^{m-1} \left( \mathbf{A}\mathbf{x}_0^k - \mathbf{b} \right)$$

$$- \frac{\beta[1 - \theta_{1,0} + (m-1)\theta_2]}{\theta_{1,0}} \left( \mathbf{A}\mathbf{x}_0^0 - \mathbf{b} \right). \tag{5.90}$$

Summing (5.89) with $s$ from 1 to $S$ and adding (5.90), we have

$$\hat{\lambda}_S^m - \boldsymbol{\lambda}^*$$

$$= \hat{\lambda}_S^m - \hat{\lambda}_0^0 + \hat{\lambda}_0^0 - \boldsymbol{\lambda}^*$$

$$= \frac{\beta}{\theta_{1,S}} \left( \mathbf{A}\mathbf{x}_S^m - \mathbf{b} \right) + \frac{\beta(\theta_2 + \theta_{1,S})}{\theta_{1,S}} \sum_{k=1}^{m-1} \left( \mathbf{A}\mathbf{x}_S^k - \mathbf{b} \right)$$

$$- \frac{\beta\left[1 - \theta_{1,0} + (m-1)\theta_2\right]}{\theta_{1,0}} \left( \mathbf{A}\mathbf{x}_0^0 - \mathbf{b} \right) + \tilde{\lambda}_0^0$$

$$+ \frac{\beta(1 - \theta_{1,0})}{\theta_{1,0}} \left( \mathbf{A}\mathbf{x}_0^0 - \mathbf{b} \right) - \boldsymbol{\lambda}^*$$

$$\overset{a}{=} \frac{(m-1)(\theta_2 + \theta_{1,S})\beta + \beta}{\theta_{1,S}} \left( \mathbf{A}\hat{\mathbf{x}}_S - \mathbf{b} \right) + \tilde{\lambda}_0^0$$

$$- \frac{\beta(m-1)\theta_2}{\theta_{1,0}} \left( \mathbf{A}\mathbf{x}_0^0 - \mathbf{b} \right) - \boldsymbol{\lambda}^*, \tag{5.91}$$

where the equality $\overset{a}{=}$ uses the definition of $\hat{\mathbf{x}}_S$. Substituting (5.91) into (5.88) and using that $L(\mathbf{x}, \boldsymbol{\lambda})$ is convex in $\mathbf{x}$, we have

$$\mathbb{E}\left[ \frac{1}{2\beta} \left\| \frac{\beta(m-1)(\theta_2 + \theta_{1,S}) + \beta}{\theta_{1,S}} \left( \mathbf{A}\hat{\mathbf{x}}_S - \mathbf{b} \right) \right. \right.$$

$$\left. \left. - \frac{\beta(m-1)\theta_2}{\theta_{1,0}} \left( \mathbf{A}\mathbf{x}_0^0 - \mathbf{b} \right) + \tilde{\lambda}_0^0 - \boldsymbol{\lambda}^* \right\|^2 \right]$$

$$+ \mathbb{E}\left[\frac{(m-1)(\theta_2 + \theta_{1,s}) + 1}{\theta_{1,s}}\left(L(\hat{\mathbf{x}}_S, \boldsymbol{\lambda}^*) - L(\mathbf{x}^*, \boldsymbol{\lambda}^*)\right)\right]$$

$$\leq \frac{1 - \theta_{1,0} + (m-1)\theta_2}{\theta_{1,0}}\left(L\left(\mathbf{x}_0^0, \boldsymbol{\lambda}^*\right) - L\left(\mathbf{x}^*, \boldsymbol{\lambda}^*\right)\right) + \frac{1}{2\beta}\left\|\hat{\boldsymbol{\lambda}}_0^0 - \boldsymbol{\lambda}^*\right\|^2$$

$$+ \frac{1}{2}\left\|\mathbf{x}_{0,1}^0 - \mathbf{x}_1^*\right\|^2_{\left(\theta_{1,0}L_1 + \beta\|\mathbf{A}_1\|_2^2\right)\mathbf{I} - \beta\mathbf{A}_1^T\mathbf{A}_1}$$

$$+ \frac{1}{2}\left\|\mathbf{x}_{0,2}^0 - \mathbf{x}_2^*\right\|^2_{\left(\alpha\theta_{1,0}L_2 + \beta\|\mathbf{A}_2\|_2^2\right)\mathbf{I}}.$$

By the definitions of $L(\mathbf{x}, \boldsymbol{\lambda})$ and $\hat{\boldsymbol{\lambda}}_0^0$ we can obtain (5.76). $\qquad\square$

With Theorem 5.3 in hand, together with the definition of $\theta_{1,s}$ we can easily obtain the convergence rate of Algorithm 5.4 to solve Problem (5.45).

**Theorem 5.4** *The convergence rate of Algorithm 5.4 to solve Problem (5.45) is* $O(1/S)$. *Specifically, we have*

$$\mathbb{E}\left(L(\hat{\mathbf{x}}_S, \boldsymbol{\lambda}^*) - L(\mathbf{x}^*, \boldsymbol{\lambda}^*)\right) \leq \frac{\theta_{1,S}}{(m-1)(\theta_2 + \theta_{1,S}) + 1}C_1$$

*and*

$$\mathbb{E}\left\|\mathbf{A}\hat{\mathbf{x}}_S - \mathbf{b}\right\| \leq \frac{\theta_{1,S}}{\beta(m-1)(\theta_2 + \theta_{1,S}) + \beta}C_2,$$

*where*

$$C_1 = \frac{1 - \theta_{1,0} + (m-1)\theta_2}{\theta_{1,0}}\left(L\left(\mathbf{x}_0^0, \boldsymbol{\lambda}^*\right) - L\left(\mathbf{x}^*, \boldsymbol{\lambda}^*\right)\right) + \frac{1}{2\beta}\left\|\hat{\boldsymbol{\lambda}}_0^0 - \boldsymbol{\lambda}^*\right\|^2$$

$$+ \frac{1}{2}\left\|\mathbf{x}_{0,1}^0 - \mathbf{x}_1^*\right\|^2_{\left(\theta_{1,0}L_1 + \beta\|\mathbf{A}_1\|_2^2\right)\mathbf{I} - \beta\mathbf{A}_1^T\mathbf{A}_1}$$

$$+ \frac{1}{2}\left\|\mathbf{x}_{0,2}^0 - \mathbf{x}_2^*\right\|^2_{\left(\alpha\theta_{1,0}L_2 + \beta\|\mathbf{A}_2\|_2^2\right)\mathbf{I}}$$

*and*

$$C_2 = \sqrt{2\beta C_1} + \left\|\frac{\beta(m-1)\theta_2}{\theta_{1,0}}\left(\mathbf{A}\mathbf{x}_0^0 - \mathbf{b}\right) - \tilde{\boldsymbol{\lambda}}_0^0 + \boldsymbol{\lambda}^*\right\|.$$

## 5.4 Nonconvex Stochastic ADMM and Its Acceleration

In this section, we consider stochastic ADMM in the nonconvex setting. We will first extend the multi-block Bregman ADMM algorithm (Algorithm 4.1) introduced in Sect. 4.1 and then propose acceleration by the VR technique.

### 5.4.1 Nonconvex SADMM

We consider stochastic ADMM in the nonconvex setting. We first study a two-block linearly constrained problem shown as follows:

$$\min_{\mathbf{x}, \mathbf{y}} \ (f(\mathbf{x}) + g(\mathbf{y})), \quad s.t. \quad \mathbf{A}\mathbf{x} + \mathbf{B}\mathbf{y} = \mathbf{b}, \tag{5.92}$$

where we allow $f(\mathbf{x})$ to be an infinite-sum of individual terms, i.e., $f(\mathbf{x}) = \mathbb{E}_\xi F(\mathbf{x}; \xi)$. We consider the following assumption.

**Assumption 5.1** $f$ and $g$ are $L_1$-smooth and $L_2$-smooth, respectively. Moreover, the variance of stochastic gradients for $f$ is uniformly bounded by $\sigma^2$, i.e.,

$$\mathbb{E}_\xi \|\nabla F(\mathbf{x}, \xi) - \nabla f(\mathbf{x})\|^2 \le \sigma^2 \text{ for all given } \mathbf{x}.$$

We consider the following update rule to solve (5.92):

$$\mathbf{x}^{k+1} = \mathbf{x}^k - \eta \left[ \tilde{\nabla} f(\mathbf{x}^k) + \beta \mathbf{A}^T \left( \mathbf{A}\mathbf{x}^k + \mathbf{B}\mathbf{y}^k - \mathbf{b} + \frac{\lambda^k}{\beta} \right) \right], \tag{5.93a}$$

$$\mathbf{y}^{k+1} = \underset{\mathbf{y}}{\operatorname{argmin}} \left( g(\mathbf{y}) + \left\langle \lambda^k, \mathbf{B}\mathbf{y} \right\rangle + \frac{\beta}{2} \left\| \mathbf{A}\mathbf{x}^{k+1} + \mathbf{B}\mathbf{y} - \mathbf{b} \right\|^2 + D_\phi(\mathbf{y}, \mathbf{y}^k) \right), \tag{5.93b}$$

$$\lambda^{k+1} = \lambda^k + \beta \left( \mathbf{A}\mathbf{x}^{k+1} + \mathbf{B}\mathbf{y}^{k+1} - \mathbf{b} \right), \tag{5.93c}$$

where $\tilde{\nabla} f(\mathbf{x}^k)$ is a stochastic estimator of $\nabla f(\mathbf{x}^k)$ given $\mathbf{x}^k$ and has the form that

$$\tilde{\nabla} f(\mathbf{x}^k) = \frac{1}{S} \sum_{\xi \in \mathcal{I}_k} \nabla F(\mathbf{x}^k, \xi),$$

$\mathcal{I}_k$ with a size of $S$ is a mini-batch of indices randomly drawn, and $D_\phi$ is a certain Bregman distance. We summarize the above algorithm in Algorithm 5.5.

---

**Algorithm 5.5** Nonconvex Stochastic ADMM (Nonconvex SADMM)

---

Initialize $\mathbf{x}^0, \mathbf{y}^0, \boldsymbol{\lambda}^0$.
**for** $k = 0, 1, 2, 3, \cdots$ **do**
    Update $\mathbf{x}^{k+1}, \mathbf{y}^{k+1}$, and $\boldsymbol{\lambda}^{k+1}$ by (5.93a), (5.93b), and (5.93c), respectively.
**end for**

---

Because the indices in $\mathcal{I}_k$ are drawn independently, we have

$$\mathbb{E}_k \tilde{\nabla} f(\mathbf{x}^k) = \nabla f(\mathbf{x}^k) \quad \text{and} \quad \mathbb{E}_k \left\| \tilde{\nabla} f(\mathbf{x}^k) - \nabla f(\mathbf{x}^k) \right\|^2 \leq \frac{\sigma^2}{S}, \tag{5.94}$$

where the expectation is taken under the condition that the previous $k$ iterates are known.

The augmented Lagrangian function is

$$L_\beta(\mathbf{x}, \mathbf{y}, \boldsymbol{\lambda}) = f(\mathbf{x}) + g(\mathbf{y}) + \langle \boldsymbol{\lambda}, \mathbf{Ax} + \mathbf{By} - \mathbf{b} \rangle + \frac{\beta}{2} \|\mathbf{Ax} + \mathbf{By} - \mathbf{b}\|^2.$$

Then $L_\beta$ is $\tilde{L}_1$-smooth with respect to $\mathbf{x}$ and $\tilde{L}_2$-smooth with respect to $\mathbf{y}$, respectively, where $\tilde{L}_1 = L_1 + \beta \|\mathbf{A}\|_2^2$ and $\tilde{L}_2 = L_2 + \beta \|\mathbf{B}\|_2^2$.

We show that under Assumption 5.1 and the surjectiveness of $\mathbf{B}$, the above nonconvex SADMM algorithm can find an $\epsilon$-approximate KKT point in $O(\epsilon^{-4})$ stochastic accesses of gradient in expectation.

**Theorem 5.5** *Assume that Assumption 5.1 holds and there exists $\mu > 0$ such that $\|\mathbf{B}^T \boldsymbol{\lambda}\| \geq \mu \|\boldsymbol{\lambda}\|$ for all $\boldsymbol{\lambda}$. Set*

$$\eta \in [\Theta(\epsilon^2), 1/\tilde{L}_1] \quad \text{and} \quad S = \eta \cdot \Theta(\epsilon^{-2}) \in \mathbb{Z}^+.$$

*Pick $\phi$ to be $\rho = \Theta(1)$-strongly convex and $L = \Theta(1)$-smooth, set $\beta \geq \frac{24(L_2^2 + 2L^2)}{\mu^2 \rho} = \Theta(1)$, and define the Lyapunov function:*

$$\Phi^k = L_\beta(\mathbf{x}^k, \mathbf{y}^k, \boldsymbol{\lambda}^k) + \frac{6L^2}{\mu^2 \beta} \|\mathbf{y}^k - \mathbf{y}^{k-1}\|^2.$$

*Then after running Algorithm 5.5 by $K = \eta^{-1}\epsilon^{-2}$ iterations, we find an $O(\epsilon)$-approximate KKT point in expectation. Specifically, letting $(\tilde{\mathbf{x}}, \tilde{\mathbf{y}}, \tilde{\boldsymbol{\lambda}})$ uniformly randomly taken from $\{\mathbf{x}^k, \mathbf{y}^k, \boldsymbol{\lambda}^k\}_{k=1}^K$, defining $D = \Phi^0 - \min_{k \geq 0} \mathbb{E}\Phi^k$, and* *assuming that D is finite, we have*

$$\tilde{\mathbb{E}}\left\|\mathbf{A}\tilde{\mathbf{x}} + \mathbf{B}\tilde{\mathbf{y}} - \mathbf{b}\right\|^2 \le \frac{1}{\beta}\left(\frac{D}{K} + \frac{\tilde{L}_1\eta^2\sigma^2}{2S}\right) = O(\epsilon^2),$$

$$\tilde{\mathbb{E}}\|\nabla g(\tilde{\mathbf{y}}) + \mathbf{B}^T\tilde{\boldsymbol{\lambda}}\|^2 \le \frac{4L^2}{\rho}\left(\frac{D}{K} + \frac{\tilde{L}_1\eta^2\sigma^2}{2S}\right) = O(\epsilon^2), \quad and$$

$$\tilde{\mathbb{E}}\|\nabla f(\tilde{\mathbf{x}}) + \mathbf{A}^T\tilde{\boldsymbol{\lambda}}\|^2 \le 2\frac{K+1}{K}\left(2 + \eta\beta\|\mathbf{A}\|_2^2\right)\left(\frac{D}{\eta(K+1)} + \frac{\tilde{L}_1\eta\sigma^2}{2S}\right)$$

$$= O(\epsilon^2),$$

*where $\tilde{\mathbb{E}}$ denotes taking expectation for all the randomness in Algorithm 5.5 and the selection of $(\tilde{\mathbf{x}}, \tilde{\mathbf{y}}, \tilde{\boldsymbol{\lambda}})$.*

**Proof** By the $\tilde{L}_1$-smoothness of $L_\beta$ with respect to $\mathbf{x}$ and Proposition A.6 we have

$$L_\beta(\mathbf{x}^{k+1}, \mathbf{y}^k, \boldsymbol{\lambda}^k)$$

$$\le L_\beta(\mathbf{x}^k, \mathbf{y}^k, \boldsymbol{\lambda}^k) + \left\langle\nabla_{\mathbf{x}}L_\beta(\mathbf{x}^k, \mathbf{y}^k, \boldsymbol{\lambda}^k), \mathbf{x}^{k+1} - \mathbf{x}^k\right\rangle + \frac{\tilde{L}_1}{2}\left\|\mathbf{x}^{k+1} - \mathbf{x}^k\right\|^2$$

$$\overset{a}{\le} L_\beta(\mathbf{x}^k, \mathbf{y}^k, \boldsymbol{\lambda}^k)$$

$$- \eta\left\langle\nabla_{\mathbf{x}}L_\beta(\mathbf{x}^k, \mathbf{y}^k, \boldsymbol{\lambda}^k), \nabla_{\mathbf{x}}L_\beta(\mathbf{x}^k, \mathbf{y}^k, \boldsymbol{\lambda}^k) + \tilde{\nabla}f(\mathbf{x}^k) - \nabla f(\mathbf{x}^k)\right\rangle$$

$$+ \frac{\tilde{L}_1\eta^2}{2}\left\|\nabla_{\mathbf{x}}L_\beta(\mathbf{x}^k, \mathbf{y}^k, \boldsymbol{\lambda}^k) + \tilde{\nabla}f(\mathbf{x}^k) - \nabla f(\mathbf{x}^k)\right\|^2,$$

where in $\overset{a}{\le}$ we plug in the update rule of $\mathbf{x}^k$. Taking expectation for $\tilde{\nabla}f(\mathbf{x}^k)$ conditionally on the previous $k$ iterations and using (5.94) and $\eta \le \frac{1}{\tilde{L}_1}$, we have

$$\mathbb{E}_k L_\beta(\mathbf{x}^{k+1}, \mathbf{y}^k, \boldsymbol{\lambda}^k)$$

$$\le L_\beta(\mathbf{x}^k, \mathbf{y}^k, \boldsymbol{\lambda}^k) - \frac{\eta}{2}\left\|\nabla_{\mathbf{x}}L_\beta(\mathbf{x}^k, \mathbf{y}^k, \boldsymbol{\lambda}^k)\right\|^2$$

$$+ \frac{\tilde{L}_1\eta^2}{2}\mathbb{E}_k\left\|\tilde{\nabla}f(\mathbf{x}^k) - \nabla f(\mathbf{x}^k)\right\|^2$$

$$\le L_\beta(\mathbf{x}^k, \mathbf{y}^k, \boldsymbol{\lambda}^k) - \frac{\eta}{2}\left\|\nabla_{\mathbf{x}}L_\beta(\mathbf{x}^k, \mathbf{y}^k, \boldsymbol{\lambda}^k)\right\|^2 + \frac{\tilde{L}_1\eta^2\sigma^2}{2S}. \tag{5.95}$$

From the update of $\mathbf{y}$, we have

$$\frac{\rho}{2}\|\mathbf{y}^{k+1} - \mathbf{y}^k\|^2 \leq L_\beta(\mathbf{x}^{k+1}, \mathbf{y}^k, \boldsymbol{\lambda}^k) - L_\beta(\mathbf{x}^{k+1}, \mathbf{y}^{k+1}, \boldsymbol{\lambda}^k). \tag{5.96}$$

From the update of $\boldsymbol{\lambda}$, we have

$$-\frac{1}{\beta}\|\boldsymbol{\lambda}^{k+1} - \boldsymbol{\lambda}^k\|^2 = L_\beta(\mathbf{x}^{k+1}, \mathbf{y}^{k+1}, \boldsymbol{\lambda}^k) - L_\beta(\mathbf{x}^{k+1}, \mathbf{y}^{k+1}, \boldsymbol{\lambda}^{k+1}). \tag{5.97}$$

Adding (5.95)–(5.97) and taking full expectation, we have

$$\frac{\eta}{2}\mathbb{E}\left\|\nabla_{\mathbf{x}}L_\beta(\mathbf{x}^k, \mathbf{y}^k, \boldsymbol{\lambda}^k)\right\|^2 + \frac{\rho}{2}\mathbb{E}\|\mathbf{y}^{k+1} - \mathbf{y}^k\|^2 - \frac{1}{\beta}\mathbb{E}\|\boldsymbol{\lambda}^{k+1} - \boldsymbol{\lambda}^k\|^2$$

$$\leq \mathbb{E}L_\beta(\mathbf{x}^k, \mathbf{y}^k, \boldsymbol{\lambda}^k) - \mathbb{E}L_\beta(\mathbf{x}^{k+1}, \mathbf{y}^{k+1}, \boldsymbol{\lambda}^{k+1}) + \frac{\tilde{L}_1\eta^2\sigma^2}{2S}. \tag{5.98}$$

On the other hand, using the same argument as (4.4) and (4.5), we have

$$\mu^2\|\boldsymbol{\lambda}^{k+1} - \boldsymbol{\lambda}^k\|^2 \leq \|\mathbf{B}^T(\boldsymbol{\lambda}^{k+1} - \boldsymbol{\lambda}^k)\|^2$$

$$\leq 3\left(L_2^2 + L^2\right)\|\mathbf{y}^{k+1} - \mathbf{y}^k\|^2 + 3L^2\|\mathbf{y}^k - \mathbf{y}^{k-1}\|^2. \tag{5.99}$$

Taking full expectation on (5.99), multiplying its both sides with $2/(\mu^2\beta)$ and adding it with (5.98), we have

$$\frac{\eta}{2}\mathbb{E}\left\|\nabla_{\mathbf{x}}L_\beta(\mathbf{x}^k, \mathbf{y}^k, \boldsymbol{\lambda}^k)\right\|^2 + \left(\frac{\rho}{2} - \frac{6L_2^2 + 12L^2}{\mu^2\beta}\right)\mathbb{E}\|\mathbf{y}^{k+1} - \mathbf{y}^k\|^2$$

$$+ \frac{1}{\beta}\mathbb{E}\|\boldsymbol{\lambda}^{k+1} - \boldsymbol{\lambda}^k\|^2 \leq \mathbb{E}\Phi^k - \mathbb{E}\Phi^{k+1} + \frac{\tilde{L}_1\eta^2\sigma^2}{2S}. \tag{5.100}$$

Summing (5.100) from $k = 0, \cdots, K - 1$, and using $\beta \geq \frac{24(L_2^2 + 2L^2)}{\mu^2\rho}$, we have

$$\frac{1}{K}\frac{\eta}{2}\sum_{k=0}^{K-1}\mathbb{E}\left\|\nabla_{\mathbf{x}}L_\beta(\mathbf{x}^k, \mathbf{y}^k, \boldsymbol{\lambda}^k)\right\|^2 + \frac{\rho}{4}\frac{1}{K}\sum_{k=0}^{K-1}\mathbb{E}\|\mathbf{y}^{k+1} - \mathbf{y}^k\|^2$$

$$+ \frac{1}{\beta}\frac{1}{K}\sum_{k=0}^{K-1}\mathbb{E}\|\boldsymbol{\lambda}^{k+1} - \boldsymbol{\lambda}^k\|^2 \leq \frac{D}{K} + \frac{\tilde{L}_1\eta^2\sigma^2}{2S}. \tag{5.101}$$

Using the same argument as (4.6) and recalling that $(\tilde{\mathbf{x}}, \tilde{\mathbf{y}}, \tilde{\boldsymbol{\lambda}})$ are uniformly randomly taken from $\left\{(\mathbf{x}^k, \mathbf{y}^k, \boldsymbol{\lambda}^k)\right\}_{k=1}^K$, we obtain

$$
\begin{aligned}
\tilde{\mathbb{E}}\left\|\nabla g(\tilde{\mathbf{y}}) + \mathbf{B}^T \tilde{\boldsymbol{\lambda}}\right\|^2 &= \frac{1}{K}\sum_{k=0}^{K-1}\mathbb{E}\left\|\nabla g(\mathbf{y}^{k+1}) + \mathbf{B}^T\boldsymbol{\lambda}^{k+1}\right\|^2 \\
&\le \frac{L^2}{K}\sum_{k=0}^{K-1}\mathbb{E}\left\|\mathbf{y}^{k+1} - \mathbf{y}^k\right\|^2 \le \frac{4L^2}{\rho}\left(\frac{D}{K} + \frac{\tilde{L}_1\eta^2\sigma^2}{2S}\right).
\end{aligned}
$$
(5.102)

From the update of $\boldsymbol{\lambda}$, we have

$$
\begin{aligned}
\tilde{\mathbb{E}}\left\|\mathbf{A}\tilde{\mathbf{x}} + \mathbf{B}\tilde{\mathbf{y}} - \mathbf{b}\right\|^2 &= \frac{1}{K}\sum_{k=0}^{K-1}\mathbb{E}\left\|\mathbf{A}\mathbf{x}^{k+1} + \mathbf{B}\mathbf{y}^{k+1} - \mathbf{b}\right\|^2 \\
&= \frac{1}{\beta^2 K}\sum_{k=0}^{K-1}\mathbb{E}\|\boldsymbol{\lambda}^{k+1} - \boldsymbol{\lambda}^k\|^2 \le \frac{1}{\beta}\left(\frac{D}{K} + \frac{\tilde{L}_1\eta^2\sigma^2}{2S}\right).
\end{aligned}
$$
(5.103)

By the update rule of $\mathbf{x}$, we have

$$
\begin{aligned}
&\tilde{\mathbb{E}}\left\|\nabla f(\tilde{\mathbf{x}}) + \mathbf{A}^T\tilde{\boldsymbol{\lambda}}\right\|^2 \\
&= \frac{1}{K}\sum_{k=1}^{K}\mathbb{E}\left\|\nabla f(\mathbf{x}^k) + \mathbf{A}^T\boldsymbol{\lambda}^k\right\|^2 \\
&= \frac{1}{K}\sum_{k=1}^{K}\mathbb{E}\left\|\nabla_{\mathbf{x}}L_\beta(\mathbf{x}^k, \mathbf{y}^k, \boldsymbol{\lambda}^k) - \beta\mathbf{A}^T(\mathbf{A}\mathbf{x}^k + \mathbf{B}\mathbf{y}^k - \mathbf{b})\right\|^2 \\
&\le \frac{2}{K}\sum_{k=1}^{K}\mathbb{E}\left\|\nabla_{\mathbf{x}}L_\beta(\mathbf{x}^k, \mathbf{y}^k, \boldsymbol{\lambda}^k)\right\|^2 + 2\|\mathbf{A}\|_2^2\frac{1}{K}\sum_{k=0}^{K-1}\mathbb{E}\left\|\boldsymbol{\lambda}^{k+1} - \boldsymbol{\lambda}^k\right\|^2 \\
&\le \frac{2(K+1)}{K}\frac{1}{K+1}\sum_{k=0}^{K}\mathbb{E}\left\|\nabla_{\mathbf{x}}L_\beta(\mathbf{x}^k, \mathbf{y}^k, \boldsymbol{\lambda}^k)\right\|^2 \\
&\quad + 2\|\mathbf{A}\|_2^2\frac{K+1}{K}\frac{1}{K+1}\sum_{k=0}^{K}\mathbb{E}\left\|\boldsymbol{\lambda}^{k+1} - \boldsymbol{\lambda}^k\right\|^2 \\
&\le 2\frac{K+1}{K}\left(2 + \eta\beta\|\mathbf{A}\|_2^2\right)\left[\frac{D}{\eta(K+1)} + \frac{\tilde{L}_1\eta\sigma^2}{2S}\right],
\end{aligned}
$$

where the last inequality is obtained by summing (5.100) from $k = 0$ to $K$, analogously to (5.101). So the proof concludes.                                                    □

From Theorem 5.5, we know that the total access of the stochastic gradient of $f$ and the updates by $g$ will be no more than $O(\epsilon^{-4})$ in expectation.

### 5.4.2  SPIDER Acceleration

The Stochastic Path-Integrated Differential Estimator (SPIDER) [4, 5, 8] technique is a radical VR method that is used to track quantities using reduced stochastic oracles. For generic $L$-smooth stochastic nonconvex optimization, SPIDER can achieve the optimal $O(\epsilon^{-3})$ expected complexity to find an $\epsilon$-approximate first-order stationary point. This result is different from variance reduction methods in the convex case, as the latter can only accelerate the convergence rate for the finite-sum problems. We also note that for the finite-sum problem with $n$ individual functions, SPIDER can improve the complexity to $O(\min(n + n^{1/2}\epsilon^{-2}, \epsilon^{-3}))$.

In this section, we apply the SPIDER technique to accelerate the nonconvex SADMM algorithm. We consider a multi-block linearly constrained problem shown as below:

$$\min_{\mathbf{x}_1,\cdots,\mathbf{x}_m,\mathbf{y}} \left( \sum_{i=1}^{m} f_i(\mathbf{x}_i) + g(\mathbf{y}) \right),$$

$$s.t. \quad \sum_{i=1}^{m} \mathbf{A}_i \mathbf{x}_i + \mathbf{B}\mathbf{y} = \mathbf{b}, \tag{5.104}$$

where $f_i(\mathbf{x}_i) = \mathbb{E}_{\xi_i} F_i(\mathbf{x}_i; \xi_i)$ for $i \in [m]$, under the following assumption.

**Assumption 5.2** $g$ is $L_0$-smooth. For each $i \in [m]$, $F_i(\mathbf{x}_i; \xi_i)$ is $L_i$-smooth with respect to $\mathbf{x}_i$ for all $\xi_i$. Moreover, the variance of stochastic gradients of $f_i$ is uniformly bounded by $\sigma^2$, i.e.,

$$\mathbb{E}_{\xi_i} \|\nabla F_i(\mathbf{x}_i, \xi_i) - \nabla f_i(\mathbf{x}_i)\|^2 \leq \sigma^2 \text{ for all given } \mathbf{x}_i.$$

The iterations to solve (5.104) go as follows:

$$\mathbf{x}_i^{k+1} = \mathbf{x}_i^k - \eta \left[ \tilde{\nabla} f_i\left(\mathbf{x}_i^k\right) \right.$$

$$\left. + \beta \mathbf{A}_i^T \left( \sum_{j<i} \mathbf{A}_j \mathbf{x}_j^{k+1} + \sum_{j\geq i} \mathbf{A}_j \mathbf{x}_j^k + \mathbf{B}\mathbf{y}^k - \mathbf{b} + \frac{\lambda^k}{\beta} \right) \right], \ i \in [m],$$

$$\tag{5.105a}$$

$$
\mathbf{y}^{k+1} = \underset{\mathbf{y}}{\operatorname{argmin}} \left( g(\mathbf{y}) + \left\langle \boldsymbol{\lambda}^k, \mathbf{B}\mathbf{y} \right\rangle + \frac{\beta}{2} \left\| \mathbf{A}\mathbf{x}^{k+1} + \mathbf{B}\mathbf{y} - \mathbf{b} \right\|^2 \right.
$$

$$
\left. + D_\phi(\mathbf{y}, \mathbf{y}^k) \right), \tag{5.105b}
$$

$$
\boldsymbol{\lambda}^{k+1} = \boldsymbol{\lambda}^k + \beta \left( \mathbf{A}\mathbf{x}^{k+1} + \mathbf{B}\mathbf{y}^{k+1} - \mathbf{b} \right), \tag{5.105c}
$$

where $D_\phi$ is a certain Bregman distance and for simplicity, we define

$$
\mathbf{x} = \left( \mathbf{x}_1^T, \cdots, \mathbf{x}_m^T \right)^T \quad \text{and} \quad \mathbf{A} = [\mathbf{A}_1, \cdots, \mathbf{A}_m].
$$

We summarize the above algorithm in Algorithm 5.6.

---

**Algorithm 5.6** SPIDER Accelerated Nonconvex Stochastic ADMM (SPIDER-ADMM)

---

Initialize $\mathbf{x}_i^0, i \in [m], \mathbf{y}^0, \boldsymbol{\lambda}^0$.
**for** $k = 0, 1, 2, 3, \cdots$ **do**
    Update $\mathbf{x}_i^{k+1}, i \in [m], \mathbf{y}^{k+1}$, and $\boldsymbol{\lambda}^{k+1}$ by (5.105a), (5.105b), and (5.105c), respectively.
**end for**

---

We will choose a more sophisticated gradient estimator $\tilde{\nabla} f_i(\mathbf{x}_i^k)$ as follows:

- For a certain hyper-parameter $q$, if the iteration $k$ is divisible by $q$, then

$$
\tilde{\nabla} f_i \left( \mathbf{x}_i^k \right) = \frac{1}{S_1} \sum_{\xi_i \in \mathcal{I}_{k,i}} \nabla F_i \left( \mathbf{x}_i^k, \xi_i \right),
$$

where $\mathcal{I}_{k,i}$ with a size of $S_1$ is a mini-batch of indices randomly drawn.
- Otherwise,

$$
\tilde{\nabla} f_i \left( \mathbf{x}_i^k \right) = \frac{1}{S_2} \sum_{\xi_i \in \mathcal{I}_{k,i}} \left[ \nabla F_i \left( \mathbf{x}_i^k, \xi_i \right) - \nabla F_i \left( \mathbf{x}_i^{k-1}, \xi_i \right) \right] + \tilde{\nabla} f_i(\mathbf{x}_i^{k-1}),
$$

where $\mathcal{I}_{k,i}$ with a size of $S_2$ is a mini-batch of indices randomly drawn.

We have the following lemma that bounds the variance of the estimator $\tilde{\nabla} f_i(\mathbf{x}_i^k)$.

**Lemma 5.7** *Under Assumption 5.2, letting $k_0 = \lfloor k/q \rfloor q$, we have that for all $k = 0, \cdots, K$,*

$$
\mathbb{E} \left\| \tilde{\nabla} f_i \left( \mathbf{x}_i^k \right) - \nabla f_i \left( \mathbf{x}_i^k \right) \right\|^2 \leq \frac{L_i^2}{S_2} \sum_{\ell=k_0+1}^{k} \| \mathbf{x}_i^\ell - \mathbf{x}_i^{\ell-1} \|^2 + \frac{\sigma^2}{S_1}. \tag{5.106}
$$

**Proof** When $k$ is divisible by $q$, because the indices in $\mathcal{I}_{k,i}$ are drawn independently, by the bounded variance assumption, (5.106) is true. For $k \neq \lfloor k/q \rfloor q$, by taking expectation conditionally on previous all updates before $\mathbf{x}_i^k$, we have

$$
\mathbb{E}_{k,i} \left\| \tilde{\nabla} f_i \left( \mathbf{x}_i^k \right) - \nabla f_i \left( \mathbf{x}_i^k \right) \right\|^2
$$

$$
= \mathbb{E}_{k,i} \left\| \frac{1}{S_2} \sum_{\xi_i \in \mathcal{I}_{k,i}} \left[ \nabla F_i \left( \mathbf{x}_i^k, \xi_i \right) - \nabla F_i \left( \mathbf{x}_i^{k-1}, \xi_i \right) - \left( \nabla f_i \left( \mathbf{x}_i^k \right) - \nabla f_i \left( \mathbf{x}_i^{k-1} \right) \right) \right] \right.
$$

$$
\left. + \left[ \tilde{\nabla} f_i \left( \mathbf{x}_i^{k-1} \right) - \nabla f_i \left( \mathbf{x}_i^{k-1} \right) \right] \right\|^2
$$

$$
\overset{a}{=} \mathbb{E}_{k,i} \left\| \frac{1}{S_2} \sum_{\xi_i \in \mathcal{I}_{k,i}} \left[ \nabla F_i \left( \mathbf{x}_i^k, \xi_i \right) - \nabla F_i \left( \mathbf{x}_i^{k-1}, \xi_i \right) - \left( \nabla f_i \left( \mathbf{x}_i^k \right) - \nabla f_i \left( \mathbf{x}_i^{k-1} \right) \right) \right] \right\|^2
$$

$$
+ \left\| \tilde{\nabla} f_i \left( \mathbf{x}_i^{k-1} \right) - \nabla f_i \left( \mathbf{x}_i^{k-1} \right) \right\|^2
$$

$$
= \frac{1}{S_2} \mathbb{E}_{k,i} \left\| \nabla F_i \left( \mathbf{x}_i^k, \xi_i \right) - \nabla F_i \left( \mathbf{x}_i^{k-1}, \xi_i \right) - \left( \nabla f_i \left( \mathbf{x}_i^k \right) - \nabla f_i \left( \mathbf{x}_i^{k-1} \right) \right) \right\|^2
$$

$$
+ \left\| \tilde{\nabla} f_i \left( \mathbf{x}_i^{k-1} \right) - \nabla f_i \left( \mathbf{x}_i^{k-1} \right) \right\|^2
$$

$$
\overset{b}{\leq} \frac{1}{S_2} \mathbb{E}_{k,i} \left\| \nabla F_i \left( \mathbf{x}_i^k, \xi_i \right) - \nabla F_i \left( \mathbf{x}_i^{k-1}, \xi_i \right) \right\|^2 + \left\| \tilde{\nabla} f_i \left( \mathbf{x}_i^{k-1} \right) - \nabla f_i \left( \mathbf{x}_i^{k-1} \right) \right\|^2
$$

$$
\overset{c}{\leq} \frac{L_i^2}{S_2} \mathbb{E}_{k,i} \left\| \mathbf{x}_i^k - \mathbf{x}_i^{k-1} \right\|^2 + \left\| \tilde{\nabla} f_i \left( \mathbf{x}_i^{k-1} \right) - \nabla f_i \left( \mathbf{x}_i^{k-1} \right) \right\|^2,
$$

where $\mathbb{E}_{k,i}$ stands for taking expectation conditionally on all previous updates before $\mathbf{x}_i^k$, in $\overset{a}{=}$ we use

$$
\mathbb{E}_{k,i} \left[ \nabla F_i \left( \mathbf{x}_i^k, \xi_i \right) - \nabla F_i \left( \mathbf{x}_i^{k-1}, \xi_i \right) - \left( \nabla f_i \left( \mathbf{x}_i^k \right) - \nabla f_i \left( \mathbf{x}_i^{k-1} \right) \right) \right] = \mathbf{0},
$$

in $\overset{b}{\leq}$ we use Proposition A.3, and in $\overset{c}{\leq}$ we use the assumption that individual $F_i$ is $L_i$-smooth. Then by taking full expectation and recursively using the above equality from $k_0 + 1$ to $k$, we have (5.106).    □

Moreover, the augmented Lagrangian function of Problem (5.104) is

$$
L_\beta(\mathbf{x}, \mathbf{y}, \lambda) = \sum_{i=1}^m f_i(\mathbf{x}_i) + g(\mathbf{y}) + \left\langle \lambda, \sum_{i=1}^m \mathbf{A}_i \mathbf{x}_i + \mathbf{B}\mathbf{y} - \mathbf{b} \right\rangle
$$

$$
+ \frac{\beta}{2} \left\| \sum_{i=1}^m \mathbf{A}_i \mathbf{x}_i + \mathbf{B}\mathbf{y} - \mathbf{b} \right\|^2.
$$

We have that $L_\beta$ is $\tilde{L}_i$-smooth with respect to $\mathbf{x}_i$ and $\tilde{L}_0$-smooth with respect to $\mathbf{y}$, where $\tilde{L}_i = L_i + \beta \|\mathbf{A}_i\|_2^2$ and $\tilde{L}_0 = L_0 + \beta \|\mathbf{B}\|_2^2$. We show that under Assumption 5.2 and the surjectiveness of $\mathbf{B}$, the above SPIDER-ADMM algorithm can find an $\epsilon$-approximate KKT point in $O(\epsilon^{-3})$ stochastic accesses of gradient for $f_i$ and $O(\epsilon^{-2})$ accesses of gradient for $g$ in expectation. (Note that we should sample $S_2$ functions in each of the $O(\epsilon^{-2})$ iterations.)

**Theorem 5.6** *Assume that Assumption 5.2 holds and there exists $\mu > 0$ such that $\|\mathbf{B}^T \boldsymbol{\lambda}\| \geq \mu \|\boldsymbol{\lambda}\|$ for all $\boldsymbol{\lambda}$. Set*

$$S_1 = \Theta(\epsilon^{-2}), \quad S_2 = \Theta(\epsilon^{-1}), \quad q = \Theta(\epsilon^{-1}), \text{ and}$$

$$\eta = \min\left\{ \frac{1}{2\max_{i\in[m]}\{\tilde{L}_i\}}, \frac{1}{2\max_{i\in[m]}\{L_i\}\sqrt{q/S_2}} \right\} = \Theta(1).$$

*Pick $\phi$ to be $\rho = \Theta(1)$-strongly convex and $L = \Theta(1)$-smooth, set $\beta \geq \frac{24(L_0^2+2L^2)}{\mu^2\rho} = \Theta(1)$, and define the Lyapunov function:*

$$\Phi^k = L_\beta(\mathbf{x}^k, \mathbf{y}^k, \boldsymbol{\lambda}^k) + \frac{6L^2}{\mu^2\beta}\|\mathbf{y}^k - \mathbf{y}^{k-1}\|^2.$$

*Then after running Algorithm 5.6 by $K = \Theta(\epsilon^{-2})$ iterations, we can find an $O(\epsilon)$-approximate KKT point in expectation. Specifically, letting $(\tilde{\mathbf{x}}, \tilde{\mathbf{y}}, \tilde{\boldsymbol{\lambda}})$ uniformly randomly taken from $\{\mathbf{x}^k, \mathbf{y}^k, \boldsymbol{\lambda}^k\}_{k=1}^K$, defining $D = \Phi^0 - \min_{k\geq 0} \mathbb{E}\Phi^k$, and assuming that $D$ is finite, we have*

$$\tilde{\mathbb{E}}\left\|\mathbf{A}\tilde{\mathbf{x}} + \mathbf{B}\tilde{\mathbf{y}} - \mathbf{b}\right\|^2 \leq \frac{1}{\beta}\left(\frac{D}{K} + \frac{m\sigma^2\eta}{2S_1}\right) = O(\epsilon^2),$$

$$\tilde{\mathbb{E}}\|\nabla g(\tilde{\mathbf{y}}) + \mathbf{B}^T\tilde{\boldsymbol{\lambda}}\|^2 \leq \frac{4L^2}{\rho}\left(\frac{D}{K} + \frac{m\sigma^2\eta}{2S_1}\right) = O(\epsilon^2), \text{ and}$$

$$\tilde{\mathbb{E}}\|\nabla f_i(\tilde{\mathbf{x}}_i) + \mathbf{A}_i^T\tilde{\boldsymbol{\lambda}}\|^2 \leq 4\frac{K+1}{K}C_i\left(\frac{D}{K+1} + \frac{m\sigma^2\eta}{2S_1}\right) + \frac{4\sigma^2}{S_1}$$

$$= O(\epsilon^2), \quad i \in [m],$$

*where*

$$C_i = \frac{8}{\eta} + \beta \|\mathbf{A}_i\|_2^2 + 8\eta\beta^2(m+1)\|\mathbf{A}_i\|_2^2 \max_{j\in[m]}\|\mathbf{A}_j\|_2^2$$

$$+ \frac{4}{\rho}\beta^2(m+1)\|\mathbf{A}_i\|_2^2\|\mathbf{B}\|_2^2 + \frac{8\eta q L_i^2}{S_2}$$

$$= \Theta(1)$$

*and* $\tilde{\mathbb{E}}$ *denotes taking expectation for all the randomness in Algorithm 5.6 and the selection of* $(\tilde{\mathbf{x}}, \tilde{\mathbf{y}}, \tilde{\boldsymbol{\lambda}})$.

**Proof** By the $\tilde{L}_i$-smoothness of $L_\beta$ with respect to $\mathbf{x}_i$ and Proposition A.6, we have

$$
L_\beta\left(\mathbf{x}_{j\leq i}^{k+1}, \mathbf{x}_{j>i}^k, \mathbf{y}^k, \boldsymbol{\lambda}^k\right)
$$

$$
\leq L_\beta\left(\mathbf{x}_{j<i}^{k+1}, \mathbf{x}_{j\geq i}^k, \mathbf{y}^k, \boldsymbol{\lambda}^k\right)
$$

$$
+ \left\langle \nabla_{\mathbf{x}_i} L_\beta\left(\mathbf{x}_{j<i}^{k+1}, \mathbf{x}_{j\geq i}^k, \mathbf{y}^k, \boldsymbol{\lambda}^k\right), \mathbf{x}_i^{k+1} - \mathbf{x}_i^k\right\rangle + \frac{\tilde{L}_i}{2}\left\|\mathbf{x}_i^{k+1} - \mathbf{x}_i^k\right\|^2
$$

$$
\overset{a}{=} L_\beta\left(\mathbf{x}_{j<i}^{k+1}, \mathbf{x}_{j\geq i}^k, \mathbf{y}^k, \boldsymbol{\lambda}^k\right)
$$

$$
+ \left\langle -\frac{1}{\eta}\left(\mathbf{x}_i^{k+1} - \mathbf{x}_i^k\right) + \nabla f_i\left(\mathbf{x}_i^k\right) - \tilde{\nabla} f_i\left(\mathbf{x}_i^k\right), \mathbf{x}_i^{k+1} - \mathbf{x}_i^k\right\rangle + \frac{\tilde{L}_i}{2}\left\|\mathbf{x}_i^{k+1} - \mathbf{x}_i^k\right\|^2
$$

$$
\overset{b}{\leq} L_\beta\left(\mathbf{x}_{j<i}^{k+1}, \mathbf{x}_{j\geq i}^k, \mathbf{y}^k, \boldsymbol{\lambda}^k\right) - \left(\frac{1}{2\eta} - \frac{\tilde{L}_i}{2}\right)\left\|\mathbf{x}_i^{k+1} - \mathbf{x}_i^k\right\|^2
$$

$$
+ \frac{\eta}{2}\left\|\tilde{\nabla} f_i\left(\mathbf{x}_i^k\right) - \nabla f_i\left(\mathbf{x}_i^k\right)\right\|^2, \tag{5.107}
$$

where in $\overset{a}{=}$ we plug in the update rule of $\mathbf{x}_i^k$ and in $\overset{b}{\leq}$ we use the fact that

$$
\left\langle \nabla f_i\left(\mathbf{x}_i^k\right) - \tilde{\nabla} f_i\left(\mathbf{x}_i^k\right), \mathbf{x}_i^{k+1} - \mathbf{x}_i^k\right\rangle
$$

$$
\leq \frac{\eta}{2}\left\|\tilde{\nabla} f_i\left(\mathbf{x}_i^k\right) - \nabla f_i\left(\mathbf{x}_i^k\right)\right\|^2 + \frac{1}{2\eta}\left\|\mathbf{x}_i^{k+1} - \mathbf{x}_i^k\right\|^2.
$$

For the **y**'s and the $\boldsymbol{\lambda}$'s updates, we have

$$
\frac{\rho}{2}\|\mathbf{y}^{k+1} - \mathbf{y}^k\|^2 \leq L_\beta(\mathbf{x}^{k+1}, \mathbf{y}^k, \boldsymbol{\lambda}^k) - L_\beta(\mathbf{x}^{k+1}, \mathbf{y}^{k+1}, \boldsymbol{\lambda}^k) \tag{5.108}
$$

and

$$
-\frac{1}{\beta}\|\boldsymbol{\lambda}^{k+1} - \boldsymbol{\lambda}^k\|^2 = L_\beta(\mathbf{x}^{k+1}, \mathbf{y}^{k+1}, \boldsymbol{\lambda}^k) - L_\beta(\mathbf{x}^{k+1}, \mathbf{y}^{k+1}, \boldsymbol{\lambda}^{k+1}), \tag{5.109}
$$

respectively.

Summing (5.107) with $i = 1, 2, \cdots, m$, adding it with (5.108) and (5.109), and using that $\eta \leq \frac{1}{2 \max_{i \in [m]} \{\tilde{L}_i\}}$, we have

$$
\frac{1}{4\eta} \sum_{i=1}^{m} \left\| \mathbf{x}_i^{k+1} - \mathbf{x}_i^k \right\|^2 + \frac{\rho}{2} \|\mathbf{y}^{k+1} - \mathbf{y}^k\|^2 - \frac{1}{\beta} \|\boldsymbol{\lambda}^{k+1} - \boldsymbol{\lambda}^k\|^2
$$

$$
\leq L_\beta(\mathbf{x}^k, \mathbf{y}^k, \boldsymbol{\lambda}^k) - L_\beta(\mathbf{x}^{k+1}, \mathbf{y}^{k+1}, \boldsymbol{\lambda}^{k+1})
$$

$$
+ \frac{\eta}{2} \sum_{i=1}^{m} \left\| \tilde{\nabla} f_i\left(\mathbf{x}_i^k\right) - \nabla f_i\left(\mathbf{x}_i^k\right) \right\|^2. \tag{5.110}
$$

Using the same argument as (4.4) and (4.5), we have

$$
\mu^2 \|\boldsymbol{\lambda}^{k+1} - \boldsymbol{\lambda}^k\|^2 \leq \|\mathbf{B}^T (\boldsymbol{\lambda}^{k+1} - \boldsymbol{\lambda}^k)\|^2
$$

$$
\leq 3 \left( L_0^2 + L^2 \right) \|\mathbf{y}^{k+1} - \mathbf{y}^k\|^2 + 3L^2 \|\mathbf{y}^k - \mathbf{y}^{k-1}\|^2. \tag{5.111}
$$

Multiplying both sides of (5.111) with $2/(\mu^2\beta)$, adding it with (5.110), and using $\beta \geq \frac{24(L_0^2 + 2L^2)}{\mu^2 \rho}$, we have

$$
\frac{1}{4\eta} \sum_{i=1}^{m} \left\| \mathbf{x}_i^{k+1} - \mathbf{x}_i^k \right\|^2 + \frac{\rho}{4} \|\mathbf{y}^{k+1} - \mathbf{y}^k\|^2 + \frac{1}{\beta} \|\boldsymbol{\lambda}^{k+1} - \boldsymbol{\lambda}^k\|^2
$$

$$
\leq \Phi^k - \Phi^{k+1} + \frac{\eta}{2} \sum_{i=1}^{m} \left\| \tilde{\nabla} f_i\left(\mathbf{x}_i^k\right) - \nabla f_i\left(\mathbf{x}_i^k\right) \right\|^2.
$$

By taking full expectation on both sides of the above inequality and plugging in (5.106), we obtain

$$
\frac{1}{4\eta} \sum_{i=1}^{m} \mathbb{E} \left\| \mathbf{x}_i^{k+1} - \mathbf{x}_i^k \right\|^2 + \frac{\rho}{4} \mathbb{E}\|\mathbf{y}^{k+1} - \mathbf{y}^k\|^2 + \frac{1}{\beta} \mathbb{E}\|\boldsymbol{\lambda}^{k+1} - \boldsymbol{\lambda}^k\|^2
$$

$$
\leq \mathbb{E}\Phi^k - \mathbb{E}\Phi^{k+1} + \frac{\eta}{2} \frac{1}{S_2} \sum_{\ell=k_0+1}^{k} \sum_{i=1}^{m} L_i^2 \mathbb{E} \left\| \mathbf{x}_i^\ell - \mathbf{x}_i^{\ell-1} \right\|^2 + \frac{m\sigma^2 \eta}{2S_1}
$$

$$
\leq \mathbb{E}\Phi^k - \mathbb{E}\Phi^{k+1} + \frac{\eta}{2} \frac{\max_{i \in [m]} L_i^2}{S_2} \sum_{\ell=k_0+1}^{k} \sum_{i=1}^{m} \mathbb{E} \left\| \mathbf{x}_i^\ell - \mathbf{x}_i^{\ell-1} \right\|^2 + \frac{m\sigma^2 \eta}{2S_1}.
$$

$$
\tag{5.112}
$$

Summing (5.112) for $k = 0$ to $K - 1$, using $k - k_0 \leq q$ and the boundedness of $D = \Phi^0 - \min_{k \geq 0} \mathbb{E}\Phi^k$, and then dividing both sides with $K$, we have

$$
\left( \frac{1}{4\eta} - \frac{\eta q \max_{i \in [m]} L_i^2}{2S_2} \right) \frac{1}{K} \sum_{k=0}^{K-1} \sum_{i=1}^{m} \mathbb{E} \left\| \mathbf{x}_i^{k+1} - \mathbf{x}_i^k \right\|^2
$$

$$
+ \frac{\rho}{4} \frac{1}{K} \sum_{k=0}^{K-1} \mathbb{E}\|\mathbf{y}^{k+1} - \mathbf{y}^k\|^2 + \frac{1}{\beta K} \sum_{k=0}^{K-1} \mathbb{E}\|\boldsymbol{\lambda}^{k+1} - \boldsymbol{\lambda}^k\|^2
$$

$$
\leq \frac{D}{K} + \frac{m\sigma^2 \eta}{2S_1}. \tag{5.113}
$$

By the value of $\eta$ we have

$$
\frac{1}{4\eta} - \frac{\eta q \max_{i \in [m]} \{L_i^2\}}{2S_2} \geq \frac{1}{8\eta}. \tag{5.114}
$$

So using a similar argument as (5.102) and (5.103), we have

$$
\tilde{\mathbb{E}} \left\| \nabla g(\tilde{\mathbf{y}}) + \mathbf{B}^T \tilde{\boldsymbol{\lambda}} \right\|^2 = \frac{1}{K} \sum_{k=0}^{K-1} \mathbb{E} \left\| \nabla g(\mathbf{y}^{k+1}) + \mathbf{B}^T \boldsymbol{\lambda}^{k+1} \right\|^2
$$

$$
\leq \frac{L^2}{K} \sum_{k=0}^{K-1} \mathbb{E} \left\| \mathbf{y}^{k+1} - \mathbf{y}^k \right\|^2 \leq \frac{4L^2}{\rho} \left( \frac{D}{K} + \frac{m\sigma^2 \eta}{2S_1} \right)
$$

and

$$
\tilde{\mathbb{E}} \left\| \mathbf{A}\tilde{\mathbf{x}} + \mathbf{B}\tilde{\mathbf{y}} - \mathbf{b} \right\|^2 = \frac{1}{K} \sum_{k=0}^{K-1} \mathbb{E} \left\| \mathbf{A}\mathbf{x}^{k+1} + \mathbf{B}\mathbf{y}^{k+1} - \mathbf{b} \right\|^2
$$

$$
= \frac{1}{\beta^2 K} \sum_{k=0}^{K-1} \mathbb{E}\|\boldsymbol{\lambda}^{k+1} - \boldsymbol{\lambda}^k\|^2 \leq \frac{1}{\beta} \left( \frac{D}{K} + \frac{m\sigma^2 \eta}{2S_1} \right).
$$

Finally, by (5.113) and (5.114) we have

$$
\frac{1}{K} \sum_{k=0}^{K-1} \sum_{i=1}^{m} \mathbb{E} \left\| \mathbf{x}_i^{k+1} - \mathbf{x}_i^k \right\|^2 \leq 8\eta \left( \frac{D}{K} + \frac{m\sigma^2 \eta}{2S_1} \right). \tag{5.115}
$$

By the update rule of $\mathbf{x}_i^k$, we have

$$
\tilde{\nabla} f_i \left( \mathbf{x}_i^k \right) + \mathbf{A}_i^T \left[ \boldsymbol{\lambda}^k + \beta \left( \sum_{j<i} \mathbf{A}_j \mathbf{x}_j^{k+1} + \sum_{j \geq i} \mathbf{A}_j \mathbf{x}_j^k + \mathbf{B} \mathbf{y}^k - \mathbf{b} \right) \right]
$$
$$
= -\frac{1}{\eta} \left( \mathbf{x}_i^{k+1} - \mathbf{x}_i^k \right),
$$

which implies that

$$
\nabla f_i \left( \mathbf{x}_i^k \right) + \mathbf{A}_i^T \boldsymbol{\lambda}^k
$$
$$
= -\frac{1}{\eta} \left( \mathbf{x}_i^{k+1} - \mathbf{x}_i^k \right) - \mathbf{A}_i^T \left( \boldsymbol{\lambda}^{k+1} - \boldsymbol{\lambda}^k \right)
$$
$$
+ \beta \mathbf{A}_i^T \left[ \sum_{j \geq i} \mathbf{A}_j \left( \mathbf{x}_j^{k+1} - \mathbf{x}_j^k \right) + \mathbf{B} \left( \mathbf{y}^{k+1} - \mathbf{y}^k \right) \right]
$$
$$
- \left[ \tilde{\nabla} f_i \left( \mathbf{x}_i^k \right) - \nabla f_i \left( \mathbf{x}_i^k \right) \right].
$$

So we obtain

$$
\left\| \nabla f_i \left( \mathbf{x}_i^k \right) + \mathbf{A}_i^T \boldsymbol{\lambda}^k \right\|^2
$$
$$
\leq \frac{4}{\eta^2} \left\| \mathbf{x}_i^{k+1} - \mathbf{x}_i^k \right\|^2 + 4 \|\mathbf{A}_i\|_2^2 \left\| \boldsymbol{\lambda}^{k+1} - \boldsymbol{\lambda}^k \right\|^2
$$
$$
+ 4 \beta^2 \left\| \mathbf{A}_i^T \left[ \sum_{j \geq i} \mathbf{A}_j \left( \mathbf{x}_j^{k+1} - \mathbf{x}_j^k \right) + \mathbf{B}(\mathbf{y}^{k+1} - \mathbf{y}^k) \right] \right\|^2
$$
$$
+ 4 \left\| \tilde{\nabla} f_i \left( \mathbf{x}_i^k \right) - \nabla f_i \left( \mathbf{x}_i^k \right) \right\|^2
$$
$$
\leq \frac{4}{\eta^2} \left\| \mathbf{x}_i^{k+1} - \mathbf{x}_i^k \right\|^2 + 4 \|\mathbf{A}_i\|_2^2 \left\| \boldsymbol{\lambda}^{k+1} - \boldsymbol{\lambda}^k \right\|^2
$$
$$
+ 4 \beta^2 (m+1) \|\mathbf{A}_i\|_2^2 \sum_{j=1}^{m} \|\mathbf{A}_j\|_2^2 \left\| \mathbf{x}_j^{k+1} - \mathbf{x}_j^k \right\|^2
$$
$$
+ 4 \beta^2 (m+1) \|\mathbf{A}_i\|_2^2 \|\mathbf{B}\|_2^2 \left\| \mathbf{y}^{k+1} - \mathbf{y}^k \right\|^2
$$
$$
+ 4 \left\| \tilde{\nabla} f_i \left( \mathbf{x}_i^k \right) - \nabla f_i \left( \mathbf{x}_i^k \right) \right\|^2. \tag{5.116}
$$

Taking full expectation on (5.116) and summing the result from $k = 1$ to $K$, we obtain

$$\tilde{\mathbb{E}} \left\| \nabla f_i \left( \tilde{\mathbf{x}}_i^k \right) + \mathbf{A}_i^T \tilde{\boldsymbol{\lambda}}^k \right\|^2$$

$$\overset{a}{=} \frac{1}{K} \sum_{k=1}^{K} \mathbb{E} \left\| \nabla f_i \left( \mathbf{x}_i^k \right) + \mathbf{A}_i^T \boldsymbol{\lambda}^k \right\|^2$$

$$\overset{b}{\leq} \frac{4}{\eta^2} \frac{K+1}{K} \frac{1}{K+1} \sum_{k=0}^{K} \mathbb{E} \left\| \mathbf{x}_i^{k+1} - \mathbf{x}_i^k \right\|^2$$

$$+ 4 \|\mathbf{A}_i\|_2^2 \frac{K+1}{K} \frac{1}{K+1} \sum_{k=0}^{K} \mathbb{E} \left\| \boldsymbol{\lambda}^{k+1} - \boldsymbol{\lambda}^k \right\|^2$$

$$+ 4\beta^2(m+1)\|\mathbf{A}_i\|_2^2 \max_{j \in [m]} \|\mathbf{A}_j\|_2^2 \frac{K+1}{K} \frac{1}{K+1} \sum_{k=0}^{K} \sum_{j=1}^{m} \mathbb{E} \left\| \mathbf{x}_j^{k+1} - \mathbf{x}_j^k \right\|^2$$

$$+ 4\beta^2(m+1)\|\mathbf{A}_i\|_2^2 \|\mathbf{B}\|_2^2 \frac{K+1}{K} \frac{1}{K+1} \sum_{k=0}^{K} \mathbb{E} \left\| \mathbf{y}^{k+1} - \mathbf{y}^k \right\|^2$$

$$+ \frac{4q L_i^2}{S_2} \frac{K+1}{K} \frac{1}{K+1} \sum_{k=0}^{K} \mathbb{E} \left\| \mathbf{x}_i^{k+1} - \mathbf{x}_i^k \right\|^2 + \frac{4\sigma^2}{S_1}$$

$$\overset{c}{\leq} 4 \frac{K+1}{K} C_i \left( \frac{D}{K+1} + \frac{m\sigma^2 \eta}{2S_1} \right) + \frac{4\sigma^2}{S_1},$$

where the first equality $\overset{a}{=}$ uses that $(\tilde{\mathbf{x}}, \tilde{\mathbf{y}}, \tilde{\boldsymbol{\lambda}})$ are uniformly randomly taken from $\{\mathbf{x}^k, \mathbf{y}^k, \boldsymbol{\lambda}^k\}_{k=1}^K$, $\overset{b}{\leq}$ uses Lemma 5.7, and $\overset{c}{\leq}$ utilizes (5.113) (similar to the proof of (5.115)). $\qquad\square$

From Theorem 5.6, we know that the number of access to the stochastic gradients of $f_i$ and updates for $\mathbf{y}$ using $g$ are $O(\epsilon^{-3})$ and $O(\epsilon^{-2})$, respectively, to find an $O(\epsilon)$-approximate KKT point in expectation. Compared with nonconvex SADMM, both Theorems 5.5 and 5.6 require $K = O(\epsilon^{-2})$ iterations to achieve an $\epsilon$-approximate solution. However, SPIDER-ADMM only samples $O(\epsilon^{-1})$ stochastic gradients, instead of $O(\epsilon^{-2})$, for each iteration. We note that by using the same argument in Sect. 4.1.1, the surjectiveness assumption can also be relaxed.

At the end of the chapter, we would like to remind the readers that when $f_i$ can be written as a sum of finite $n$ individual functions for $i \in [m]$, SPIDER-ADMM can also find an $\epsilon$-approximate KKT point in expectation in $O\left(n + n^{1/2}\epsilon^{-2}\right)$ complexity. Moreover, we note that for generic nonconvex optimization, SPIDER is a more efficient technique compared with the traditional VR methods, such as SVRG [6], in the sense that the latter can only achieve a complexity of $O\left(\min\left(\epsilon^{-10/3}, n + n^{2/3}\epsilon^{-2}\right)\right)$.

# References

1. Z. Allen-Zhu, Katyusha: the first truly accelerated stochastic gradient method, in *ACM Symposium on the Theory of Computing* (2017), pp. 1200–1205
2. A. Defazio, F. Bach, S. Lacoste-Julien, SAGA: a fast incremental gradient method with support for non-strongly convex composite objectives, in *Advances in Neural Information Processing Systems* (2014), pp. 1646–1654
3. C. Fang, F. Cheng, Z. Lin, Faster and non-ergodic $O(1/k)$ stochastic alternating direction method of multipliers, in *Advances in Neural Information Processing Systems* (2017), pp. 4476–4485
4. C. Fang, C.J. Li, Z. Lin, T. Zhang, SPIDER: Near-optimal non-convex optimization via stochastic path-integrated differential estimator, in *Advances in Neural Information Processing Systems* (2018), pp. 689–699
5. F. Huang, S. Chen, H. Huang, Faster stochastic alternating direction method of multipliers for nonconvex optimization, in *International Conference on Machine Learning* (2019), pp. 2839–2848
6. R. Johnson, T. Zhang, Accelerating stochastic gradient descent using predictive variance reduction, in *Advances in Neural Information Processing Systems* (2013), pp. 315–323
7. J. Mairal, Optimization with first-order surrogate functions, in *International Conference on Machine Learning* (2013), pp. 783–791
8. L.M. Nguyen, J. Liu, K. Scheinberg, M. Takac, SARAH: a novel method for machine learning problems using stochastic recursive gradient, in *International Conference on Machine Learning* (2017), pp. 2613–2621
9. H. Ouyang, N. He, L. Tran, A. Gray, Stochastic alternating direction method of multipliers, in *International Conference on Machine Learning* (2013), pp. 80–88
10. M. Schmidt, N. Le Roux, F. Bach, Minimizing finite sums with the stochastic average gradient. *Math. Program.* **162**(1–2), 83–112 (2017)
11. S. Shalev-Shwartz, T. Zhang, Stochastic dual coordinate ascent methods for regularized loss minimization. *J. Mach. Learn. Res.* **14**(Feb), 567–599 (2013)
12. S. Zheng, J.T. Kwok, Fast-and-light stochastic ADMM, in *International Joint Conference on Artificial Intelligence* (2016), pp. 2407–2613

# Chapter 6
# ADMM for Distributed Optimization

In this chapter, we introduce the application of ADMM to distributed optimization. We first introduce how to use ADMM, linearized ADMM, and accelerated linearized ADMM to centralized distributed optimization, and give the corresponding convergence rates. Then, we focus on decentralized distributed optimization and show that the corresponding ADMM is equivalent to the linearized augmented Lagrangian method, and give its accelerated version. Next, we introduce the asynchronous ADMM. At last, we end this chapter by the nonconvex and generally linearly constrained distributed ADMM.

Consider the following problem in a distributed environment:

$$\min_{\mathbf{x} \in \mathbb{R}^d} f(\mathbf{x}) \equiv \sum_{i=1}^{m} f_i(\mathbf{x}), \qquad (6.1)$$

where $m$ agents form a connected and undirected network and the local function $f_i$ is only accessible by agent $i$ due to storage or privacy reasons. We consider two kinds of networks. The first one is the centralized network with one centralized master agent and $m$ worker agents. Each worker agent is connected to the master agent. We will introduce this kind of network in Sect. 6.1. The second one is the decentralized network, which does not have the centralized agent and each agent only communicates with its neighbors. This kind of network will be introduced in Sect. 6.2. All the agents cooperate to solve Problem (6.1).

Z. Lin et al., *Alternating Direction Method of Multipliers for Machine Learning*, https://doi.org/10.1007/978-981-16-9840-8_6

## 6.1   Centralized Optimization

In the centralized network, we reformulate Problem (6.1) as the following linearly constrained one:

$$\min_{\{\mathbf{x}_i\}, \mathbf{z}} \sum_{i=1}^{m} f_i(\mathbf{x}_i),$$

$$s.t. \ \ \mathbf{x}_i = \mathbf{z}, \quad i \in [m], \tag{6.2}$$

so that we can use the ADMM type methods to solve it.

### 6.1.1   ADMM

Introduce the augmented Lagrangian function

$$L(\mathbf{x}, \mathbf{z}, \boldsymbol{\lambda}) = \sum_{i=1}^{m} \left( f_i(\mathbf{x}_i) + \langle \boldsymbol{\lambda}_i, \mathbf{x}_i - \mathbf{z} \rangle + \frac{\beta}{2} \|\mathbf{x}_i - \mathbf{z}\|^2 \right). \tag{6.3}$$

ADMM can be used to solve problem (6.2) with the following iterations (for example, see [2, 3]):

$$\mathbf{z}^{k+1} = \underset{\mathbf{z}}{\operatorname{argmin}} \sum_{i=1}^{m} \left( \langle \boldsymbol{\lambda}_i^k, \mathbf{x}_i^k - \mathbf{z} \rangle + \frac{\beta}{2} \left\| \mathbf{x}_i^k - \mathbf{z} \right\|^2 \right)$$

$$= \frac{1}{m} \sum_{i=1}^{m} \left( \mathbf{x}_i^k + \frac{1}{\beta} \boldsymbol{\lambda}_i^k \right), \tag{6.4a}$$

$$\mathbf{x}_i^{k+1} = \underset{\mathbf{x}_i}{\operatorname{argmin}} \left( f_i(\mathbf{x}_i) + \langle \boldsymbol{\lambda}_i^k, \mathbf{x}_i - \mathbf{z}^{k+1} \rangle + \frac{\beta}{2} \left\| \mathbf{x}_i - \mathbf{z}^{k+1} \right\|^2 \right)$$

$$= \operatorname{Prox}_{\beta^{-1} f_i} \left( \mathbf{z}^{k+1} - \frac{1}{\beta} \boldsymbol{\lambda}_i^k \right), \quad i \in [m], \tag{6.4b}$$

$$\boldsymbol{\lambda}_i^{k+1} = \boldsymbol{\lambda}_i^k + \beta \left( \mathbf{x}_i^{k+1} - \mathbf{z}^{k+1} \right), \quad i \in [m]. \tag{6.4c}$$

In the above method, the master agent is responsible for updating $\mathbf{z}$ while each worker agent is responsible for $\mathbf{x}_i$ and $\boldsymbol{\lambda}_i$. Steps (6.4b) and (6.4c) are carried out independently at each worker agent, while step (6.4a) is performed at the master agent. At each iteration, the master agent collects $\mathbf{x}_i^k$ and $\boldsymbol{\lambda}_i^k$ from each worker agent, computes the average, and sends $\mathbf{z}^{k+1}$ back to each worker agent. Then each worker agent computes $\mathbf{x}_i^{k+1}$ and $\boldsymbol{\lambda}_i^{k+1}$ in parallel. We present the above method in Algorithms 6.1 and 6.2.

---

**Algorithm 6.1** Centralized ADMM of the master

---

    **for** $k = 0, 1, 2, \cdots$ **do**

        Wait until receiving $\mathbf{x}_i^k$ and $\boldsymbol{\lambda}_i^k$ from all the workers $i \in [m]$.

        Update $\mathbf{z}^{k+1}$ by (6.4a).

        Send $\mathbf{z}^{k+1}$ to all the workers.

    **end for**

---

**Algorithm 6.2** Centralized ADMM of the $i$th worker

---

    Initialize: $\mathbf{x}_i^0, \boldsymbol{\lambda}_i^0, i \in [m]$.

    **for** $k = 0, 1, 2, \cdots$ **do**

        Send $(\mathbf{x}_i^k, \boldsymbol{\lambda}_i^k)$ to the master.

        Wait until receiving $\mathbf{z}^{k+1}$ from the master.

        Update $\mathbf{x}_i^{k+1}$ and $\boldsymbol{\lambda}_i^{k+1}$ by (6.4b) and (6.4c), respectively.

    **end for**

---

Now we discuss the convergence of Algorithms 6.1–6.2. Denote

$$(\mathbf{x}_1^*, \cdots, \mathbf{x}_m^*, \mathbf{z}^*, \boldsymbol{\lambda}_1^*, \cdots, \boldsymbol{\lambda}_m^*)$$

as a KKT point of Problem (6.2). From Theorem 3.3 we have the following convergence result.

**Theorem 6.1** *Suppose that each $f_i(\mathbf{x}_i)$ is convex, $i \in [m]$. Then for Algorithms 6.1–6.2, we have*

$$\left| \sum_{i=1}^m f_i(\hat{\mathbf{x}}_i^{K+1}) - \sum_{i=1}^m f_i(\mathbf{x}_i^*) \right| \leq \frac{C}{2(K+1)} + \frac{2\sqrt{C}\sqrt{\sum_{i=1}^m \|\boldsymbol{\lambda}_i^*\|^2}}{\sqrt{\beta}(K+1)},$$

$$\sqrt{\sum_{i=1}^m \|\hat{\mathbf{x}}_i^{K+1} - \hat{\mathbf{z}}^{K+1}\|^2} \leq \frac{2\sqrt{C}}{\sqrt{\beta}(K+1)},$$

*where*

$$\hat{\mathbf{x}}_i^{K+1} = \frac{1}{K+1} \sum_{k=1}^{K+1} \mathbf{x}_i^k, \ i \in [m], \quad \hat{\mathbf{z}}^{K+1} = \frac{1}{K+1} \sum_{k=1}^{K+1} \mathbf{z}^k, \ and$$

$$C = \frac{1}{\beta} \sum_{i=1}^m \|\boldsymbol{\lambda}_i^0 - \boldsymbol{\lambda}_i^*\|^2 + \beta \sum_{i=1}^m \|\mathbf{x}_i^0 - \mathbf{x}_i^*\|^2.$$

***Proof*** Algorithms 6.1–6.2 are a direct application of the original ADMM (Algorithm 2.1) to Problem (6.2) by setting

$$\mathbf{x} = \mathbf{z}, \quad \mathbf{y} = (\mathbf{x}_1^T, \cdots, \mathbf{x}_m^T)^T, \quad \mathbf{A} = \mathbf{1}_m \otimes \mathbf{I}_d, \quad \mathbf{B} = -\mathbf{I}_{md},$$

$$\mathbf{b} = \mathbf{0}, \quad f(\mathbf{x}) = 0, \quad \text{and} \quad g(\mathbf{y}) = \sum_i f_i(\mathbf{x}_i)$$

in (2.13), where $d$ is the dimension of $\mathbf{x}_i$, $\mathbf{1}_m$ is the vector of $m$ ones, and $\otimes$ is the Kronecker product. $\qquad\qquad\square$

Similarly, from Theorem 3.4 we have the following linear convergence result.

**Theorem 6.2** *Suppose that each $f_i(\mathbf{x}_i)$ is $\mu$-strongly convex and $L$-smooth, $i \in [m]$. Let $\beta = \sqrt{\mu L}$. Then for Algorithms 6.1–6.2, we have*

$$\sum_{i=1}^{m} \left( \frac{1}{2\beta} \|\boldsymbol{\lambda}_i^{k+1} - \boldsymbol{\lambda}_i^*\|^2 + \frac{\beta}{2} \|\mathbf{x}_i^{k+1} - \mathbf{x}_i^*\|^2 \right)$$

$$\leq \left( 1 + \frac{1}{2}\sqrt{\frac{\mu}{L}} \right)^{-1} \sum_{i=1}^{m} \left( \frac{1}{2\beta} \|\boldsymbol{\lambda}_i^k - \boldsymbol{\lambda}_i^*\|^2 + \frac{\beta}{2} \|\mathbf{x}_i^k - \mathbf{x}_i^*\|^2 \right).$$

## 6.1.2   Linearized ADMM

When each $f_i$ is $L$-smooth, we can also linearize $f_i$ in step (6.4b) to simplify the computation, if the proximal mapping of $f_i$ is not easily computable. The iterations of resulting linearized ADMM are as follows:

$$\mathbf{z}^{k+1} = \underset{\mathbf{z}}{\operatorname{argmin}} \sum_{i=1}^{m} \left( \langle \boldsymbol{\lambda}_i^k, \mathbf{x}_i^k - \mathbf{z} \rangle + \frac{\beta}{2} \left\| \mathbf{x}_i^k - \mathbf{z} \right\|^2 \right)$$

$$= \frac{1}{m} \sum_{i=1}^{m} \left( \mathbf{x}_i^k + \frac{1}{\beta} \boldsymbol{\lambda}_i^k \right), \qquad\qquad (6.5a)$$

$$\mathbf{x}_i^{k+1} = \underset{\mathbf{x}_i}{\operatorname{argmin}} \left( f_i(\mathbf{x}_i) + \langle \boldsymbol{\lambda}_i^k, \mathbf{x}_i - \mathbf{z}^{k+1} \rangle + \frac{\beta}{2} \left\| \mathbf{x}_i - \mathbf{z}^{k+1} \right\|^2 + D_{\psi_i}(\mathbf{x}_i, \mathbf{x}_i^k) \right)$$

$$= \underset{\mathbf{x}_i}{\operatorname{argmin}} \left( \langle \nabla f_i(\mathbf{x}_i^k), \mathbf{x}_i - \mathbf{x}_i^k \rangle + \frac{L}{2} \left\| \mathbf{x}_i - \mathbf{x}_i^k \right\|^2 \right.$$

$$\left. + \langle \boldsymbol{\lambda}_i^k, \mathbf{x}_i - \mathbf{z}^{k+1} \rangle + \frac{\beta}{2} \left\| \mathbf{x}_i - \mathbf{z}^{k+1} \right\|^2 \right)$$

$$= \frac{1}{L+\beta}\left(L\mathbf{x}_i^k + \beta\mathbf{z}^{k+1} - \nabla f_i(\mathbf{x}_i^k) - \boldsymbol{\lambda}_i^k\right), \quad i \in [m], \tag{6.5b}$$

$$\boldsymbol{\lambda}_i^{k+1} = \boldsymbol{\lambda}_i^k + \beta\left(\mathbf{x}_i^{k+1} - \mathbf{z}^{k+1}\right), \quad i \in [m], \tag{6.5c}$$

by choosing

$$\psi_i(\mathbf{x}_i) = \frac{L}{2}\|\mathbf{x}_i\|^2 - f_i(\mathbf{x}_i).$$

We summarize the method in Algorithms 6.3 and 6.4.

---

**Algorithm 6.3** Centralized linearized ADMM of the master

---

**for** $k = 0, 1, 2, \cdots$ **do**
    Wait until receiving $\mathbf{x}_i^k$ and $\boldsymbol{\lambda}_i^k$ from all the workers $i \in [m]$.
    Update $\mathbf{z}^{k+1}$ by (6.5a).
    Send $\mathbf{z}^{k+1}$ to all the workers.
**end for**

---

**Algorithm 6.4** Centralized linearized ADMM of the $i$th worker

---

Initialize: $\mathbf{x}_i^0, \boldsymbol{\lambda}_i^0, i \in [m]$.
**for** $k = 0, 1, 2, \cdots$ **do**
    Send $(\mathbf{x}_i^k, \boldsymbol{\lambda}_i^k)$ to the master.
    Wait until receiving $\mathbf{z}^{k+1}$ from the master.
    Update $\mathbf{x}_i^{k+1}$ and $\boldsymbol{\lambda}_i^{k+1}$ by (6.5b) and (6.5c), respectively.
**end for**

---

Similar to Theorem 6.1, from Theorem 3.6 we can also have the $O(1/K)$ convergence rate. We omit the details and mainly discuss the linear convergence rate under stronger conditions. From Theorem 3.8 and using $L_\psi \leq L - \mu$, where $\psi(\mathbf{x}) = \sum_{i=1}^m \psi_i(\mathbf{x}_i)$, we have the following linear convergence result.

**Theorem 6.3** *Suppose that each $f_i(\mathbf{x}_i)$ is $\mu$-strongly convex and $L$-smooth, $i \in [m]$. Let $\beta = \sqrt{\mu(2L - \mu)}$. Then for Algorithm 6.3–6.4, we have*

$$\sum_{i=1}^m \left(\frac{1}{2\beta}\|\boldsymbol{\lambda}_i^{k+1} - \boldsymbol{\lambda}_i^*\|^2 + \frac{\beta}{2}\|\mathbf{x}_i^{k+1} - \mathbf{x}_i^*\|^2 + D_{\psi_i}(\mathbf{x}_i^*, \mathbf{x}_i^{k+1})\right)$$

$$\leq \left[1 + \frac{1}{3}\min\left(\sqrt{\frac{\mu}{2L - \mu}}, \frac{\mu}{L - \mu}\right)\right]^{-1}$$

$$\times \sum_{i=1}^m \left(\frac{1}{2\beta}\|\boldsymbol{\lambda}_i^k - \boldsymbol{\lambda}_i^*\|^2 + \frac{\beta}{2}\|\mathbf{x}_i^k - \mathbf{x}_i^*\|^2 + D_{\psi_i}(\mathbf{x}_i^*, \mathbf{x}_i^k)\right).$$

### 6.1.3   Accelerated Linearized ADMM

Motivated by the results in Sect. 3.3.2, we can also use the accelerated linearized ADMM to solve Problem (6.2) to further improve the convergence rate of the linearized ADMM. From Algorithm 3.6 given in Sect. 3.3.2, we have the following iterations:

$$\mathbf{w}_i^k = \theta \mathbf{x}_i^k + (1-\theta)\widetilde{\mathbf{x}}_i^k, \tag{6.6a}$$

$$\mathbf{z}^{k+1} = \operatorname*{argmin}_{\mathbf{z}} \sum_{i=1}^m \left( \left\langle \boldsymbol{\lambda}_i^k, \mathbf{x}_i^k - \mathbf{z} \right\rangle + \frac{\beta\theta}{2} \|\mathbf{x}_i^k - \mathbf{z}\|^2 \right)$$

$$= \frac{1}{m} \sum_{i=1}^m \left( \mathbf{x}_i^k + \frac{1}{\beta\theta} \boldsymbol{\lambda}_i^k \right), \tag{6.6b}$$

$$\mathbf{x}_i^{k+1} = \frac{1}{\frac{\theta}{\alpha} + \mu} \left\{ \mu \mathbf{w}_i^k + \frac{\theta}{\alpha} \mathbf{x}_i^k - \left[ \nabla f_i(\mathbf{x}_i^k) + \boldsymbol{\lambda}_i^k + \beta\theta \left( \mathbf{x}_i^k - \mathbf{z}^{k+1} \right) \right] \right\}, \tag{6.6c}$$

$$\widetilde{\mathbf{z}}^{k+1} = \theta \mathbf{z}^{k+1} + (1-\theta)\widetilde{\mathbf{z}}^k, \tag{6.6d}$$

$$\widetilde{\mathbf{x}}_i^{k+1} = \theta \mathbf{x}_i^{k+1} + (1-\theta)\widetilde{\mathbf{x}}_i^k, \tag{6.6e}$$

$$\boldsymbol{\lambda}_i^{k+1} = \boldsymbol{\lambda}_i^k + \beta\theta \left( \mathbf{x}_i^{k+1} - \mathbf{z}^{k+1} \right). \tag{6.6f}$$

We summarize the method in Algorithms 6.5 and 6.6.

---

**Algorithm 6.5** Accelerated centralized linearized ADMM of the master

---

Initialize: $\widetilde{\mathbf{z}}^0$.
**for** $k = 0, 1, 2, \cdots$ **do**
    Wait until receiving $\mathbf{x}_i^k$ and $\boldsymbol{\lambda}_i^k$ from all the workers, $i \in [m]$.
    Update $\mathbf{z}^{k+1}$ and $\widetilde{\mathbf{z}}^{k+1}$ by (6.6b) and (6.6d), respectively.
    Send $\mathbf{z}^{k+1}$ to all the workers.
**end for**

---

**Algorithm 6.6** Accelerated centralized linearized ADMM of the $i$th worker

---

Initialize: $\mathbf{x}_i^0, \boldsymbol{\lambda}_i^0, i \in [m]$, and $\widetilde{\mathbf{x}}_i^0$.
**for** $k = 0, 1, 2, \cdots$ **do**
    Send $(\mathbf{x}_i^k, \boldsymbol{\lambda}_i^k)$ to the master.
    Wait until receiving $\mathbf{z}^{k+1}$ from the master.
    Update $\mathbf{x}_i^{k+1}, \widetilde{\mathbf{x}}_i^{k+1}, \boldsymbol{\lambda}_i^{k+1}$, and $\mathbf{w}_i^{k+1}$ by (6.6c), (6.6e), (6.6f), and (6.6a), respectively.
**end for**

---

**Table 6.1** Complexity comparisons between centralized ADMM, centralized linearized ADMM (LADMM), and its accelerated version

| Centralized ADMM | Centralized LADMM | Accelerated centralized LADMM |
|---|---|---|
| $O\left(\sqrt{\frac{L}{\mu}}\log\frac{1}{\epsilon}\right)$ | $O\left(\frac{L}{\mu}\log\frac{1}{\epsilon}\right)$ | $O\left(\sqrt{\frac{L}{\mu}}\log\frac{1}{\epsilon}\right)$ |

Denote

$$\ell_k = (1-\theta)\sum_{i=1}^{m}\left(f_i(\widetilde{\mathbf{x}}_i^k) - f_i(\mathbf{x}_i^*) + \left\langle \boldsymbol{\lambda}_i^*, \widetilde{\mathbf{x}}_i^k - \widetilde{\mathbf{z}}^k\right\rangle\right)$$

$$+ \frac{\theta^2}{2\alpha}\sum_{i=1}^{m}\|\mathbf{x}_i^k - \mathbf{x}_i^*\|^2 + \frac{1}{2\beta}\sum_{i=1}^{m}\|\boldsymbol{\lambda}_i^k - \boldsymbol{\lambda}_i^*\|^2.$$

From Theorem 3.12 we have the following linear convergence result.

**Theorem 6.4** *Suppose that each $f_i(\mathbf{x}_i)$ is $\mu$-strongly convex and $L$-smooth, $i \in [m]$. Let*

$$\alpha = \frac{1}{4L}, \quad \beta = L, \quad and \quad \theta = \sqrt{\frac{\mu}{L}}.$$

*Then for the accelerated linearized ADMM (Algorithms 6.5–6.6), we have*

$$\ell_{k+1} \le \left(1 - \sqrt{\frac{\mu}{L}}\right)\ell_k.$$

We list the convergence rate comparisons of different centralized ADMM methods in Table 6.1. Similar to the comparisons in Table 3.2, we see that the accelerated linearized ADMM is faster than the linearized ADMM with a better dependence on the condition number $L/\mu$. The original ADMM has the same convergence rate as the accelerated linearized ADMM. However, the original ADMM may need to solve a subproblem iteratively at each iteration, while the accelerated linearized ADMM only performs a gradient descent type update.

## 6.2 Decentralized Optimization

In this section we consider the decentralized topology. In this case, we cannot use the constraints in (6.2) since there is no central node to compute $\mathbf{z}$. Denote $\mathcal{E}$ as the set of edges. Assume that all the nodes are ordered from 1 to $m$. For any two nodes $i$ and $j$, if $i$ and $j$ are directly connected in the network and $i < j$, we say $(i, j) \in \mathcal{E}$. To simplify the presentation, we order the edges from 1 to $|\mathcal{E}|$. For each node $i$, we denote $\mathcal{N}_i$ as its neighborhood:

$$\mathcal{N}_i = \{j | (i, j) \in \mathcal{E} \text{ or } (j, i) \in \mathcal{E}\},$$

and $d_i = |\mathcal{N}_i|$ as its degree.

Introduce auxiliary variables $\mathbf{z}_{ij}$ if $(i, j) \in \mathcal{E}$. Then we can reformulate Problem (6.1) as follows (for example, see [1, 10, 12, 14]):

$$\min_{\mathbf{x}_i, \mathbf{z}_{ij}} \sum_{i=1}^{m} f_i(\mathbf{x}_i),$$

$$s.t. \ \mathbf{x}_i = \mathbf{z}_{ij}, \quad \mathbf{x}_j = \mathbf{z}_{ij}, \quad \forall (i, j) \in \mathcal{E}. \tag{6.7}$$

That is to say, each variable $\mathbf{x}_i$ corresponds to one node, while each variable $\mathbf{z}_{ij}$ ($i < j$) corresponds to one edge. The augmented Lagrangian function of Problem (6.7)

$$L(\mathbf{x}, \mathbf{z}, \boldsymbol{\lambda}) = \sum_{i=1}^{m} f_i(\mathbf{x}_i) + \sum_{(i,j) \in \mathcal{E}} \left( \langle \boldsymbol{\lambda}_{ij}, \mathbf{x}_i - \mathbf{z}_{ij} \rangle + \langle \boldsymbol{\gamma}_{ij}, \mathbf{x}_j - \mathbf{z}_{ij} \rangle \right.$$

$$\left. + \frac{\beta}{2} \| \mathbf{x}_i - \mathbf{z}_{ij} \|^2 + \frac{\beta}{2} \| \mathbf{x}_j - \mathbf{z}_{ij} \|^2 \right).$$

### 6.2.1  ADMM

We can use ADMM to solve Problem (6.7), which consists of the following iterations:

$$\mathbf{x}_i^{k+1} = \underset{\mathbf{x}_i}{\text{argmin}} \left[ f_i(\mathbf{x}_i) + \sum_{j:(i,j) \in \mathcal{E}} \left( \langle \boldsymbol{\lambda}_{ij}^k, \mathbf{x}_i - \mathbf{z}_{ij}^k \rangle + \frac{\beta}{2} \left\| \mathbf{x}_i - \mathbf{z}_{ij}^k \right\|^2 \right) \right.$$

$$\left. + \sum_{j:(j,i) \in \mathcal{E}} \left( \langle \boldsymbol{\gamma}_{ji}^k, \mathbf{x}_i - \mathbf{z}_{ji}^k \rangle + \frac{\beta}{2} \left\| \mathbf{x}_i - \mathbf{z}_{ji}^k \right\|^2 \right) \right], \tag{6.8a}$$

$$\mathbf{z}_{ij}^{k+1} = \underset{\mathbf{z}_{ij}}{\text{argmin}} \left( -\langle \boldsymbol{\lambda}_{ij}^k + \boldsymbol{\gamma}_{ij}^k, \mathbf{z}_{ij} \rangle + \frac{\beta}{2} \left\| \mathbf{x}_i^{k+1} - \mathbf{z}_{ij} \right\|^2 + \frac{\beta}{2} \left\| \mathbf{x}_j^{k+1} - \mathbf{z}_{ij} \right\|^2 \right)$$

$$= \frac{1}{2\beta} \left( \boldsymbol{\lambda}_{ij}^k + \boldsymbol{\gamma}_{ij}^k \right) + \frac{1}{2} \left( \mathbf{x}_i^{k+1} + \mathbf{x}_j^{k+1} \right), \tag{6.8b}$$

$$\boldsymbol{\lambda}_{ij}^{k+1} = \boldsymbol{\lambda}_{ij}^k + \beta \left( \mathbf{x}_i^{k+1} - \mathbf{z}_{ij}^{k+1} \right), \tag{6.8c}$$

$$\boldsymbol{\gamma}_{ij}^{k+1} = \boldsymbol{\gamma}_{ij}^k + \beta \left( \mathbf{x}_j^{k+1} - \mathbf{z}_{ij}^{k+1} \right). \tag{6.8d}$$

Next, we introduce the result in [10] to simplify the above method by eliminating variables $\mathbf{z}_{ij}$, $\boldsymbol{\lambda}_{ij}$, and $\boldsymbol{\gamma}_{ij}$.

Summing (6.8c) and (6.8d) and using (6.8b), we have

$$\boldsymbol{\lambda}_{ij}^{k+1} + \boldsymbol{\gamma}_{ij}^{k+1} = \mathbf{0}, \quad \forall k \geq 0.$$

Initialize $\boldsymbol{\lambda}_{ij}^0 = \boldsymbol{\gamma}_{ij}^0 = \mathbf{0}$, we have

$$\boldsymbol{\lambda}_{ij}^{k} + \boldsymbol{\gamma}_{ij}^{k} = \mathbf{0}, \quad \forall k \geq 0.$$

Plugging it into (6.8b), we have

$$\mathbf{z}_{ij}^{k+1} = \frac{1}{2} \left( \mathbf{x}_i^{k+1} + \mathbf{x}_j^{k+1} \right), \quad \forall k \geq 0. \tag{6.9}$$

We may initialize

$$\mathbf{z}_{ij}^{0} = \frac{1}{2} \left( \mathbf{x}_i^{0} + \mathbf{x}_j^{0} \right).$$

From (6.9) and (6.8c), we have

$$\boldsymbol{\lambda}_{ij}^{k+1} = \boldsymbol{\lambda}_{ij}^{k} + \frac{\beta}{2} \left( \mathbf{x}_i^{k+1} - \mathbf{x}_j^{k+1} \right). \tag{6.10}$$

So we have

$$\boldsymbol{\lambda}_{ij}^{k+1} = \beta \sum_{t=1}^{k+1} \frac{1}{2} \left( \mathbf{x}_i^{t} - \mathbf{x}_j^{t} \right).$$

Similarly, we can have

$$\boldsymbol{\gamma}_{ij}^{k+1} = \beta \sum_{t=1}^{k+1} \frac{1}{2} \left( \mathbf{x}_j^{t} - \mathbf{x}_i^{t} \right).$$

Note that we only define $\boldsymbol{\lambda}_{ij}$, $\boldsymbol{\gamma}_{ij}$, and $\mathbf{z}_{ij}$ for $i < j$. Now we define

$$\boldsymbol{\lambda}_{ij} \equiv \boldsymbol{\gamma}_{ji} \quad \text{and} \quad \mathbf{z}_{ij} \equiv \mathbf{z}_{ji} \quad \text{for } i > j.$$

Then

$$\boldsymbol{\lambda}_{ij}^{k+1} = \beta \sum_{t=1}^{k+1} \frac{1}{2} \left( \mathbf{x}_i^{t} - \mathbf{x}_j^{t} \right) \quad \text{and} \quad \mathbf{z}_{ij}^{k+1} = \frac{1}{2} \left( \mathbf{x}_i^{k+1} + \mathbf{x}_j^{k+1} \right)$$

for both $i < j$ and $i > j$. So is (6.10). Thus (6.8a) can be simplified to

$$
\begin{aligned}
\mathbf{x}_i^{k+1} &= \underset{\mathbf{x}_i}{\operatorname{argmin}} \left[ f_i(\mathbf{x}_i) + \sum_{j:(i,j)\in\mathcal{E}} \left( \left\langle \boldsymbol{\lambda}_{ij}^k - \beta\mathbf{z}_{ij}^k, \mathbf{x}_i \right\rangle + \frac{\beta}{2}\|\mathbf{x}_i\|^2 \right) \right. \\
&\qquad \left. + \sum_{j:(j,i)\in\mathcal{E}} \left( \left\langle \boldsymbol{\gamma}_{ji}^k - \beta\mathbf{z}_{ji}^k, \mathbf{x}_i \right\rangle + \frac{\beta}{2}\|\mathbf{x}_i\|^2 \right) \right] \\
&= \underset{\mathbf{x}_i}{\operatorname{argmin}} \left[ f_i(\mathbf{x}_i) + \sum_{j\in\mathcal{N}_i} \left( \left\langle \boldsymbol{\lambda}_{ij}^k - \beta\mathbf{z}_{ij}^k, \mathbf{x}_i \right\rangle + \frac{\beta}{2}\|\mathbf{x}_i\|^2 \right) \right] \\
&= \underset{\mathbf{x}_i}{\operatorname{argmin}} \left[ f_i(\mathbf{x}_i) + \sum_{j\in\mathcal{N}_i} \left( \left\langle \boldsymbol{\lambda}_{ij}^k - \beta\mathbf{z}_{ij}^k + \beta\mathbf{x}_i^k, \mathbf{x}_i \right\rangle + \frac{\beta}{2}\|\mathbf{x}_i - \mathbf{x}_i^k\|^2 \right) \right] \\
&= \underset{\mathbf{x}_i}{\operatorname{argmin}} \left[ f_i(\mathbf{x}_i) + \sum_{j\in\mathcal{N}_i} \left( \left\langle \boldsymbol{\lambda}_{ij}^k + \frac{\beta}{2}\left(\mathbf{x}_i^k - \mathbf{x}_j^k\right), \mathbf{x}_i \right\rangle + \frac{\beta}{2}\|\mathbf{x}_i - \mathbf{x}_i^k\|^2 \right) \right].
\end{aligned}
$$
(6.11)

Denote $\mathbf{L} \in \mathbb{R}^{m\times m}$ as the Laplacian matrix (Definition A.2) and $\mathbf{D}$ as the diagonal degree matrix with $\mathbf{D}_{ii} = d_i$. It is well known that $\mathbf{L}$ is symmetric and satisfies $\mathbf{0} \preceq \mathbf{L} \preceq 2\mathbf{D}$.[1]

Define

$$
\mathbf{X} = \begin{pmatrix} \mathbf{x}_1^T \\ \vdots \\ \mathbf{x}_m^T \end{pmatrix} \in \mathbb{R}^{m\times d}, \quad f(\mathbf{X}) = \sum_{i=1}^m f_i(\mathbf{x}_i),
$$

$$
\boldsymbol{v}_i = \sum_{j\in\mathcal{N}_i} \boldsymbol{\lambda}_{ij}, \quad \text{and} \quad \boldsymbol{\Upsilon} = \begin{pmatrix} \boldsymbol{v}_1^T \\ \vdots \\ \boldsymbol{v}_m^T \end{pmatrix} \in \mathbb{R}^{m\times d}.
$$

Then we have

$$
\mathbf{L}_i^T \mathbf{X} = d_i \mathbf{x}_i^T - \sum_{j\in\mathcal{N}_i} \mathbf{x}_j^T,
$$

where $\mathbf{L}_i$ is the $i$-th column of $\mathbf{L}$.

---

[1] $0 \leq \boldsymbol{\alpha}^T \mathbf{L}\boldsymbol{\alpha} = \frac{1}{2}\sum_{(i,j)\in\mathcal{E}}(\alpha_i - \alpha_j)^2 \leq \sum_{(i,j)\in\mathcal{E}} \left(\alpha_i^2 + \alpha_j^2\right) = 2\boldsymbol{\alpha}^T \mathbf{D}\boldsymbol{\alpha}$.

With the Laplacian matrix $\mathbf{L}$ and $\boldsymbol{v}_i$ introduced, (6.11) can be written as

$$\mathbf{x}_i^{k+1} = \underset{\mathbf{x}_i}{\text{argmin}} \left[ f_i(\mathbf{x}_i) + \left\langle \boldsymbol{v}_i^k, \mathbf{x}_i \right\rangle + \frac{\beta}{2} \left\langle \sum_{j \in \mathcal{N}_i} \mathbf{L}_{ij} \mathbf{x}_j^k, \mathbf{x}_i \right\rangle + \frac{\beta d_i}{2} \left\| \mathbf{x}_i - \mathbf{x}_i^k \right\|^2 \right]$$

$$= \text{Prox}_{(\beta d_i)^{-1} f_i} \left( \mathbf{x}_i^k - \frac{1}{\beta d_i} \left( \boldsymbol{v}_i^k + \frac{\beta}{2} \sum_{j \in \mathcal{N}_i} \mathbf{L}_{ij} \mathbf{x}_j^k \right) \right), \quad i \in [m]. \quad (6.12)$$

Summing (6.10) over $j \in \mathcal{N}_i$, we have that (6.10) gives

$$\boldsymbol{v}_i^{k+1} = \boldsymbol{v}_i^k + \frac{\beta}{2} \sum_{j \in \mathcal{N}_i} \mathbf{L}_{ij} \mathbf{x}_j^{k+1}, \quad i \in [m]. \quad (6.13)$$

(6.12)–(6.13) can be written in a compact form:

$$\mathbf{X}^{k+1} = \underset{\mathbf{X}}{\text{argmin}} \left( f(\mathbf{X}) + \left\langle \boldsymbol{\Upsilon}^k + \frac{\beta}{2} \mathbf{L} \mathbf{X}^k, \mathbf{X} \right\rangle + \frac{\beta}{2} \left\| \sqrt{\mathbf{D}} (\mathbf{X} - \mathbf{X}^k) \right\|^2 \right), \quad (6.14)$$

$$\boldsymbol{\Upsilon}^{k+1} = \boldsymbol{\Upsilon}^k + \frac{\beta}{2} \mathbf{L} \mathbf{X}^{k+1}. \quad (6.15)$$

Denoting $\mathbf{W} = \sqrt{\mathbf{L}/2}$, (6.15) can be rewritten as

$$\boldsymbol{\Upsilon}^{k+1} = \boldsymbol{\Upsilon}^k + \beta \mathbf{W}^2 \mathbf{X}^{k+1}.$$

Letting $\boldsymbol{\Upsilon}^0 \in \text{Span}(\mathbf{W}^2)$, we know that

$$\boldsymbol{\Upsilon}^k \in \text{Span}(\mathbf{W}^2), \quad \forall k \geq 0,$$

and there exists $\boldsymbol{\Omega}^k$ such that $\boldsymbol{\Upsilon}^k = \mathbf{W} \boldsymbol{\Omega}^k$.[2] Then (6.14) and (6.15) can be rewritten as[3]

$$\mathbf{X}^{k+1} = \underset{\mathbf{X}}{\text{argmin}} \left( f(\mathbf{X}) + \left\langle \boldsymbol{\Omega}^k, \mathbf{W} \mathbf{X} \right\rangle + \beta \left\langle \mathbf{W}^2 \mathbf{X}^k, \mathbf{X} \right\rangle + \frac{\beta}{2} \left\| \sqrt{\mathbf{D}} (\mathbf{X} - \mathbf{X}^k) \right\|^2 \right)$$

$$= \underset{\mathbf{X}}{\text{argmin}} \left( f(\mathbf{X}) + \left\langle \boldsymbol{\Omega}^k, \mathbf{W} \mathbf{X} \right\rangle + \frac{\beta}{2} \| \mathbf{W} \mathbf{X} \|^2 + D_\psi(\mathbf{X}, \mathbf{X}^k) \right), \quad (6.16a)$$

$$\boldsymbol{\Omega}^{k+1} = \boldsymbol{\Omega}^k + \beta \mathbf{W} \mathbf{X}^{k+1}, \quad (6.16b)$$

---

[2] Denote $\mathbf{U} \boldsymbol{\Lambda} \mathbf{U}^T$ to be the eigen-decomposition of $\mathbf{W}$ with $\mathbf{U} \in \mathbb{R}^{m \times (m-1)}$ and $\boldsymbol{\Lambda} \in \mathbb{R}^{(m-1) \times (m-1)}$, then $\boldsymbol{\Lambda}$ is invertible. Since there exists $\mathbf{R}^k$ such that $\boldsymbol{\Upsilon}^k = \mathbf{U} \mathbf{R}^k$, we can choose $\boldsymbol{\Omega}^k = \mathbf{U} \boldsymbol{\Lambda}^{-1} \mathbf{R}^k$ such that $\boldsymbol{\Upsilon}^k = \mathbf{W} \boldsymbol{\Omega}^k$.

[3] From (6.15), we have $\mathbf{R}^{k+1} = \mathbf{R}^k + \beta \boldsymbol{\Lambda}^2 \mathbf{U}^T \mathbf{X}^{k+1}$. Multiplying both sides by $\mathbf{U} \boldsymbol{\Lambda}^{-1}$, we get (6.16b).

with

$$\psi(\mathbf{X}) = \frac{\beta}{2} \left\| \sqrt{\mathbf{D}} \mathbf{X} \right\|^2 - \frac{\beta}{2} \|\mathbf{W}\mathbf{X}\|^2.$$

Thus, algorithm (6.8a)–(6.8d) is equivalent to using the linearized augmented Lagrangian method to solve problem

$$\min_{\mathbf{X}} f(\mathbf{X}), \quad s.t. \quad \mathbf{W}\mathbf{X} = \mathbf{0}.$$

Algorithm (6.16a)–(6.16b) is not implementable in the distributed manner due to $\mathbf{W} = \sqrt{\mathbf{L}/2}$, which is only used for analysis. In practice, we implement the original (6.12)–(6.13) instead. We present algorithm (6.12)–(6.13) in Algorithm 6.7.

---

**Algorithm 6.7** Decentralized ADMM of the $i$th node

---

Initialize: $\mathbf{x}_i^0$ and $\boldsymbol{v}_i^0 = \mathbf{0}$, $i \in [m]$.
Send $\mathbf{x}_i^0$ to its neighbors.
Wait until receiving $\mathbf{x}_j^0$ from all its neighbors, $j \in \mathcal{N}_i$.
**for** $k = 0, 1, 2, \cdots$ **do**
  Update $\mathbf{x}_i^{k+1}$ by (6.12).
  Send $\mathbf{x}_i^{k+1}$ to its neighbors.
  Wait until receiving $\mathbf{x}_j^{k+1}$ from all its neighbors, $j \in \mathcal{N}_i$.
  Update $\boldsymbol{v}_i^{k+1}$ by (6.13).
**end for**

---

#### 6.2.1.1   Convergence Analysis

We consider the linearized augmented Lagrangian method (6.16a)–(6.16b) with a general $\psi$. From Theorem 3.14 or 3.8, we have the following convergence result.

**Theorem 6.5** *Assume that each $f_i$ is $\mu$-strongly convex and $L$-smooth, $i \in [m]$, and $\psi(\mathbf{y})$ is convex and $L_\psi$-smooth. Initialize $\boldsymbol{\Omega}^0 = \mathbf{0}$. Then for algorithm (6.16a)–(6.16b) we have*

$$\frac{1}{2\beta} \|\boldsymbol{\Omega}^{k+1} - \boldsymbol{\Omega}^*\|^2 + \frac{\beta}{2} \|\mathbf{W}\mathbf{X}^{k+1} - \mathbf{W}\mathbf{X}^*\|^2 + D_\psi(\mathbf{X}^*, \mathbf{X}^{k+1})$$

$$\leq \left(1 + \frac{1}{3} \min \left\{ \frac{\beta\sigma_{\mathbf{L}}}{2(L + L_\psi)}, \frac{\mu}{\beta\|\mathbf{W}\|_2^2}, \frac{\mu}{L_\psi} \right\}\right)^{-1}$$

$$\times \left( \frac{1}{2\beta} \|\boldsymbol{\Omega}^k - \boldsymbol{\Omega}^*\|^2 + \frac{\beta}{2} \|\mathbf{W}\mathbf{X}^k - \mathbf{W}\mathbf{X}^*\|^2 + D_\psi(\mathbf{X}^*, \mathbf{X}^k) \right),$$

*where $\sigma_{\mathbf{L}}$ is the smallest positive eigenvalue of $\mathbf{L}$.*

***Proof*** From the proof of Theorem 3.8, to prove this theorem we only need to check

$$\|\mathbf{W}(\mathbf{\Omega}^k - \mathbf{\Omega}^*)\| \geq \sqrt{\sigma_L/2}\|\mathbf{\Omega}^k - \mathbf{\Omega}^*\|.$$

Note that $\mathbf{B} = \mathbf{W}$ and $\sigma^2 = \frac{\sigma_L}{2}$ in Theorem 3.8.

Since the network has to be connected, the rank of the Laplacian matrix $\mathbf{L}$ is $m-1$ (Proposition A.2). Let $\mathbf{V}\mathbf{\Sigma}\mathbf{V}^T = \mathbf{L}$ be its economical SVD with $\mathbf{V} \in \mathbb{R}^{m \times (m-1)}$. For any $\mathbf{\Omega}$ belonging to the column space of $\mathbf{W}$, we have

$$\|\mathbf{W}\mathbf{\Omega}\|^2 = \sum_{i=1}^{d} \mathbf{\Omega}_i^T \mathbf{W}^2 \mathbf{\Omega}_i$$

$$= \frac{1}{2}\sum_{i=1}^{d} \mathbf{\Omega}_i^T \mathbf{L}\mathbf{\Omega}_i$$

$$= \frac{1}{2}\sum_{i=1}^{d} (\mathbf{V}^T \mathbf{\Omega}_i)^T \mathbf{\Sigma} (\mathbf{V}^T \mathbf{\Omega}_i)$$

$$\geq \frac{\sigma_L}{2}\sum_{i=1}^{d} \|\mathbf{V}^T \mathbf{\Omega}_i\|^2 = \frac{\sigma_L}{2}\|\mathbf{V}^T \mathbf{\Omega}\|^2 \overset{a}{=} \frac{\sigma_L}{2}\|\mathbf{\Omega}\|^2,$$

where we denote $\mathbf{\Omega}_i$ to be the $i$th column of $\mathbf{\Omega}$, and $\overset{a}{=}$ follows from the fact that $\mathbf{\Omega}$ belongs to the column space of $\mathbf{W}$, i.e., there exists $\alpha \in \mathbb{R}^{(m-1) \times d}$ such that $\mathbf{\Omega} = \mathbf{V}\alpha$.

From (6.16b) and the KKT condition, we know that both $\mathbf{\Omega}^k$ and $\mathbf{\Omega}^*$ belong to the column space of $\mathbf{W}$. So we have

$$\|\mathbf{W}(\mathbf{\Omega}^k - \mathbf{\Omega}^*)\| \geq \sqrt{\sigma_L/2}\|\mathbf{\Omega}^k - \mathbf{\Omega}^*\|.$$

From Theorem 3.8, we get the conclusion. □

Now, we discuss algorithm (6.16a)–(6.16b) with the special

$$\psi(\mathbf{X}) = \frac{\beta}{2}\left\|\sqrt{\mathbf{D}}\mathbf{X}\right\|^2 - \frac{\beta}{2}\|\mathbf{W}\mathbf{X}\|^2 \quad \text{and} \quad L_\psi = \beta d_{max},$$

where $d_{max} = \max\{d_i\}$. Then algorithm (6.16a)–(6.16b) reduces to Algorithm 6.7. From Remark 3.4 and

$$\|\mathbf{W}\|_2^2 = \frac{1}{2}\|\mathbf{L}\|_2 \leq \|\mathbf{D}\|_2 \leq d_{max}$$

(that is, $\|\mathbf{B}\|_2^2 \leq d_{max}$ and $\sigma^2 = \frac{\sigma_L}{2}$ in Remark 3.4), we have the following theorem.

**Theorem 6.6** *Assume that each $f_i$ is $\mu$-strongly convex and $L$-smooth, $i \in$ [m]. Initialize $\Omega^0 = \mathbf{0}$ and let $\beta = O\left(\sqrt{\frac{\mu L}{\sigma_\mathbf{L} d_{max}}}\right)$. Then Algorithm 6.7 needs $O\left(\left(\sqrt{\frac{L d_{max}}{\mu \sigma_\mathbf{L}}} + \frac{d_{max}}{\sigma_\mathbf{L}}\right) \log \frac{1}{\epsilon}\right)$ iterations to find an $\epsilon$-approximate solution $(\mathbf{X}, \Omega)$, i.e.,*

$$\frac{1}{2\beta}\|\Omega - \Omega^*\|^2 + \frac{\beta}{2}\|\mathbf{WX} - \mathbf{WX}^*\|^2 + D_\psi(\mathbf{X}^*, \mathbf{X}) \leq \epsilon.$$

We see that the complexity depends on the condition number $\frac{L}{\mu}$ of the objective function and $\frac{d_{max}}{\sigma_\mathbf{L}}$. The latter one can be regarded as the condition number of the Laplacian matrix $\mathbf{L}$.

## 6.2.2   Linearized ADMM

The subproblem in (6.8a) is a proximal mapping of $f_i$ (c.f. (6.12)). When the proximal mapping of $f_i$ is not easily computable, as in Sect. 3.2 we may linearize the objective $f_i$, which leads to the following step [10]:

$$\mathbf{x}_i^{k+1} = \underset{\mathbf{x}_i}{\operatorname{argmin}}\left[\left\langle \nabla f_i(\mathbf{x}_i^k), \mathbf{x}_i - \mathbf{x}_i^k\right\rangle + \frac{L}{2}\left\|\mathbf{x}_i - \mathbf{x}_i^k\right\|^2\right.$$
$$+ \sum_{j:(i,j)\in\mathcal{E}}\left(\left\langle \boldsymbol{\lambda}_{ij}^k, \mathbf{x}_i - \mathbf{z}_{ij}^k\right\rangle + \frac{\beta}{2}\left\|\mathbf{x}_i - \mathbf{z}_{ij}^k\right\|^2\right)$$
$$\left.+ \sum_{j:(j,i)\in\mathcal{E}}\left(\left\langle \boldsymbol{\gamma}_{ji}^k, \mathbf{x}_i - \mathbf{z}_{ji}^k\right\rangle + \frac{\beta}{2}\left\|\mathbf{x}_i - \mathbf{z}_{ji}^k\right\|^2\right)\right].$$

Steps (6.8b)–(6.8d) remain unchanged. Similar to (6.11), we have

$$\mathbf{x}_i^{k+1} = \underset{\mathbf{x}_i}{\operatorname{argmin}}\left[\left\langle \nabla f_i(\mathbf{x}_i^k), \mathbf{x}_i - \mathbf{x}_i^k\right\rangle + \frac{L}{2}\left\|\mathbf{x}_i - \mathbf{x}_i^k\right\|^2\right.$$
$$\left.+ \sum_{j\in\mathcal{N}_i}\left(\left\langle \boldsymbol{\lambda}_{ij}^k + \frac{\beta}{2}\left(\mathbf{x}_i^k - \mathbf{x}_j^k\right), \mathbf{x}_i\right\rangle + \frac{\beta}{2}\left\|\mathbf{x}_i - \mathbf{x}_i^k\right\|^2\right)\right]$$
$$= \mathbf{x}_i^k - \frac{1}{L + \beta d_i}\left\{\nabla f_i(\mathbf{x}_i^k) + \sum_{j\in\mathcal{N}_i}\left[\boldsymbol{\lambda}_{ij}^k + \frac{\beta}{2}\left(\mathbf{x}_i^k - \mathbf{x}_j^k\right)\right]\right\}.$$

Similar to the deductions in Sect. 6.2.1, the resultant linearized ADMM can be rewritten as

$$
\mathbf{X}^{k+1} = \underset{\mathbf{X}}{\operatorname{argmin}} \left( \left\langle \nabla f(\mathbf{X}^k), \mathbf{X} \right\rangle + \frac{L}{2} \|\mathbf{X} - \mathbf{X}^k\|^2 \right.
$$

$$
\left. + \left\langle \boldsymbol{\Omega}^k, \mathbf{W}\mathbf{X} \right\rangle + \beta \left\langle \mathbf{W}^2\mathbf{X}^k, \mathbf{X} \right\rangle + \frac{\beta}{2} \left\| \sqrt{\mathbf{D}}(\mathbf{X} - \mathbf{X}^k) \right\|^2 \right)
$$

$$
= \mathbf{X}^k - (L\mathbf{I} + \beta\mathbf{D})^{-1} \left( \beta\mathbf{W}^2\mathbf{X}^k + \nabla f(\mathbf{X}^k) + \mathbf{W}\boldsymbol{\Omega}^k \right), \tag{6.17a}
$$

$$
\boldsymbol{\Omega}^{k+1} = \boldsymbol{\Omega}^k + \beta\mathbf{W}\mathbf{X}^{k+1}, \tag{6.17b}
$$

which is also a special case of algorithm (6.16a)–(6.16b) with

$$
\psi(\mathbf{X}) = \frac{L}{2}\|\mathbf{X}\|^2 - f(\mathbf{X}) + \frac{\beta}{2} \left\| \sqrt{\mathbf{D}}\mathbf{X} \right\|^2 - \frac{\beta}{2}\|\mathbf{W}\mathbf{X}\|^2 \quad \text{and} \quad L_\psi = L + \beta d_{\max}.
$$

We present the method in Algorithm 6.8, which is a distributed version of (6.17a)–(6.17b).

---

**Algorithm 6.8** Decentralized linearized ADMM of the $i$th node

---

Initialize: $\mathbf{x}_i^0$ and $\boldsymbol{v}_i^0 = \mathbf{0}$, $i \in [m]$.
Send $\mathbf{x}_i^0$ to its neighbors.
Wait until receiving $\mathbf{x}_j^0$ from all its neighbors, $j \in \mathcal{N}_i$.
**for** $k = 0, 1, 2, \cdots$ **do**

$$
\mathbf{x}_i^{k+1} = \underset{\mathbf{x}_i}{\operatorname{argmin}} \left( \left\langle \nabla f_i(\mathbf{x}_i^k), \mathbf{x}_i \right\rangle + \left\langle \boldsymbol{v}_i^k, \mathbf{x}_i \right\rangle + \frac{\beta}{2} \left\langle \sum_{j \in \mathcal{N}_i} \mathbf{L}_{ij}\mathbf{x}_j^k, \mathbf{x}_i \right\rangle + \frac{\beta d_i + L}{2} \left\| \mathbf{x}_i - \mathbf{x}_i^k \right\|^2 \right)
$$

$$
= \mathbf{x}_i^k - \frac{1}{\beta d_i + L} \left( \nabla f_i(\mathbf{x}_i^k) + \boldsymbol{v}_i^k + \frac{\beta}{2} \sum_{j \in \mathcal{N}_i} \mathbf{L}_{ij}\mathbf{x}_j^k \right).
$$

Send $\mathbf{x}_i^{k+1}$ to its neighbors.
Wait until receiving $\mathbf{x}_j^{k+1}$ from all its neighbors, $j \in \mathcal{N}_i$.
$\boldsymbol{v}_i^{k+1} = \boldsymbol{v}_i^k + \frac{\beta}{2} \sum_{j \in \mathcal{N}_i} \mathbf{L}_{ij}\mathbf{x}_j^{k+1}$.
**end for**

---

From Remark 3.4, we have the following theorem.

**Theorem 6.7** *Assume that each $f_i$ is $\mu$-strongly convex and $L$-smooth, $i \in [m]$. Initialize $\boldsymbol{\Omega}^0 = \mathbf{0}$ and let $\beta = O\left( \sqrt{\frac{\mu L}{\sigma_{\mathrm{L}} d_{\max}}} \right)$. Then Algorithm 6.8 needs*

$O\left(\left(\frac{L}{\mu} + \frac{d_{\max}}{\sigma_L}\right) \log \frac{1}{\epsilon}\right)$ *iterations to find* $(\mathbf{X}, \mathbf{\Omega})$ *such that*

$$\frac{1}{2\beta}\|\mathbf{\Omega} - \mathbf{\Omega}^*\|^2 + \frac{\beta}{2}\|\mathbf{WX} - \mathbf{WX}^*\|^2 + D_\psi(\mathbf{X}^*, \mathbf{X}) \le \epsilon.$$

### 6.2.3   Accelerated Linearized ADMM

In this section, we accelerate algorithm (6.17a)–(6.17b) using Algorithm 3.6. The resultant algorithm has the following iterations [8]:

$$\mathbf{Y}^k = \theta\mathbf{X}^k + (1 - \theta)\widetilde{\mathbf{X}}^k, \tag{6.18a}$$

$$\mathbf{X}^{k+1} = \frac{1}{\frac{\theta}{\alpha} + \mu}\left[\mu\mathbf{Y}^k + \frac{\theta}{\alpha}\mathbf{X}^k - \left(\nabla f(\mathbf{Y}^k) + \mathbf{W}\mathbf{\Omega}^k + \beta\theta\mathbf{W}^2\mathbf{X}^k\right)\right], \tag{6.18b}$$

$$\widetilde{\mathbf{X}}^{k+1} = \theta\mathbf{X}^{k+1} + (1 - \theta)\widetilde{\mathbf{X}}^k, \tag{6.18c}$$

$$\mathbf{\Omega}^{k+1} = \mathbf{\Omega}^k + \beta\theta\mathbf{WX}^{k+1}, \tag{6.18d}$$

and it is presented in Algorithm 6.9 in the distributed manner.

---

**Algorithm 6.9** Accelerated decentralized linearized ADMM of the $i$th node

---

Initialize: $\mathbf{x}_i^0 = \widetilde{\mathbf{x}}_i^0$ and $\boldsymbol{v}_i^0 = \mathbf{0}$, $i \in [m]$.
Send $\mathbf{x}_i^0$ to its neighbors.
Wait until receiving $\mathbf{x}_j^0$ from all its neighbors, $j \in \mathcal{N}_i$.
**for** $k = 0, 1, 2, \cdots$ **do**
  $\mathbf{y}_i^k = \theta\mathbf{x}_i^k + (1 - \theta)\widetilde{\mathbf{x}}_i^k$.
  $\mathbf{x}_i^{k+1} = \frac{1}{\frac{\theta}{\alpha}+\mu}\left[\mu\mathbf{y}_i^k + \frac{\theta}{\alpha}\mathbf{x}_i^k - \left(\nabla f_i(\mathbf{y}_i^k) + \boldsymbol{v}_i^k + \frac{\beta\theta}{2}\sum_{j \in \mathcal{N}_i}\mathbf{L}_{ij}\mathbf{x}_j^k\right)\right]$.
  $\widetilde{\mathbf{x}}_i^{k+1} = \theta\mathbf{x}_i^{k+1} + (1 - \theta)\widetilde{\mathbf{x}}_i^k$.
  Send $\mathbf{x}_i^{k+1}$ to its neighbors.
  Wait until receiving $\mathbf{x}_j^{k+1}$ from all its neighbors, $j \in \mathcal{N}_i$.
  $\boldsymbol{v}_i^{k+1} = \boldsymbol{v}_i^k + \frac{\beta\theta}{2}\sum_{j \in \mathcal{N}_i}\mathbf{L}_{ij}\mathbf{x}_j^{k+1}$.
**end for**

---

Denote

$$\ell_k = (1 - \theta)\left(f(\widetilde{\mathbf{X}}^k) - f(\mathbf{X}^*) + \left\langle\mathbf{\Omega}^*, \mathbf{W}\widetilde{\mathbf{X}}^k\right\rangle\right)$$

$$+ \frac{\theta^2}{2\alpha}\|\mathbf{X}^k - \mathbf{X}^*\|^2 + \frac{1}{2\beta}\|\mathbf{\Omega}^k - \mathbf{\Omega}^*\|^2.$$

**Table 6.2** Complexity comparisons between decentralized ADMM, decentralized linearized ADMM (LADMM), and its accelerated version

| Decentralized ADMM | Decentralized LADMM | Accelerated decentralized LADMM |
|---|---|---|
| $O\left(\left(\sqrt{\frac{Ld_{\max}}{\mu\sigma_{\mathbf{L}}}} + \frac{d_{\max}}{\sigma_{\mathbf{L}}}\right) \log \frac{1}{\epsilon}\right)$ | $O\left(\left(\frac{L}{\mu} + \frac{d_{\max}}{\sigma_{\mathbf{L}}}\right) \log \frac{1}{\epsilon}\right)$ | $O\left(\sqrt{\frac{Ld_{\max}}{\mu\sigma_{\mathbf{L}}}} \log \frac{1}{\epsilon}\right)$ |

From Theorem 3.15 (note that $\|\mathbf{B}\|_2^2 \leq d_{\max}$ and $\sigma^2 = \frac{\sigma_{\mathbf{L}}}{2}$ in Theorem 3.15), we have the following convergence result.

**Theorem 6.8** *Suppose that each $f_i$ is $\mu$-strongly convex and $L$-smooth, $i \in [m]$. Assume that $\frac{2d_{\max}}{\sigma_{\mathbf{L}}} \leq \frac{L}{\mu}$, where $\sigma_{\mathbf{L}}$ is the smallest non-zero singular value of $\mathbf{L}$. Let*

$$\alpha = \frac{1}{4L}, \quad \beta = \frac{L}{d_{\max}}, \quad and \quad \theta = \sqrt{\frac{2\mu d_{\max}}{L\sigma_{\mathbf{L}}}}.$$

*Then for algorithm (6.18a)–(6.18d) (Algorithm 6.9), we have*

$$\ell_{k+1} \leq O\left(1 - \sqrt{\frac{\mu\sigma_{\mathbf{L}}}{2Ld_{\max}}}\right)\ell_k.$$

We list the convergence rates comparisons in Table 6.2.

## 6.3 Asynchronous Distributed ADMM

Algorithms 6.1–6.2 proceed in a synchronous manner. That is, the master needs to wait for all the workers to finish their updates before it can proceed. When the workers have different delays, the master has to wait for the slowest worker before the next iteration, i.e., the system proceeds at the pace of the slowest worker. In this section, we introduce the asynchronous ADMM proposed in [4, 5] to reduce the waiting time.

In the asynchronous ADMM, the master does not wait for all the workers, but proceeds as long as it receives information from a partial set of workers instead. We denote the partial set at iteration $k$ as $\mathcal{A}^k$, and $\mathcal{A}_c^k$ as the complementary set of $\mathcal{A}^k$, which means the set of workers whose information does not arrive at iteration $k$. We use $\alpha$ to lower bound the size of $\mathcal{A}^k$. In the asynchronous ADMM, we often require that the master has to receive the updates from every worker at least once in every $\tau$ iterations. That is, we do not allow some workers to be absent for a long time. So we make the following bounded delay assumption.

**Assumption 3** The maximum tolerable delay for all $i$ and $k$ is upper bounded.

---

**Algorithm 6.10** Asynchronous ADMM of the master

---

Initialize: $\tilde{d}_1^1 = \cdots = \tilde{d}_m^1 = 0$.

**for** $k = 1, 2, \cdots$ **do**

Wait until receiving $\hat{\mathbf{x}}_i^k$ and $\hat{\boldsymbol{\lambda}}_i^k$ from workers $i \in \mathcal{A}^k$ such that $|\mathcal{A}^k| \geq \alpha$ and $\tilde{d}_j^k < \tau - 1$ for all $j \in \mathcal{A}_c^k$.

$$\mathbf{x}_i^{k+1} = \begin{cases} \hat{\mathbf{x}}_i^k, & \forall i \in \mathcal{A}^k, \\ \mathbf{x}_i^k, & \forall i \in \mathcal{A}_c^k. \end{cases}$$

$$\boldsymbol{\lambda}_i^{k+1} = \begin{cases} \hat{\boldsymbol{\lambda}}_i^k, & \forall i \in \mathcal{A}^k, \\ \boldsymbol{\lambda}_i^k, & \forall i \in \mathcal{A}_c^k. \end{cases}$$

$$\tilde{d}_i^{k+1} = \begin{cases} 0, & \forall i \in \mathcal{A}^k, \\ \tilde{d}_i^k + 1, & \forall i \in \mathcal{A}_c^k. \end{cases}$$

$$\mathbf{z}^{k+1} = \operatorname*{argmin}_{\mathbf{z}} \left[ \sum_{i=1}^m \left( \left\langle \boldsymbol{\lambda}_i^{k+1}, \mathbf{x}_i^{k+1} - \mathbf{z} \right\rangle + \frac{\beta}{2} \|\mathbf{x}_i^{k+1} - \mathbf{z}\|^2 \right) + \frac{\rho}{2} \|\mathbf{z} - \mathbf{z}^k\|^2 \right]$$

$$= \frac{1}{\rho + m\beta} \left[ \rho \mathbf{z}^k + \sum_{i=1}^m \left( \boldsymbol{\lambda}_i^{k+1} + \beta \mathbf{x}_i^{k+1} \right) \right].$$

Broadcast $\mathbf{z}^{k+1}$ to the workers in $\mathcal{A}^k$.

**end for**

---

Denote the upper bound as $\tau$, then it must be that for every $i$,

$$i \in \mathcal{A}^k \cup \mathcal{A}^{k-1} \cdots \cup \mathcal{A}^{\max\{k-\tau+1,0\}}.$$

We describe the asynchronous ADMM in Algorithms 6.10–6.11. It has several differences from the synchronous ADMM:

1. The master only updates $(\mathbf{x}_i^{k+1}, \boldsymbol{\lambda}_i^{k+1})$ with $i \in \mathcal{A}^k$.
2. $\mathbf{z}$ is updated by solving a subproblem with an additional proximal term.
3. We introduce $\tilde{d}_i$, the amount of delay, for each worker such that the bounded delay assumption holds. The master must wait if there exists one worker with $\tilde{d}_i = \tau - 1$.
4. The master only broadcasts the up-to-date $\mathbf{z}$ to the arrived workers in $\mathcal{A}^k$.

### 6.3.1   Convergence

To simplify the analysis, we rewrite the method from the master's point of view:

$$\mathbf{x}_i^{k+1} = \begin{cases} \operatorname*{argmin}_{\mathbf{x}_i} \left( f_i(\mathbf{x}_i) + \left\langle \boldsymbol{\lambda}_i^{\bar{k}_i+1}, \mathbf{x}_i \right\rangle + \frac{\beta}{2} \left\| \mathbf{x}_i - \mathbf{z}^{\bar{k}_i+1} \right\|^2 \right), & \forall i \in \mathcal{A}^k, \\ \mathbf{x}_i^k, & \forall i \in \mathcal{A}_c^k. \end{cases}$$

(6.19a)

---

**Algorithm 6.11** Asynchronous ADMM of the $i$th worker

---

Initialize: $\hat{\mathbf{x}}_i^0$ and $\hat{\boldsymbol{\lambda}}_i^0$, $i \in [m]$.
**for** $k_i = 1, 2, \cdots$ **do**
    Wait until receiving $\mathbf{z}$ from the master.

$$\hat{\mathbf{x}}_i^{k_i+1} = \operatorname*{argmin}_{\mathbf{x}_i} \left( f_i(\mathbf{x}_i) + \left\langle \hat{\boldsymbol{\lambda}}_i^{k_i}, \mathbf{x}_i - \mathbf{z} \right\rangle + \frac{\beta}{2} \|\mathbf{x}_i - \mathbf{z}\|^2 \right)$$

$$= \operatorname{Prox}_{\beta^{-1} f_i} \left( \mathbf{z} - \frac{1}{\beta} \hat{\boldsymbol{\lambda}}_i^{k_i} \right).$$

$$\hat{\boldsymbol{\lambda}}_i^{k_i+1} = \hat{\boldsymbol{\lambda}}_i^{k_i} + \beta \left( \hat{\mathbf{x}}_i^{k_i+1} - \mathbf{z} \right).$$

    Send $(\hat{\mathbf{x}}_i^{k_i+1}, \hat{\boldsymbol{\lambda}}_i^{k_i+1})$ to the master.
**end for**

---

$$\boldsymbol{\lambda}_i^{k+1} = \begin{cases} \boldsymbol{\lambda}_i^{\bar{k}_i+1} + \beta \left( \mathbf{x}_i^{k+1} - \mathbf{z}^{\bar{k}_i+1} \right), & \forall i \in \mathcal{A}^k, \\ \boldsymbol{\lambda}_i^k, & \forall i \in \mathcal{A}_c^k. \end{cases} \tag{6.19b}$$

$$\mathbf{z}^{k+1} = \operatorname*{argmin}_{\mathbf{z}} \left[ \sum_{i=1}^m \left( \left\langle \boldsymbol{\lambda}_i^{k+1}, \mathbf{x}_i^{k+1} - \mathbf{z} \right\rangle + \frac{\beta}{2} \left\| \mathbf{x}_i^{k+1} - \mathbf{z} \right\|^2 \right) + \frac{\rho}{2} \left\| \mathbf{z} - \mathbf{z}^k \right\|^2 \right], \tag{6.19c}$$

where we denote $\bar{k}_i$ as the last iteration before iteration $k$ for which worker $i \in \mathcal{A}^k$ arrives, i.e., $i \in \mathcal{A}^{\bar{k}_i}$. Thus, for all workers $i \in \mathcal{A}^k$, we have

$$\mathbf{x}_i^{\bar{k}_i+1} = \mathbf{x}_i^{\bar{k}_i+2} = \cdots = \mathbf{x}_i^k,$$

$$\boldsymbol{\lambda}_i^{\bar{k}_i+1} = \boldsymbol{\lambda}_i^{\bar{k}_i+2} = \cdots = \boldsymbol{\lambda}_i^k, \text{ and} \tag{6.20}$$

$$\max\{k - \tau, 0\} \leq \bar{k}_i < k.$$

For each $i \in \mathcal{A}_c^k$, we denote $\tilde{k}_i$ as the last iteration before iteration $k$ for which worker $i$ arrives, i.e., $i \in \mathcal{A}^{\tilde{k}_i}$. Under the bounded delay assumption, we have

$$\max\{k - \tau + 1, 0\} \leq \tilde{k}_i < k.$$

Thus, for all workers $i \in \mathcal{A}_c^k$, we have

$$\mathbf{x}_i^{\tilde{k}_i+1} = \mathbf{x}_i^{\tilde{k}_i+2} = \cdots = \mathbf{x}_i^k = \mathbf{x}_i^{k+1} \text{ and}$$

$$\boldsymbol{\lambda}_i^{\tilde{k}_i+1} = \boldsymbol{\lambda}_i^{\tilde{k}_i+2} = \cdots = \boldsymbol{\lambda}_i^k = \boldsymbol{\lambda}_i^{k+1}.$$

We also denote $\hat{k}_i$ as the last iteration before $\widetilde{k}_i$ for which $i \in \mathcal{A}^{\widetilde{k}_i}$ arrives, i.e., $i \in \mathcal{A}^{\hat{k}_i}$. We also have

$$\max\{\widetilde{k}_i - \tau, 0\} \le \hat{k}_i < \widetilde{k}_i.$$

Thus, for all workers $i \in \mathcal{A}_c^k$, we have

$$\mathbf{x}_i^{k+1} = \mathbf{x}_i^{\widetilde{k}_i+1} = \operatorname*{argmin}_{\mathbf{x}_i} \left( f_i(\mathbf{x}_i) + \left\langle \lambda_i^{\hat{k}_i+1}, \mathbf{x}_i \right\rangle + \frac{\beta}{2} \left\| \mathbf{x}_i - \mathbf{z}^{\hat{k}_i+1} \right\|^2 \right), \qquad (6.21)$$

$$\lambda_i^{k+1} = \lambda_i^{\widetilde{k}_i+1} = \lambda_i^{\hat{k}_i+1} + \beta \left( \mathbf{x}_i^{\widetilde{k}_i+1} - \mathbf{z}^{\hat{k}_i+1} \right), \qquad (6.22)$$

$$\mathbf{x}_i^{\hat{k}_i+1} = \mathbf{x}_i^{\hat{k}_i+2} = \cdots = \mathbf{x}_i^{\widetilde{k}_i}, \quad \text{and}$$

$$\lambda_i^{\hat{k}_i+1} = \lambda_i^{\hat{k}_i+2} = \cdots = \lambda_i^{\widetilde{k}_i}. \qquad (6.23)$$

Denote $(\mathbf{x}_1^*, \cdots, \mathbf{x}_m^*, \mathbf{z}^*, \lambda_1^*, \cdots, \lambda_m^*)$ to be a KKT point. We have

$$\sum_{i=1}^m \lambda_i^* = \mathbf{0}, \quad \mathbf{z}^* = \mathbf{x}_i^*, \quad \text{and} \quad \nabla f_i(\mathbf{x}_i^*) + \lambda_i^* = \mathbf{0}, \quad i \in [m].$$

Also denote $f^* = \sum_{i=1}^m f_i(\mathbf{z}^*)$.

**Theorem 6.9** *Assume that each $f_i$ is convex and $L$-smooth, $i \in [m]$, and Assumption 3 holds true. Let*

$$\beta > \frac{1 + L^2 + \sqrt{(1 + L^2)^2 + 8L^2}}{2} \quad \text{and} \quad \rho > \frac{1}{2}\left[ m(1 + \beta^2)(\tau - 1)^2 - m\beta \right].$$

*Suppose that $(\mathbf{x}_1^k, \cdots, \mathbf{x}_m^k, \mathbf{z}^k, \lambda_1^k, \cdots, \lambda_m^k)$ generated by (6.19a)–(6.19c) are bounded, then $(\mathbf{x}_1^k, \cdots, \mathbf{x}_m^k, \mathbf{z}^k, \lambda_1^k, \cdots, \lambda_m^k)$ converge to the set of KKT points of Problem (6.2) in the sense of*

$$\sum_{i=1}^m \lambda_i^k \to \mathbf{0}, \quad \mathbf{x}_i^{k+1} - \mathbf{z}^{k+1} \to \mathbf{0}, \quad \text{and} \quad \nabla f_i(\mathbf{x}_i^{k+1}) + \lambda_i^{k+1} = \mathbf{0}, \quad i \in [m].$$

**Proof** Recall the augmented Lagrangian function in (6.3). Notice that

$$L(\mathbf{x}^{k+1}, \mathbf{z}^{k+1}, \lambda^{k+1}) - L(\mathbf{x}^k, \mathbf{z}^k, \lambda^k)$$

$$= \left( L(\mathbf{x}^{k+1}, \mathbf{z}^{k+1}, \lambda^{k+1}) - L(\mathbf{x}^{k+1}, \mathbf{z}^k, \lambda^{k+1}) \right)$$

$$+ \left( L(\mathbf{x}^{k+1}, \mathbf{z}^k, \boldsymbol{\lambda}^{k+1}) - L(\mathbf{x}^{k+1}, \mathbf{z}^k, \boldsymbol{\lambda}^k) \right)$$

$$+ \left( L(\mathbf{x}^{k+1}, \mathbf{z}^k, \boldsymbol{\lambda}^k) - L(\mathbf{x}^k, \mathbf{z}^k, \boldsymbol{\lambda}^k) \right).$$

We bound the three terms one by one.

For the first term, from the $(m\beta + \rho)$-strong convexity of $L(\mathbf{x}, \mathbf{z}, \boldsymbol{\lambda}) + \frac{\rho}{2} \|\mathbf{z} - \mathbf{z}^k\|^2$ with respect to $\mathbf{z}$, (6.19c), and (A.7), we have

$$L(\mathbf{x}^{k+1}, \mathbf{z}^k, \boldsymbol{\lambda}^{k+1}) - \left( L(\mathbf{x}^{k+1}, \mathbf{z}^{k+1}, \boldsymbol{\lambda}^{k+1}) + \frac{\rho}{2} \|\mathbf{z}^{k+1} - \mathbf{z}^k\|^2 \right)$$

$$\geq \frac{m\beta + \rho}{2} \|\mathbf{z}^{k+1} - \mathbf{z}^k\|^2.$$

Therefore,

$$L(\mathbf{x}^{k+1}, \mathbf{z}^{k+1}, \boldsymbol{\lambda}^{k+1}) - L(\mathbf{x}^{k+1}, \mathbf{z}^k, \boldsymbol{\lambda}^{k+1}) \leq - \left( \frac{m\beta}{2} + \rho \right) \|\mathbf{z}^{k+1} - \mathbf{z}^k\|^2.$$

For the second term, from the augmented Lagrangian function in (6.3), we have

$$L(\mathbf{x}^{k+1}, \mathbf{z}^k, \boldsymbol{\lambda}^{k+1}) - L(\mathbf{x}^{k+1}, \mathbf{z}^k, \boldsymbol{\lambda}^k)$$

$$= \sum_{i=1}^{m} \left\langle \boldsymbol{\lambda}_i^{k+1} - \boldsymbol{\lambda}_i^k, \mathbf{x}_i^{k+1} - \mathbf{z}^k \right\rangle$$

$$\overset{a}{=} \sum_{i \in \mathcal{A}^k} \left\langle \boldsymbol{\lambda}_i^{k+1} - \boldsymbol{\lambda}_i^k, \mathbf{x}_i^{k+1} - \mathbf{z}^k \right\rangle$$

$$= \sum_{i \in \mathcal{A}^k} \left( \left\langle \boldsymbol{\lambda}_i^{k+1} - \boldsymbol{\lambda}_i^k, \mathbf{x}_i^{k+1} - \mathbf{z}^{\bar{k}_i+1} \right\rangle + \left\langle \boldsymbol{\lambda}_i^{k+1} - \boldsymbol{\lambda}_i^k, \mathbf{z}^{\bar{k}_i+1} - \mathbf{z}^k \right\rangle \right)$$

$$\overset{b}{=} \sum_{i \in \mathcal{A}^k} \left( \frac{1}{\beta} \left\| \boldsymbol{\lambda}_i^{k+1} - \boldsymbol{\lambda}_i^k \right\|^2 + \left\langle \boldsymbol{\lambda}_i^{k+1} - \boldsymbol{\lambda}_i^k, \mathbf{z}^{\bar{k}_i+1} - \mathbf{z}^k \right\rangle \right),$$

where we use $\boldsymbol{\lambda}_i^{k+1} = \boldsymbol{\lambda}_i^k$ for $i \in \mathcal{A}_i^k$ in $\overset{a}{=}$, and (6.19b) and (6.20) in $\overset{b}{=}$.

For the third term, from the $\beta$-strong convexity of $L(\mathbf{x}, \mathbf{z}, \boldsymbol{\lambda})$ with respect to $\mathbf{x}_i$, we have

$$L(\mathbf{x}^{k+1}, \mathbf{z}^k, \boldsymbol{\lambda}^k) - L(\mathbf{x}^k, \mathbf{z}^k, \boldsymbol{\lambda}^k)$$

$$\overset{c}{=} \sum_{i \in \mathcal{A}^k} \left[ \left( f_i(\mathbf{x}_i^{k+1}) + \left\langle \boldsymbol{\lambda}_i^k, \mathbf{x}_i^{k+1} - \mathbf{z}^k \right\rangle + \frac{\beta}{2} \left\| \mathbf{x}_i^{k+1} - \mathbf{z}^k \right\|^2 \right) \right.$$

$$-\left(f_i(\mathbf{x}_i^k) + \left\langle \boldsymbol{\lambda}_i^k, \mathbf{x}_i^k - \mathbf{z}^k \right\rangle + \frac{\beta}{2} \left\| \mathbf{x}_i^k - \mathbf{z}^k \right\|^2 \right)\Bigg]$$

$$\leq \sum_{i \in \mathcal{A}^k} \left( \left\langle \nabla f_i(\mathbf{x}_i^{k+1}) + \boldsymbol{\lambda}_i^k + \beta(\mathbf{x}_i^{k+1} - \mathbf{z}^k), \mathbf{x}_i^{k+1} - \mathbf{x}_i^k \right\rangle - \frac{\beta}{2} \left\| \mathbf{x}_i^{k+1} - \mathbf{x}_i^k \right\|^2 \right)$$

$$\stackrel{d}{=} \sum_{i \in \mathcal{A}^k} \left( \beta \left\langle \mathbf{z}^{\bar{k}_i+1} - \mathbf{z}^k, \mathbf{x}_i^{k+1} - \mathbf{x}_i^k \right\rangle - \frac{\beta}{2} \left\| \mathbf{x}_i^{k+1} - \mathbf{x}_i^k \right\|^2 \right), \tag{6.24}$$

where we use $\mathbf{x}_i^{k+1} = \mathbf{x}_i^k$ for $i \in \mathcal{A}_c^k$ in $\stackrel{c}{=}$, and the optimality condition of (6.19a) and (6.20) in $\stackrel{d}{=}$.

Thus, we have

$$L(\mathbf{x}^{k+1}, \mathbf{z}^{k+1}, \boldsymbol{\lambda}^{k+1}) - L(\mathbf{x}^k, \mathbf{z}^k, \boldsymbol{\lambda}^k)$$

$$\leq -\left(\frac{m\beta}{2} + \rho\right) \left\| \mathbf{z}^{k+1} - \mathbf{z}^k \right\|^2 - \frac{\beta}{2} \sum_{i \in \mathcal{A}^k} \left\| \mathbf{x}_i^{k+1} - \mathbf{x}_i^k \right\|^2$$

$$+ \sum_{i \in \mathcal{A}^k} \left( \frac{1}{\beta} \left\| \boldsymbol{\lambda}_i^{k+1} - \boldsymbol{\lambda}_i^k \right\|^2 + \left\langle \boldsymbol{\lambda}_i^{k+1} - \boldsymbol{\lambda}_i^k, \mathbf{z}^{\bar{k}_i+1} - \mathbf{z}^k \right\rangle \right)$$

$$+ \beta \left\langle \mathbf{z}^{\bar{k}_i+1} - \mathbf{z}^k, \mathbf{x}_i^{k+1} - \mathbf{x}_i^k \right\rangle \Bigg).$$

From (6.19a)–(6.19b) and (6.21)–(6.22), for any $i$, we have

$$\mathbf{0} = \nabla f_i(\mathbf{x}_i^{k+1}) + \boldsymbol{\lambda}_i^{k+1}. \tag{6.25}$$

From the $L$-smoothness of $f_i$ and $\langle \mathbf{a}, \mathbf{b} \rangle \leq \frac{\alpha}{2}\|\mathbf{a}\|^2 + \frac{1}{2\alpha}\|\mathbf{b}\|^2$ for $\alpha > 0$, we have

$$\|\boldsymbol{\lambda}_i^{k+1} - \boldsymbol{\lambda}_i^k\| \leq L\|\mathbf{x}_i^{k+1} - \mathbf{x}_i^k\|$$

and

$$L(\mathbf{x}^{k+1}, \mathbf{z}^{k+1}, \boldsymbol{\lambda}^{k+1}) - L(\mathbf{x}^k, \mathbf{z}^k, \boldsymbol{\lambda}^k)$$

$$\leq -\left(\frac{m\beta}{2} + \rho\right) \left\| \mathbf{z}^{k+1} - \mathbf{z}^k \right\|^2 - \sum_{i \in \mathcal{A}^k} \left( \frac{\beta}{2} - \frac{L^2}{\beta} - \frac{L^2}{2} - \frac{1}{2} \right) \left\| \mathbf{x}_i^{k+1} - \mathbf{x}_i^k \right\|^2$$

$$+ \sum_{i \in \mathcal{A}^k} \frac{1 + \beta^2}{2} \left\| \mathbf{z}^{\bar{k}_i+1} - \mathbf{z}^k \right\|^2. \tag{6.26}$$

Now, we bound the last term in (6.26). It is easy to show that

$$\sum_{k=0}^{K} \sum_{i \in \mathcal{A}^k} \left\| \mathbf{z}^{\bar{k}_i + 1} - \mathbf{z}^k \right\|^2 = \sum_{k=0}^{K} \sum_{i \in \mathcal{A}^k} \left\| \sum_{t=\bar{k}_i+1}^{k-1} \left( \mathbf{z}^t - \mathbf{z}^{t+1} \right) \right\|^2$$

$$\leq \sum_{k=0}^{K} \sum_{i \in \mathcal{A}^k} (k - \bar{k}_i - 1) \sum_{t=\bar{k}_i+1}^{k-1} \left\| \mathbf{z}^t - \mathbf{z}^{t+1} \right\|^2$$

$$\leq \sum_{k=0}^{K} \sum_{i \in \mathcal{A}^k} (\tau - 1) \sum_{t=\max\{k-\tau+1,1\}}^{k-1} \left\| \mathbf{z}^t - \mathbf{z}^{t+1} \right\|^2$$

$$\leq m(\tau - 1) \sum_{k=0}^{K} \sum_{t=\max\{k-\tau+1,1\}}^{k-1} \left\| \mathbf{z}^t - \mathbf{z}^{t+1} \right\|^2$$

$$\leq m(\tau - 1)^2 \sum_{k=0}^{K} \left\| \mathbf{z}^k - \mathbf{z}^{k+1} \right\|^2 \tag{6.27}$$

due to

$$\max\{k - \tau, 0\} \leq \bar{k}_i < k \quad \text{and} \quad |\mathcal{A}^k| \leq m.$$

Thus we have

$$L(\mathbf{x}^{K+1}, \mathbf{z}^{K+1}, \boldsymbol{\lambda}^{K+1}) - L(\mathbf{x}^0, \mathbf{z}^0, \boldsymbol{\lambda}^0)$$

$$\leq -\sum_{k=0}^{K} \left[ \left( \frac{m\beta}{2} + \rho \right) - \frac{(1 + \beta^2)m(\tau - 1)^2}{2} \right] \left\| \mathbf{z}^{k+1} - \mathbf{z}^k \right\|^2$$

$$- \sum_{k=0}^{K} \sum_{i \in \mathcal{A}^k} \left( \frac{\beta}{2} - \frac{L^2}{\beta} - \frac{L^2}{2} - \frac{1}{2} \right) \left\| \mathbf{x}_i^{k+1} - \mathbf{x}_i^k \right\|^2.$$

Letting $\rho$ and $\beta$ be large enough such that

$$\frac{\beta}{2} - \frac{L^2}{\beta} - \frac{L^2}{2} - \frac{1}{2} > 0 \quad \text{and} \quad \left( \frac{m\beta}{2} + \rho \right) - \frac{(1 + \beta^2)m(\tau - 1)^2}{2} > 0,$$

from the assumption that $(\mathbf{x}^{K+1}, \mathbf{z}^{K+1}, \boldsymbol{\lambda}^{K+1})$ is bounded, we have

$$\mathbf{z}^{k+1} - \mathbf{z}^k \to \mathbf{0} \quad \text{and} \quad \mathbf{x}_i^{k+1} - \mathbf{x}_i^k \to \mathbf{0}, \quad \forall i \in \mathcal{A}^k.$$

From (6.25) and the smoothness of $f_i$, we have

$$\lambda_i^{k+1} - \lambda_i^k \to \mathbf{0}, \quad \forall i \in \mathcal{A}^k.$$

From (6.19b), we have

$$\mathbf{x}_i^{k+1} - \mathbf{z}^{\bar{k}_i+1} \to \mathbf{0}, \quad \forall i \in \mathcal{A}^k,$$

which further gives

$$\mathbf{x}_i^{k+1} - \mathbf{z}^{k+1} \to \mathbf{0}, \quad \forall i \in \mathcal{A}^k,$$

due to

$$\max\{k - \tau, 0\} \le \bar{k}_i < k \quad \text{and} \quad \mathbf{z}^{\bar{k}_i+1} - \mathbf{z}^{k+1} \to \mathbf{0}.$$

For any $i \in \mathcal{A}_c^k$, we have $i \in \mathcal{A}^{\widetilde{k}_i}$ and

$$
\begin{aligned}
\left\| \mathbf{z}^{k+1} - \mathbf{x}_i^{k+1} \right\| &= \left\| \mathbf{z}^{k+1} - \mathbf{x}_i^{\widetilde{k}_i+1} \right\| \\
&\le \left\| \mathbf{z}^{k+1} - \mathbf{z}^{\hat{k}_i+1} \right\| + \left\| \mathbf{z}^{\hat{k}_i+1} - \mathbf{x}_i^{\widetilde{k}_i+1} \right\| \\
&\overset{a}{=} \left\| \mathbf{z}^{k+1} - \mathbf{z}^{\hat{k}_i+1} \right\| + \frac{1}{\beta} \left\| \lambda_i^{\widetilde{k}_i} - \lambda_i^{\widetilde{k}_i+1} \right\| \to 0,
\end{aligned}
$$

where $\overset{a}{=}$ uses (6.22) and (6.23). So we have

$$\mathbf{x}_i^{k+1} - \mathbf{z}^{k+1} \to \mathbf{0}, \quad \forall i.$$

Then from the optimality condition of (6.19c), we have

$$\sum_{i=1}^m \lambda_i^k \to \mathbf{0}.$$

$\square$

## 6.3.2   Linear Convergence Rate

When we further assume that each $f_i$ is strongly convex, we have the linear convergence rate.

**Theorem 6.10** *Assume that each $f_i$ is $\mu$-strongly convex and L-smooth, $i \in [m]$, and Assumption 3 holds true. Let $\beta$ and $\rho$ be large enough such that*

$$8m(\beta - \mu) \le \rho,$$

$$\frac{m\beta + 2\rho}{2} - 1 - \tau 2^{2\tau} - \left( \frac{1 + \beta^2}{2} + \frac{1}{2m} \right) m\tau 2^\tau > 0, \text{ and}$$

$$\frac{\beta}{2} - \frac{L^2}{\beta} - \frac{L^2}{2} - \frac{1}{2} - \frac{L^2}{4m\beta^2} - \frac{L^2}{4m\beta^2} 2^{\tau-1} \tau > 0.$$

*Then we have*

$$L(\mathbf{x}^{K+1}, \mathbf{z}^{K+1}, \boldsymbol{\lambda}^{K+1}) - f^* \le \left( 1 + \frac{1}{\delta\rho} \right)^{-(K+1)} \left( L(\mathbf{x}^0, \mathbf{z}^0, \boldsymbol{\lambda}^0) - f^* \right),$$

*where* $\delta \ge \max \left\{ 1, \frac{1}{\rho}, \frac{\rho + m\beta}{m\mu} - 1 \right\}$.

**Proof** From the strong convexity of $f_i$ and (6.25), we have

$$f_i(\mathbf{z}^*) - f_i(\mathbf{x}_i^{k+1}) \ge - \left\langle \boldsymbol{\lambda}_i^{k+1}, \mathbf{z}^* - \mathbf{x}_i^{k+1} \right\rangle + \frac{\mu}{2} \left\| \mathbf{z}^* - \mathbf{x}_i^{k+1} \right\|^2.$$

From the optimality condition of (6.19c), we have

$$- \sum_{i=1}^m \left[ \boldsymbol{\lambda}_i^{k+1} + \beta(\mathbf{x}_i^{k+1} - \mathbf{z}^{k+1}) \right] + \rho(\mathbf{z}^{k+1} - \mathbf{z}^k) = \mathbf{0}.$$

So we have

$$\sum_{i=1}^m \left\langle \boldsymbol{\lambda}_i^{k+1} + \beta(\mathbf{x}_i^{k+1} - \mathbf{z}^{k+1}), \mathbf{z}^{k+1} - \mathbf{z}^* \right\rangle = \rho \left\langle \mathbf{z}^{k+1} - \mathbf{z}^k, \mathbf{z}^{k+1} - \mathbf{z}^* \right\rangle$$

and

$$\sum_{i=1}^m f_i(\mathbf{z}^*) - \sum_{i=1}^m f_i(\mathbf{x}_i^{k+1})$$

$$\ge - \sum_{i=1}^m \left\langle \boldsymbol{\lambda}_i^{k+1}, \mathbf{z}^{k+1} - \mathbf{x}_i^{k+1} \right\rangle + \frac{\mu}{2} \sum_{i=1}^m \left\| \mathbf{z}^* - \mathbf{x}_i^{k+1} \right\|^2$$

$$+ \rho \left\langle \mathbf{z}^{k+1} - \mathbf{z}^k, \mathbf{z}^{k+1} - \mathbf{z}^* \right\rangle - \beta \sum_{i=1}^m \left\langle \mathbf{x}_i^{k+1} - \mathbf{z}^{k+1}, \mathbf{z}^{k+1} - \mathbf{z}^* \right\rangle$$

$$= -\sum_{i=1}^{m} \left\langle \boldsymbol{\lambda}_i^{k+1}, \mathbf{z}^{k+1} - \mathbf{x}_i^{k+1} \right\rangle + \frac{\mu}{2} \sum_{i=1}^{m} \left\| \mathbf{z}^* - \mathbf{x}_i^{k+1} \right\|^2$$

$$+ \frac{\rho + m\beta}{2} \left\| \mathbf{z}^{k+1} - \mathbf{z}^* \right\|^2 - \frac{\rho}{2} \left\| \mathbf{z}^k - \mathbf{z}^* \right\|^2 + \frac{\rho}{2} \left\| \mathbf{z}^{k+1} - \mathbf{z}^k \right\|^2$$

$$- \frac{\beta}{2} \sum_{i=1}^{m} \left\| \mathbf{x}_i^{k+1} - \mathbf{z}^* \right\|^2 + \frac{\beta}{2} \sum_{i=1}^{m} \left\| \mathbf{x}_i^{k+1} - \mathbf{z}^{k+1} \right\|^2.$$

Thus we have

$$L(\mathbf{x}^{k+1}, \mathbf{z}^{k+1}, \boldsymbol{\lambda}^{k+1}) - f^*$$

$$\leq \frac{\beta - \mu}{2} \sum_{i=1}^{m} \left\| \mathbf{x}_i^{k+1} - \mathbf{z}^* \right\|^2 + \frac{\rho}{2} \left\| \mathbf{z}^k - \mathbf{z}^* \right\|^2$$

$$- \frac{\rho + m\beta}{2} \left\| \mathbf{z}^{k+1} - \mathbf{z}^* \right\|^2 - \frac{\rho}{2} \left\| \mathbf{z}^{k+1} - \mathbf{z}^k \right\|^2.$$

We want to eliminate the first three terms. Since

$$\frac{\beta - \mu}{2} \sum_{i=1}^{m} \left\| \mathbf{x}_i^{k+1} - \mathbf{z}^* \right\|^2 \leq \frac{(\beta - \mu)(1 + \delta)}{2} \sum_{i=1}^{m} \left\| \mathbf{x}_i^{k+1} - \mathbf{z}^{k+1} \right\|^2$$

$$+ \frac{(\beta - \mu)m}{2} \left( 1 + \frac{1}{\delta} \right) \left\| \mathbf{z}^{k+1} - \mathbf{z}^* \right\|^2 \quad \text{and}$$

$$\frac{\rho}{2} \left\| \mathbf{z}^k - \mathbf{z}^* \right\|^2 \leq \frac{\rho}{2}(1 + \delta) \left\| \mathbf{z}^{k+1} - \mathbf{z}^k \right\|^2 + \frac{\rho}{2} \left( 1 + \frac{1}{\delta} \right) \left\| \mathbf{z}^{k+1} - \mathbf{z}^* \right\|^2,$$

we have

$$L(\mathbf{x}^{k+1}, \mathbf{z}^{k+1}, \boldsymbol{\lambda}^{k+1}) - f^*$$

$$\leq \frac{\rho \delta}{2} \left\| \mathbf{z}^{k+1} - \mathbf{z}^k \right\|^2 + \left[ \frac{\rho + m(\beta - \mu)}{2\delta} - \frac{m\mu}{2} \right] \left\| \mathbf{z}^{k+1} - \mathbf{z}^* \right\|^2$$

$$+ (\beta - \mu)\delta \sum_{i=1}^{m} \left\| \mathbf{x}_i^{k+1} - \mathbf{z}^{k+1} \right\|^2$$

$$\leq \frac{\rho \delta}{2} \left\| \mathbf{z}^{k+1} - \mathbf{z}^k \right\|^2 + (\beta - \mu)\delta \sum_{i=1}^{m} \left\| \mathbf{x}_i^{k+1} - \mathbf{z}^{k+1} \right\|^2$$

by letting $\delta > 1$ be large enough such that

$$\frac{\rho + m(\beta - \mu)}{2\delta} - \frac{m\mu}{2} \leq 0.$$

Since

$$\sum_{i=1}^{m} \left\| \mathbf{x}_i^{k+1} - \mathbf{z}^{k+1} \right\|^2$$

$$= \sum_{i \in \mathcal{A}^k} \left( \left\| \mathbf{x}_i^{k+1} - \mathbf{z}^{\bar{k}_i+1} + \mathbf{z}^{\bar{k}_i+1} - \mathbf{z}^{k+1} \right\|^2 \right)$$

$$+ \sum_{i \in \mathcal{A}_c^k} \left( \left\| \mathbf{x}_i^{k+1} - \mathbf{z}^{\hat{k}_i+1} + \mathbf{z}^{\hat{k}_i+1} - \mathbf{z}^{k+1} \right\|^2 \right)$$

$$\leq \sum_{i \in \mathcal{A}^k} \left( 2 \left\| \mathbf{x}_i^{k+1} - \mathbf{z}^{\bar{k}_i+1} \right\|^2 + 2 \left\| \mathbf{z}^{\bar{k}_i+1} - \mathbf{z}^{k+1} \right\|^2 \right)$$

$$+ \sum_{i \in \mathcal{A}_c^k} \left( 2 \left\| \mathbf{x}_i^{k+1} - \mathbf{z}^{\hat{k}_i+1} \right\|^2 + 2 \left\| \mathbf{z}^{\hat{k}_i+1} - \mathbf{z}^{k+1} \right\|^2 \right)$$

$$\overset{a}{=} \sum_{i \in \mathcal{A}^k} \left( \frac{2}{\beta^2} \left\| \boldsymbol{\lambda}_i^{k+1} - \boldsymbol{\lambda}_i^{k} \right\|^2 + 2 \left\| \mathbf{z}^{\bar{k}_i+1} - \mathbf{z}^{k+1} \right\|^2 \right)$$

$$+ \sum_{i \in \mathcal{A}_c^k} \left( \frac{2}{\beta^2} \left\| \boldsymbol{\lambda}_i^{\tilde{k}_i+1} - \boldsymbol{\lambda}_i^{\tilde{k}_i} \right\|^2 + 2 \left\| \mathbf{z}^{\hat{k}_i+1} - \mathbf{z}^{k+1} \right\|^2 \right)$$

$$\overset{b}{\leq} \sum_{i \in \mathcal{A}^k} \left( \frac{2L^2}{\beta^2} \left\| \mathbf{x}_i^{k+1} - \mathbf{x}_i^{k} \right\|^2 + 4 \left\| \mathbf{z}^{\bar{k}_i+1} - \mathbf{z}^{k} \right\|^2 \right)$$

$$+ \sum_{i \in \mathcal{A}_c^k} \left( \frac{2L^2}{\beta^2} \left\| \mathbf{x}_i^{\tilde{k}_i+1} - \mathbf{x}_i^{\tilde{k}_i} \right\|^2 + 4 \left\| \mathbf{z}^{\hat{k}_i+1} - \mathbf{z}^{k} \right\|^2 \right) + 4m \left\| \mathbf{z}^{k+1} - \mathbf{z}^{k} \right\|^2,$$

where $\overset{a}{=}$ uses (6.19b), (6.20), (6.22), and (6.23) and $\overset{b}{\leq}$ uses (6.25) to replace $\boldsymbol{\lambda}_i^{k}$ with $-\nabla f_i(\mathbf{x}_i^{k})$ and then apply the $L$-smoothness of $f_i$. $\overset{b}{\leq}$ also uses the inequality $\|\mathbf{a} + \mathbf{b}\|^2 \leq 2(\|\mathbf{a}\|^2 + \|\mathbf{b}\|^2)$.

By letting $\rho$ be large enough such that $8m(\beta - \mu) \leq \rho$, we have

$$L(\mathbf{x}^{k+1}, \mathbf{z}^{k+1}, \boldsymbol{\lambda}^{k+1}) - f^*$$

$$\leq \left[ \frac{\rho\delta}{2} + 4m(\beta - \mu)\delta \right] \left\| \mathbf{z}^{k+1} - \mathbf{z}^{k} \right\|^2$$

$$+ \sum_{i \in \mathcal{A}^k} (\beta - \mu)\delta \left( \frac{2L^2}{\beta^2} \left\| \mathbf{x}_i^{k+1} - \mathbf{x}_i^{k} \right\|^2 + 4 \left\| \mathbf{z}^{\bar{k}_i+1} - \mathbf{z}^{k} \right\|^2 \right)$$

$$+ \sum_{i \in \mathcal{A}_c^k} (\beta - \mu)\delta \left( \frac{2L^2}{\beta^2} \left\| \mathbf{x}_i^{\tilde{k}_i+1} - \mathbf{x}_i^{\tilde{k}_i} \right\|^2 + 4 \left\| \mathbf{z}^{\hat{k}_i+1} - \mathbf{z}^k \right\|^2 \right)$$

$$\leq \rho\delta \left\| \mathbf{z}^{k+1} - \mathbf{z}^k \right\|^2 + \sum_{i \in \mathcal{A}^k} \frac{\rho\delta}{8m} \left( \frac{2L^2}{\beta^2} \left\| \mathbf{x}_i^{k+1} - \mathbf{x}_i^k \right\|^2 + 4 \left\| \mathbf{z}^{\bar{k}_i+1} - \mathbf{z}^k \right\|^2 \right)$$

$$+ \sum_{i \in \mathcal{A}_c^k} \frac{\rho\delta}{8m} \left( \frac{2L^2}{\beta^2} \left\| \mathbf{x}_i^{\tilde{k}_i+1} - \mathbf{x}_i^{\tilde{k}_i} \right\|^2 + 4 \left\| \mathbf{z}^{\hat{k}_i+1} - \mathbf{z}^k \right\|^2 \right).$$

Dividing both sides of the above inequality by $\rho\delta$ and adding it with (6.26), we have

$$\left( L(\mathbf{x}^{k+1}, \mathbf{z}^{k+1}, \boldsymbol{\lambda}^{k+1}) - f^* \right) - \frac{1}{\eta} \left( L(\mathbf{x}^k, \mathbf{z}^k, \boldsymbol{\lambda}^k) - f^* \right)$$

$$\leq \frac{1}{\eta} \left[ \sum_{i \in \mathcal{A}_c^k} \frac{L^2}{4m\beta^2} \left\| \mathbf{x}_i^{\tilde{k}_i+1} - \mathbf{x}_i^{\tilde{k}_i} \right\|^2 + \sum_{i \in \mathcal{A}_c^k} \frac{1}{2m} \left\| \mathbf{z}^{\hat{k}_i+1} - \mathbf{z}^k \right\|^2 \right.$$

$$- \left( \frac{m\beta + \rho}{2} - 1 \right) \left\| \mathbf{z}^{k+1} - \mathbf{z}^k \right\|^2 + \sum_{i \in \mathcal{A}^k} \left( \frac{1+\beta^2}{2} + \frac{1}{2m} \right) \left\| \mathbf{z}^{\bar{k}_i+1} - \mathbf{z}^k \right\|^2$$

$$\left. - \sum_{i \in \mathcal{A}^k} \left( \frac{\beta}{2} - \frac{L^2}{\beta} - \frac{L^2}{2} - \frac{1}{2} - \frac{L^2}{4m\beta^2} \right) \left\| \mathbf{x}_i^{k+1} - \mathbf{x}_i^k \right\|^2 \right]$$

$$= \frac{1}{\eta} \left[ \sum_{i \in \mathcal{A}_c^k} \frac{L^2}{4m\beta^2} \left\| \mathbf{x}_i^{\tilde{k}_i+1} - \mathbf{x}_i^{\tilde{k}_i} \right\|^2 + \sum_{i \in \mathcal{A}_c^k} \frac{1}{2m} \left\| \mathbf{z}^{\hat{k}_i+1} - \mathbf{z}^k \right\|^2 \right.$$

$$- \left( \frac{m\beta + \rho}{2} - 1 \right) \left\| \mathbf{z}^{k+1} - \mathbf{z}^k \right\|^2 + \sum_{i \in \mathcal{A}^k} \left( \frac{1+\beta^2}{2} + \frac{1}{2m} \right) \left\| \mathbf{z}^{\bar{k}_i+1} - \mathbf{z}^k \right\|^2$$

$$\left. - \sum_{i=1}^m \left( \frac{\beta}{2} - \frac{L^2}{\beta} - \frac{L^2}{2} - \frac{1}{2} - \frac{L^2}{4m\beta^2} \right) \left\| \mathbf{x}_i^{k+1} - \mathbf{x}_i^k \right\|^2 \right],$$

where we denote $\eta = 1 + \frac{1}{\rho\delta}$ and use $\mathbf{x}_i^{k+1} = \mathbf{x}_i^k$ for all $i \in \mathcal{A}_c^k$ in the last line. Telescoping the above inequality from $k = 0$ to $K$, we have

$$\left( L(\mathbf{x}^{K+1}, \mathbf{z}^{K+1}, \boldsymbol{\lambda}^{K+1}) - f^* \right) - \frac{1}{\eta^{K+1}} \left( L(\mathbf{x}^0, \mathbf{z}^0, \boldsymbol{\lambda}^0) - f^* \right)$$

$$\leq \frac{L^2}{4m\beta^2} \sum_{k=0}^K \frac{1}{\eta^{K+1-k}} \sum_{i \in \mathcal{A}_c^k} \left\| \mathbf{x}_i^{\tilde{k}_i+1} - \mathbf{x}_i^{\tilde{k}_i} \right\|^2$$

$$+ \frac{1}{2m} \sum_{k=0}^{K} \frac{1}{\eta^{K+1-k}} \sum_{i \in \mathcal{A}_c^k} \left\| \mathbf{z}^{\hat{k}_i+1} - \mathbf{z}^k \right\|^2$$

$$- \left( \frac{m\beta + \rho}{2} - 1 \right) \sum_{k=0}^{K} \frac{1}{\eta^{K+1-k}} \left\| \mathbf{z}^{k+1} - \mathbf{z}^k \right\|^2$$

$$+ \left( \frac{1+\beta^2}{2} + \frac{1}{2m} \right) \sum_{k=0}^{K} \frac{1}{\eta^{K+1-k}} \sum_{i \in \mathcal{A}^k} \left\| \mathbf{z}^{\bar{k}_i+1} - \mathbf{z}^k \right\|^2$$

$$- \left( \frac{\beta}{2} - \frac{L^2}{\beta} - \frac{L^2}{2} - \frac{1}{2} - \frac{L^2}{4m\beta^2} \right) \sum_{k=0}^{K} \frac{1}{\eta^{K+1-k}} \sum_{i=1}^{m} \left\| \mathbf{x}_i^{k+1} - \mathbf{x}_i^k \right\|^2.$$

We want to choose $\beta$ and $\rho$ large enough such that the right hand side is negative. Similar to (6.27), we have

$$\sum_{k=0}^{K} \sum_{i \in \mathcal{A}^k} \eta^k \left\| \mathbf{z}^{\bar{k}_i+1} - \mathbf{z}^k \right\|^2$$

$$= \sum_{k=0}^{K} \sum_{i \in \mathcal{A}^k} \eta^k \left\| \sum_{t=\bar{k}_i+1}^{k-1} \left( \mathbf{z}^t - \mathbf{z}^{t+1} \right) \right\|^2$$

$$\leq \sum_{k=0}^{K} \sum_{i \in \mathcal{A}^k} (k - \bar{k}_i - 1) \eta^k \sum_{t=\bar{k}_i+1}^{k-1} \left\| \mathbf{z}^t - \mathbf{z}^{t+1} \right\|^2$$

$$\leq \sum_{k=0}^{K} \sum_{i \in \mathcal{A}^k} (\tau - 1) \eta^k \sum_{t=\max\{k-\tau+1,1\}}^{k-1} \left\| \mathbf{z}^t - \mathbf{z}^{t+1} \right\|^2$$

$$\leq m(\tau - 1) \sum_{k=0}^{K} \eta^k \sum_{t=\max\{k-\tau+1,1\}}^{k-1} \left\| \mathbf{z}^t - \mathbf{z}^{t+1} \right\|^2$$

$$\leq m(\tau - 1) \sum_{k=0}^{K} \left( \eta^{k+1} + \eta^{k+2} + \cdots + \eta^{k+\tau-1} \right) \left\| \mathbf{z}^k - \mathbf{z}^{k+1} \right\|^2$$

$$\leq m(\tau - 1) \frac{\eta^\tau - \eta}{\eta - 1} \sum_{k=0}^{K} \eta^k \left\| \mathbf{z}^k - \mathbf{z}^{k+1} \right\|^2.$$

Analogously, we have

$$\sum_{k=0}^{K} \sum_{i \in \mathcal{A}_c^k} \eta^k \left\| \mathbf{z}^{\hat{k}_i+1} - \mathbf{z}^k \right\|^2$$

$$\leq m(2\tau - 1) \frac{\eta^{2\tau} - \eta}{\eta - 1} \sum_{k=0}^{K} \eta^k \left\| \mathbf{z}^k - \mathbf{z}^{k+1} \right\|^2,$$

due to

$$\max\{k - \tau + 1, 0\} \leq \tilde{k}_i < k, \quad \max\{\tilde{k}_i - \tau, 0\} \leq \hat{k}_i < \tilde{k}_i,$$

and thus

$$\max\{k - 2\tau + 1, 0\} \leq \hat{k}_i < k.$$

We also have

$$\sum_{k=0}^{K} \sum_{i \in \mathcal{A}_c^k} \eta^k \left\| \mathbf{x}_i^{\tilde{k}_i+1} - \mathbf{x}_i^{\tilde{k}_i} \right\|^2$$

$$= \sum_{k=0}^{K} \sum_{i \in \mathcal{A}_c^k} \eta^{k-\tilde{k}_i} \eta^{\tilde{k}_i} \left\| \mathbf{x}_i^{\tilde{k}_i+1} - \mathbf{x}_i^{\tilde{k}_i} \right\|^2$$

$$\leq \eta^{\tau-1} \sum_{k=0}^{K} \sum_{i \in \mathcal{A}_c^k} \eta^{\tilde{k}_i} \left\| \mathbf{x}_i^{\tilde{k}_i+1} - \mathbf{x}_i^{\tilde{k}_i} \right\|^2$$

$$\overset{a}{\leq} \eta^{\tau-1}(\tau - 1) \sum_{k=0}^{K} \sum_{i=1}^{m} \eta^k \left\| \mathbf{x}_i^{k+1} - \mathbf{x}_i^k \right\|^2,$$

where in $\overset{a}{\leq}$ we use the fact that each $\eta^{\tilde{k}_i} \left\| \mathbf{x}_i^{\tilde{k}_i+1} - \mathbf{x}_i^{\tilde{k}_i} \right\|^2$ appears no more than $\tau - 1$ times in the summation $\sum_{k=0}^{K} \sum_{i \in \mathcal{A}_c^k} \eta^{\tilde{k}_i} \left\| \mathbf{x}_i^{\tilde{k}_i+1} - \mathbf{x}_i^{\tilde{k}_i} \right\|^2$.

Thus, we have

$$\left( L(\mathbf{x}^{K+1}, \mathbf{z}^{K+1}, \boldsymbol{\lambda}^{K+1}) - f^* \right) - \frac{1}{\eta^{K+1}} \left( L(\mathbf{x}^0, \mathbf{z}^0, \boldsymbol{\lambda}^0) - f^* \right)$$

$$\leq - \left[ \frac{m\beta + \rho}{2} - 1 - \frac{1}{2m} m(2\tau - 1) \frac{\eta^{2\tau} - \eta}{\eta - 1} \right.$$

$$-\left(\frac{1+\beta^2}{2}+\frac{1}{2m}\right)m(\tau-1)\frac{\eta^\tau-\eta}{\eta-1}\right]\sum_{k=0}^{K}\frac{1}{\eta^{K+1-k}}\left\|\mathbf{z}^{k+1}-\mathbf{z}^k\right\|^2$$

$$-\left[\frac{\beta}{2}-\frac{L^2}{\beta}-\frac{L^2}{2}-\frac{1}{2}-\frac{L^2}{4m\beta^2}-\frac{L^2}{4m\beta^2}\eta^{\tau-1}(\tau-1)\right]$$

$$\times\sum_{i=1}^{m}\sum_{k=0}^{K}\frac{1}{\eta^{K+1-k}}\left\|\mathbf{x}_i^{k+1}-\mathbf{x}_i^k\right\|^2$$

$$\leq-\left[\frac{m\beta+\rho}{2}-1-\tau 2^{2\tau}-\left(\frac{1+\beta^2}{2}+\frac{1}{2m}\right)m\tau 2^\tau\right]$$

$$\times\sum_{k=0}^{K}\frac{1}{\eta^{K+1-k}}\left\|\mathbf{z}^{k+1}-\mathbf{z}^k\right\|^2$$

$$-\left(\frac{\beta}{2}-\frac{L^2}{\beta}-\frac{L^2}{2}-\frac{1}{2}-\frac{L^2}{4m\beta^2}-\frac{L^2}{4m\beta^2}2^{\tau-1}\tau\right)$$

$$\times\sum_{i=1}^{m}\sum_{k=0}^{K}\frac{1}{\eta^{K+1-k}}\left\|\mathbf{x}_i^{k+1}-\mathbf{x}_i^k\right\|^2$$

$$\leq 0,$$

where we use

$$\eta\leq 2,\quad \frac{\eta^\tau-\eta}{\eta-1}=\eta+\cdots+\eta^{\tau-1}\leq 2+\cdots+2^{\tau-1}\leq 2^\tau,\quad\text{and}$$

$$\frac{\eta^{2\tau}-\eta}{\eta-1}\leq 2^{2\tau}.$$

$\square$

From Theorem 6.2, we see that the synchronous ADMM needs $O\left(\sqrt{\frac{L}{\mu}}\log\frac{1}{\epsilon}\right)$ iterations to find an $\epsilon$-optimal solution, which has the optimal dependence on $\frac{L}{\mu}$. For the asynchronous ADMM, Theorem 6.10 only proves the linear convergence without a complexity explicitly dependent on $\frac{L}{\mu}$. We believe that in general the asynchronous ADMM needs more iterations than synchronous ADMM. It is unclear whether the time saved per iteration of the asynchronous ADMM can offset the cost of more iterations in theory, although it shows great advantages in practice.

There are some other ways to analyze asynchronous ADMM. For example, [6, 7, 13, 15] studied randomized asynchronous ADMM, which requires more assumptions than Algorithms 6.10–6.11 do, and it is also unclear whether it needs less running time than the synchronous ADMM in theory.

## 6.4   Nonconvex Distributed ADMM

Next, we introduce the nonconvex distributed ADMM. In fact, the asynchronous ADMM (Algorithm 6.11) can also be used to solve nonconvex problems. In this case, $L(\mathbf{x}, \mathbf{z}, \boldsymbol{\lambda})$ is $(\beta - L)$-strongly convex with respect to $\mathbf{x}$, and (6.24) should be replaced by the following one:

$$L(\mathbf{x}^{k+1}, \mathbf{z}^k, \boldsymbol{\lambda}^k) - L(\mathbf{x}^k, \mathbf{z}^k, \boldsymbol{\lambda}^k)$$

$$\leq \sum_{i \in \mathcal{A}^k} \left( \beta \left\langle \mathbf{z}^{\bar{k}_i+1} - \mathbf{z}^k, \mathbf{x}_i^{k+1} - \mathbf{x}_i^k \right\rangle - \frac{\beta - L}{2} \left\| \mathbf{x}_i^{k+1} - \mathbf{x}_i^k \right\|^2 \right).$$

Accordingly, we have the following convergence guarantee [4].

**Theorem 6.11** *Assume that each $f_i$ is $L$-smooth, $i \in [m]$, and Assumption 3 holds true. Let*

$$\beta > \frac{1 + L + L^2 + \sqrt{(1 + L + L^2)^2 + 8L^2}}{2} \quad and \quad \rho > \frac{m(1 + \beta^2)(\tau - 1)^2 - m\beta}{2}.$$

*Suppose that $(\mathbf{x}_1^k, \cdots, \mathbf{x}_m^k, \mathbf{z}^k, \boldsymbol{\lambda}_1^k, \cdots, \boldsymbol{\lambda}_m^k)$ generated by (6.19a)–(6.19c) are bounded, then $(\mathbf{x}_1^k, \cdots, \mathbf{x}_m^k, \mathbf{z}^k, \boldsymbol{\lambda}_1^k, \cdots, \boldsymbol{\lambda}_m^k)$ converge to the set of KKT points of Problem (6.2) in the sense of*

$$\sum_{i=1}^m \boldsymbol{\lambda}_i^k \to \mathbf{0}, \quad \mathbf{x}_i^{k+1} - \mathbf{z}^{k+1} \to \mathbf{0}, \quad and \quad \nabla f_i(\mathbf{x}_i^{k+1}) + \boldsymbol{\lambda}_i^{k+1} = \mathbf{0}, \quad i \in [m].$$

The synchronous ADMM is a special case of the asynchronous ADMM with $\mathcal{A}_c^k = \varnothing$ and $\bar{k}_i + 1 = k$. Thus, the above theorem also holds for the synchronous ADMM with a much simpler proof.

## 6.5   ADMM with Generally Linear Constraints

We end this chapter by non-consensus-based distributed ADMM. Namely, the problem is the generally linearly constrained one (3.71). The linearized ADMM with parallel splitting [9, 11] given in Algorithm 3.11 can be used to solve the problem directly. We present it in Algorithms 6.12–6.13 in the distributed manner. If the proximal mapping of $f_i$ is not easily computable, we may linearize $f_i$ as well, but since this is a straightforward modification over Algorithms 6.12–6.13, we omit the details.

---

**Algorithm 6.12** Distributed linearized ADMM with parallel splitting for the master

---

**for** $k = 0, 1, 2, \cdots$ **do**

    Wait until receiving $\mathbf{y}_i^{k+1}$ from all the workers $i \in [m]$.

    $\mathbf{s}^{k+1} = \sum_{i=1}^{m} \mathbf{y}_i^{k+1}$.

    $\boldsymbol{\lambda}^{k+1} = \boldsymbol{\lambda}^k + \beta \left( \mathbf{s}^{k+1} - \mathbf{b} \right)$.

    Send $\mathbf{s}^{k+1}$ and $\boldsymbol{\lambda}^{k+1}$ to all the workers.

**end for**

---

**Algorithm 6.13** Distributed linearized ADMM with parallel splitting for the $i$th worker

---

Initialize: $\mathbf{x}_i^0$ and $\boldsymbol{\lambda}_i^0$, $i \in [m]$.

$\mathbf{y}_i^0 = \mathbf{A}_i \mathbf{x}_i^0$.

Send $\mathbf{y}_i^0$ to the master.

Wait until receiving $\mathbf{s}^0$ and $\boldsymbol{\lambda}^0$ from the master.

**for** $k = 0, 1, 2, \cdots$ **do**

$$
\mathbf{x}_i^{k+1} = \operatorname*{argmin}_{\mathbf{x}_i} \left( f_i(\mathbf{x}_i) + \left\langle \boldsymbol{\lambda}^k, \mathbf{A}_i \mathbf{x}_i \right\rangle + \beta \left\langle \mathbf{A}_i^T \left( \mathbf{s}^k - \mathbf{b} \right), \mathbf{x}_i - \mathbf{x}_i^k \right\rangle \right.
$$

$$
\left. + \frac{m\beta \|\mathbf{A}_i\|_2^2}{2} \left\| \mathbf{x}_i - \mathbf{x}_i^k \right\|^2 \right)
$$

$$
= \operatorname{Prox}_{\left( m\beta \|\mathbf{A}_i\|_2^2 \right)^{-1} f_i} \left( \mathbf{x}_i^k - \frac{1}{m\beta \|\mathbf{A}_i\|_2^2} \mathbf{A}_i^T \left[ \boldsymbol{\lambda}^k + \beta \left( \mathbf{s}^k - \mathbf{b} \right) \right] \right).
$$

$\mathbf{y}_i^{k+1} = \mathbf{A}_i \mathbf{x}_i^{k+1}$.

Send $\mathbf{y}_i^{k+1}$ to the master.

Wait until receiving $\mathbf{s}^{k+1}$ and $\boldsymbol{\lambda}^{k+1}$ from the master.

**end for**

---

# References

1. N.S. Aybat, Z. Wang, T. Lin, S. Ma, Distributed linearized alternating direction mehod of multipliers. IEEE Trans. Automat. Contr. **63**(1), 5–20 (2018)
2. D.P. Bertsekas, J.N. Tsitsiklis, *Parallel and Distributed Computation: Numerical Methods* (Prentice Hall, Hoboken, 1989)
3. S. Boyd, N. Parikh, E. Chu, B. Peleato, J. Eckstein, Distributed optimization and statistical learning via the alternating direction method of multipliers. Found. Trends Mach. Learn. **3**(1), 1–122 (2011)
4. T.-H. Chang, M. Hong, X. Wang, Asynchronous distributed ADMM for large-scale optimization – part I: algorithm and convergence analysis. IEEE Trans. Signal Process. **64**(12), 3118–3130 (2016)
5. T.-H. Chang, W.-C. Liao, M. Hong, X. Wang, Asynchronous distributed ADMM for large-scale optimization – part II: linear convergence analysis and numerical performance. IEEE Trans. Signal Process. **64**(12), 3131–3144 (2016)
6. F. Iutzeler, P. Bianchi, P. Ciblat, W. Hachem, Asynchronous distributed optimization using a randomized alternating direction method of multipliers, in *IEEE Conference on Decision and Control* (2013), pp. 3671–3676

7. S. Kumar, R. Jain, K. Rajawat, Asynchronous optimization over heterogeneous networks via consensus ADMM. IEEE Trans. Signal Inf. Process. Netw. **3**(1), 114–129 (2017)
8. H. Li, Z. Lin, Y. Fang, Variance reduced EXTRA and DIGing and their optimal acceleration for strongly convex decentralized optimization (2020). Arxiv:2009.04373
9. Z. Lin, R. Liu, H. Li, Linearized alternating direction method with parallel splitting and adaptive penalty for separable convex programs in machine learning. Mach. Learn. **99**(2), 287–325 (2015)
10. Q. Ling, W. Shi, G. Wu, A. Ribeiro, DLM: Decentralized linearized alternating direction method of multipliers. IEEE Trans. Signal Process. **63**(15), 4051–4064 (2015)
11. R. Liu, Z. Lin, Z. Su, Linearized alternating direction method with parallel splitting and adaptive penalty for separable convex programs in machine learning, in *Asian Conference on Machine Learning* (2013), pp. 116–132
12. M. Maros, J. Jalden, On the Q-linear convergence of distributed generalized ADMM under non-strongly convex function components. IEEE Trans. Signal Inf. Process. Netw. **5**(3), 442–453 (2019)
13. Z. Peng, Y. Xu, M. Yan, W. Yin, ARock: an algorithmic framework for asynchronous parallel coordinate updates. SIAM J. Sci. Comput. **38**(5), 2851–2879 (2016)
14. W. Shi, Q. Ling, K. Yuan, G. Wu, W. Yin, On the linear convergence of the ADMM in decentralized consensus optimization. IEEE Trans. Signal Process. **62**(7), 1750–1761 (2014)
15. E. Wei, A. Ozdaglar, On the $O(1/k)$ convergence of asynchronous distributed alternating direction method of multipliers, in *IEEE Global Conference on Signal and Information Processing* (2013), pp. 551–554

# Chapter 7
# Practical Issues and Conclusions

In the previous chapters, we have introduced the major steps of various ADMMs and their convergence and convergence rate analysis. However, those are the theoretical aspects of algorithms. In real implementations, some practical issues need to be considered. In this chapter we first briefly discuss the practical issues and then give conclusions.

## 7.1 Practical Issues

In this section, we discuss several implementation issues of ADMM in practice. Since we are unable to discuss all variants, we focus on the model problem (2.13) and the original ADMM (Algorithm 2.1), where $\mathbf{A} \in \mathbb{R}^{q \times n}$ and $\mathbf{B} \in \mathbb{R}^{q \times m}$.

### 7.1.1 Stopping Criterion

When solving Problem (2.13), our purpose is to find $(\mathbf{x}^*, \mathbf{y}^*)$ satisfying the following KKT conditions:

$$- \mathbf{A}^T \boldsymbol{\lambda}^* \in \partial f(\mathbf{x}^*), \quad -\mathbf{B}^T \boldsymbol{\lambda}^* \in \partial g(\mathbf{y}^*), \quad \text{and} \quad \mathbf{A}\mathbf{x}^* + \mathbf{B}\mathbf{y}^* = \mathbf{b}, \quad (7.1)$$

where $\boldsymbol{\lambda}^*$ is the optimal dual variable. On the other hand, when we use ADMM to solve Problem (2.13), the optimality conditions of the subproblems (2.15a)–(2.15b) are

$$-\mathbf{A}^T \boldsymbol{\lambda}^{k+1} - \beta \mathbf{A}^T \mathbf{B}(\mathbf{y}^k - \mathbf{y}^{k+1}) = -\mathbf{A}^T \boldsymbol{\lambda}^k - \beta \mathbf{A}^T (\mathbf{A}\mathbf{x}^{k+1} + \mathbf{B}\mathbf{y}^k - \mathbf{b})$$

$$\in \partial f(\mathbf{x}^{k+1}),$$

© The Author(s), under exclusive license to Springer Nature Singapore Pte Ltd. 2022
Z. Lin et al., *Alternating Direction Method of Multipliers for Machine Learning*,
https://doi.org/10.1007/978-981-16-9840-8_7

$$-\mathbf{B}^T \boldsymbol{\lambda}^{k+1} = -\mathbf{B}^T \boldsymbol{\lambda}^k - \beta \mathbf{B}^T (\mathbf{A}\mathbf{x}^{k+1} + \mathbf{B}\mathbf{y}^{k+1} - \mathbf{b})$$

$$\in \partial g(\mathbf{y}^{k+1}).$$

Define the primal residual and the dual residual as

$$\mathbf{p}^{k+1} = \mathbf{A}\mathbf{x}^{k+1} + \mathbf{B}\mathbf{y}^{k+1} - \mathbf{b} \quad \text{and} \quad \mathbf{d}^{k+1} = \beta \mathbf{A}^T \mathbf{B}(\mathbf{y}^k - \mathbf{y}^{k+1}), \qquad (7.2)$$

respectively. We see that if $\mathbf{p}^{k+1} = \mathbf{0}$ and $\mathbf{d}^{k+1} = \mathbf{0}$, then $(\mathbf{x}^{k+1}, \mathbf{y}^{k+1}, \boldsymbol{\lambda}^{k+1})$ satisfies the KKT conditions in (7.1). Thus, the convergence can be monitored by $\mathbf{p}^{k+1}$ and $\mathbf{d}^{k+1}$, and we can use

$$\|\mathbf{p}^{k+1}\| \le \epsilon_p \quad \text{and} \quad \|\mathbf{d}^{k+1}\| \le \epsilon_d$$

as the stopping criterion. Boyd et al. [1] suggested using a combination of absolute error $\epsilon_{\text{abs}}$ and relative error $\epsilon_{\text{rel}}$ to prescribe the tolerances $\epsilon_p$ and $\epsilon_d$:

$$\epsilon_p = \sqrt{q}\epsilon_{\text{abs}} + \epsilon_{\text{rel}} \max\{\|\mathbf{A}\mathbf{x}^k\|, \|\mathbf{B}\mathbf{y}^k\|, \|\mathbf{b}\|\},$$

$$\epsilon_d = \sqrt{n}\epsilon_{\text{abs}} + \epsilon_{\text{rel}}\|\mathbf{A}^T \boldsymbol{\lambda}^k\|.$$

This is a relatively good strategy so that the tolerances are applicable for a wide range of problems with different dimensions and magnitudes.

For variants of ADMM, the stopping criterion can also be deduced by discerning the discrepancy between the KKT conditions and the optimality conditions of the subproblems for updating the primal variables, e.g., [7] for linearized ADMM.

### 7.1.2  Choice of Penalty Parameters

Throughout the book, a constant penalty parameter $\beta$ has been used. In this case, for many of the ADMMs introduced in the book a scaled form of the Lagrange multiplier can be used. Namely, use $\tilde{\boldsymbol{\lambda}} = \beta^{-1}\boldsymbol{\lambda}$ instead. For example, we can rewrite (2.15a)–(2.15c) as

$$\mathbf{x}^{k+1} = \underset{\mathbf{x}}{\operatorname{argmin}} \left( f(\mathbf{x}) + g(\mathbf{y}^k) + \frac{\beta}{2}\|\mathbf{A}\mathbf{x} + \mathbf{B}\mathbf{y}^k - \mathbf{b} + \tilde{\boldsymbol{\lambda}}^k\|^2 \right),$$

$$\mathbf{y}^{k+1} = \underset{\mathbf{y}}{\operatorname{argmin}} \left( f(\mathbf{x}^{k+1}) + g(\mathbf{y}) + \frac{\beta}{2}\|\mathbf{A}\mathbf{x}^{k+1} + \mathbf{B}\mathbf{y} - \mathbf{b} + \tilde{\boldsymbol{\lambda}}^k\|^2 \right),$$

$$\tilde{\boldsymbol{\lambda}}^{k+1} = \tilde{\boldsymbol{\lambda}}^k + (\mathbf{A}\mathbf{x}^{k+1} + \mathbf{B}\mathbf{y}^{k+1} - \mathbf{b}).$$

Note that in the update (2.15c) of $\boldsymbol{\lambda}$, the coefficient $\beta$ can also be chosen as other values. For example, Theorem 5.1 of [3] shows that for Algorithm 2.1 the update (2.15c) can be changed to

$$\boldsymbol{\lambda}^{k+1} = \boldsymbol{\lambda}^k + \tau\beta(\mathbf{A}\mathbf{x}^{k+1} + \mathbf{B}\mathbf{y}^{k+1} - \mathbf{b}),$$

where $\tau$ can be any fixed value in $(0, (\sqrt{5}+1)/2)$. Glowinski actually asked whether the upper bound of $\tau$ can be 2 [3] and Tao and Yuan gave an affirmative answer when the objective is quadratic [10]. So in real implementations, it is worthwhile to tune $\tau$ as well.

Actually, the choice of $\beta$ greatly affects the convergence speed, although may not change the order of convergence. When $\beta$ is fixed it is difficult to find the fixed value that fits for various problems. So in reality it is more desirable to make $\beta$ vary along iteration. Typically, there are three modes of change: $\beta_k$ being non-decreasing, non-increasing, and oscillating. When $\beta_k$ is non-decreasing, normally it has to be upper bounded[1] [8]. When $\beta_k$ is non-increasing, normally it has to be lower bounded away from 0, i.e., $\beta_k \geq \beta_{min} > 0$ [4]. In either case, all the variants of ADMM still converge because $\beta$ will eventually be fixed at the upper bound or the lower bound (i.e., the change of $\beta$ only speeds up the initial steps of iterations when the tolerances $\epsilon_p$ and $\epsilon_d$ are not too small, which suffices for many real applications).

The change of $\beta$ can be non-adaptive or adaptive. For example, Tian and Yuan [11] proposed a dynamic but non-adaptive strategy to update $\beta$:

$$\beta_k = \tilde{\beta}_{\left\lfloor \frac{k}{\gamma} \right\rfloor}, \text{ with } \tilde{\beta}_{k+1} = \frac{\tilde{\beta}_k}{\sqrt{1 + L_g^{-1}\tilde{\beta}_k}}, \tag{7.3}$$

to improve the convergence of Algorithm 2.1 from $O(k^{-1})$ to $O(k^{-2})$ in an ergodic sense, without the strong convexity assumption on $f$ and $g$, where $\lfloor x \rfloor$ is the largest integer not exceeding $x$, $\gamma > 1$, being a real number, is the frequency of adjusting the penalty parameter, and $L_g$ is the Lipschitz constant of $\nabla g$. Note that (7.3) means that $\beta$ decreases after roughly every $\gamma$ iterations.

While a non-adaptive change of $\beta$ may help speedup, it is natural to think that adaptive change of $\beta$ may work even better. A straightforward idea is to balance the primal and the dual residuals. Since

$$\mathbf{p}^{k+1} = \beta_k^{-1}(\boldsymbol{\lambda}^{k+1} - \boldsymbol{\lambda}^k)$$

and if ADMM converges $\boldsymbol{\lambda}^k$ typically has to be bounded, we may expect that a larger $\beta$ yields a smaller primal residual but a larger dual residual (see the definition of $\mathbf{d}^{k+1}$ in (7.2)). Conversely, a smaller $\beta$ leads to a larger primal residual and a

---

[1] Under some circumstances, the upper boundedness condition can be removed, e.g., when $\partial f$ and $\partial g$ are uniformly bounded, which can be satisfied when $f$ and $g$ are norms [8].

smaller dual residual. So we can adaptively tune $\beta$ to balance the two residuals, that is, increase $\beta$ when the primal residual is larger, and decrease $\beta$ when the dual residual is larger. A simple scheme for achieving this goal is (for example, see [1, 5])

$$\beta_{k+1} = \begin{cases} \eta\beta_k, & \text{if } \|\mathbf{p}^k\| \geq v\|\mathbf{d}^k\|, \\ \beta_k/\eta, & \text{if } \|\mathbf{d}^k\| \geq v\|\mathbf{p}^k\|, \\ \beta_k, & \text{otherwise}, \end{cases}$$

with $\eta > 1$ and $v > 1$. However, it is more challenging to prove the convergence of ADMM under the above adaptive penalty scheme since $\beta_k$ is oscillating.[2] Moreover, the above scheme requires two parameters $\eta$ and $v$. So it is less convenient to tune their values. To address these issues, Lin et al. [7, 8] proposed another adaptive penalty scheme that only allows to increase $\beta$ when the dual residual is less than its tolerance. The scheme in [7, 8] was originally for linearized ADMM. When it is adapted to the original ADMM, it becomes

$$\beta_{k+1} = \begin{cases} \min\{\rho\beta_k, \beta_{\max}\}, & \text{if } \|\mathbf{d}^k\| \leq \epsilon_d, \\ \beta_k, & \text{otherwise}, \end{cases}$$

where $\rho > 1$ and $\beta_{\max} > 0$ is the upper bound. As $\beta$ is non-decreasing, the convergence is guaranteed and only one parameter $\rho$ needs to be tuned, which greatly facilitates real applications. A good choice of $\rho$ should make $\beta$ increase after every several iterations, rather than being stagnant for many iterations. The corresponding scheme for decreasing $\beta$, i.e.,

$$\beta_{k+1} = \begin{cases} \max\{\rho^{-1}\beta_k, \beta_{\min}\}, & \text{if } \|\mathbf{p}^k\| \leq \epsilon_p, \\ \beta_k, & \text{otherwise}, \end{cases}$$

also works well, where $\beta_{\min} > 0$ is the lower bound.

### 7.1.3    Avoiding Excessive Auxiliary Variables

ADMM works in the principle of "divide and conquer" so that the updates for each primal variable are relatively simple. As shown in Sect. 1.1, introducing auxiliary variables is often necessary for obtaining relatively simple subproblems. However, introducing more auxiliary variables brings the side effect of slowing down the convergence (e.g., evidenced by the choice of $L_i$ in Sect. 3.5.3, which is proportional to the number of blocks of variables). Therefore, if possible we should reduce the number of auxiliary variables. Take Problem (1.6) for example, introducing the

---

[2] For example, [5] requires $\eta$ to be varying along iteration and $\sum_{k=0}^{\infty}(\eta_k - 1) < \infty$. So $\beta_k$ only changes very slightly when $k$ is large, making virtually no effect on balancing the residuals.

auxiliary variable $\mathbf{Y}$ is necessary because otherwise the subproblem for updating $\mathbf{X}$:

$$\mathbf{X}^{k+1} = \underset{\mathbf{X} \geq \mathbf{0}}{\operatorname{argmin}} \left( \|\mathbf{X}\|_* + \left\langle \boldsymbol{\lambda}^k, \mathbf{b} - \mathcal{P}_{\Omega}(\mathbf{X}) - \mathbf{e}^k \right\rangle + \frac{\beta}{2} \|\mathbf{b} - \mathcal{P}_{\Omega}(\mathbf{X}) - \mathbf{e}^k\|^2 \right)$$

is not easily solvable, even after linearizing the augmented term, due to the non-negativity constraint $\mathbf{X} \geq \mathbf{0}$. However, if $\|\mathbf{X}\|_*$ is replaced by $\|\mathbf{X}\|^2$ then introducing $\mathbf{Y}$ is unnecessary because in this case the subproblem for updating $\mathbf{X}$:

$$\mathbf{X}^{k+1} = \underset{\mathbf{X} \geq \mathbf{0}}{\operatorname{argmin}} \left( \|\mathbf{X}\|^2 + \left\langle \boldsymbol{\lambda}^k, \mathbf{b} - \mathcal{P}_{\Omega}(\mathbf{X}) - \mathbf{e}^k \right\rangle + \frac{\beta}{2} \|\mathbf{b} - \mathcal{P}_{\Omega}(\mathbf{X}) - \mathbf{e}^k\|^2 \right)$$

is easily solvable. So is the case when $\|\mathbf{X}\|_*$ is replaced by $\|\mathbf{X}\|_1$.

### 7.1.4   Solving Subproblems Inexactly

In the previous chapters, we have assumed that the subproblems for updating the primal variables are all easily solvable. For example, the proximal mappings of $f$ and $g$ have closed-form solutions, or $f$ and $g$ are $L$-smooth functions so that they can be linearized. If neither conditions can be met, we may have to solve the subproblems iteratively. Since we can only run the iterations for solving the subproblems in finite time and obtain approximate solutions, there is an issue of when to terminate the iterations. There have been some results showing that for several ADMMs (e.g., [2, 6, 9, 12]), as long as the errors in solving the subproblems are well controlled, e.g.,

$$\sum_{k=0}^{\infty} \epsilon_k < \infty,$$

where $\epsilon_k$ is the error (absolute error or relative error, whose exact definition may vary with different algorithms of solving the subproblems) in the $k$-th iteration, the resulting ADMMs can still converge to the solution of the original problem. This indicates that the subproblems need not be solved at high precision at the beginning, thus can save some time.

### 7.1.5   Other Considerations

There are other issues in real implementations, such as the initialization and the order of updating the primal variables. However, in reality they do not affect the convergence speed very much. So we need not discuss them in detail. Nonetheless, initializing with a good guess on the solution, if some prior information is available, definitely helps. Moreover, some convergence guarantees impose asymmetric conditions on $\mathbf{x}$ and $\mathbf{y}$, e.g., the full row rankness of $\mathbf{B}$ and different convexity or smoothness conditions on $g$ (e.g., see some theorems in Chap. 3). In this case, we should pay attention to which primal variable is $\mathbf{x}$ and which is $\mathbf{y}$ in order to meet the conditions.

## 7.2 Conclusions

In this book, we have introduced many variants of ADMM for various scenarios: deterministic and convex optimization, deterministic and nonconvex optimization, stochastic optimization, and distributed optimization. There has been abundant literature on ADMM, studying various aspects of ADMM. Regretfully, what we have introduced here is only the tip of an iceberg, as we organize the materials in the types of problems to solve, want to present details of proofs, and also attach practical values to the chosen algorithms.

ADMM works beautifully in the philosophy of "divide and conquer." It is easily implementable for solving real problems if one masters the trick of introducing the auxiliary variables to decouple the target problem, the augmented Lagrangian function, and the linearization technique. However, ADMM can be very slow if the penalty parameter is not chosen appropriately. For some particular problems, such as the linear program, ADMM can be extremely slow, despite its linear convergence in theory (see, e.g., Theorem 3.5 and Lemma A.3). So ADMM is a good choice if high precision is not required. This is one of the major reasons for its popularity in the machine learning community nowadays.

## References

1. S. Boyd, N. Parikh, E. Chu, B. Peleato, J. Eckstein, Distributed optimization and statistical learning via the alternating direction method of multipliers. Found. Trends Mach. Learn. **3**(1), 1–122 (2011)
2. J. Eckstein, D.P. Bertsekas, On the Douglas-Rachford splitting method and the proximal point algorithm for maximal monotone operators. Math. Program. **55**(1), 293–318 (1992)
3. R. Glowinski, *Numerical Methods for Nonlinear Variational Problems* (Springer, Berlin, 1984)
4. B. He, H. Yang, Some convergence properties of a method of multipliers for linearly constrained monotone variational inequalities. Oper. Res. Lett. **23**(3-5), 151–161 (1998)
5. B. He, H. Yang, S. Wang, Alternating direction method with self-adaptive penalty parameters for monotone variational inequalities. J. Optim. Theory Appl. **106**(2), 337–356 (2000)
6. B. He, L.-Z. Liao, D. Han, H. Yang, A new inexact alternating directions method for monotone variational inequalities. Math. Program. **92**(1), 103–118 (2002)

7. Z. Lin, R. Liu, Z. Su, Linearized alternating direction method with adaptive penalty for low-rank representation, in *Advances in Neural Information Processing Systems* (2011), pp. 612–620

8. Z. Lin, R. Liu, H. Li, Linearized alternating direction method with parallel splitting and adaptive penalty for separable convex programs in machine learning. Mach. Learn. **99**(2), 287–325 (2015)

9. M.K. Ng, F. Wang, X. Yuan, Inexact alternating direction methods for image recovery. SIAM J. Sci. Comput. **33**(4), 1643–1668 (2011)

10. M. Tao, X. Yuan, On Glowinski's open question on the alternating direction method of multipliers. J. Optim. Theory Appl. **179**(1), 163–196 (2018)

11. W. Tian, X. Yuan, An alternating direction method of multipliers with a worst-case $O(1/n^2)$ convergence rate. Math. Comput. **88**(318), 1685–1713 (2019)

12. W. Yao, Approximate Versions of the Alternating Direction Method of Multipliers. Ph.D. Thesis. The State University of New Jersey (2016)

# Appendix A
# Mathematical Preliminaries

In this appendix, we list the conventions of notations and some basic definitions and facts that are used in the book.

## A.1 Notations

| Notations | Meanings |
|---|---|
| Normal font, e.g., $s$ | A scalar. |
| Bold lowercase, e.g., $\mathbf{v}$ | A vector. |
| Bold capital, e.g., $\mathbf{M}$ | A matrix. |
| Calligraphic capital, e.g., $\mathcal{T}$ | A subspace, an operator, or a set. |
| $\mathbb{R}, \mathbb{Z}^+$ | Set of real numbers, set of non-negative integers. |
| $[n]$ | $\{1, 2, \cdots, n\}$. |
| $\mathbb{E}X$ | Expectation of random variable (or random vector) $X$. |
| $\mathbf{I}, \mathbf{0}, \mathbf{1}$ | The identity matrix, all-zero matrix or vector, and all-one vector. |
| $\mathbf{x} \geq \mathbf{y}$ | $\mathbf{x} - \mathbf{y}$ is a non-negative vector. |
| $\mathbf{X} \succeq \mathbf{Y}$ | $\mathbf{X} - \mathbf{Y}$ is a positive semidefinite matrix. |
| $f(N) = O(g(N))$ | $\exists a > 0$, such that $\frac{f(N)}{g(N)} \leq a$ for all $N \in \mathbb{Z}^+$. |
| $f(N) = \tilde{O}(g(N))$ | $\exists a > 0$, such that $\frac{\tilde{f}(N)}{g(N)} \leq a$ for all $N \in \mathbb{Z}^+$, where $\tilde{f}(N)$ is |

Z. Lin et al., *Alternating Direction Method of Multipliers for Machine Learning*, https://doi.org/10.1007/978-981-16-9840-8

| Notations | Meanings |
|---|---|
| | the function ignoring poly-logarithmic factors in $f(N)$. |
| $f(N) = \Omega(g(N))$ | $\exists a > 0$, such that $\frac{f(N)}{g(N)} \geq a$ for all $N \in \mathbb{Z}^+$. |
| $f(N) = \Theta(g(N))$ | $\exists b \geq a > 0$, such that $a \leq \frac{f(N)}{g(N)} \leq b$ for all $N \in \mathbb{Z}^+$. |
| $\nabla f(\mathbf{x})$ | Gradient of $f$ at $\mathbf{x}$. |
| $\nabla_i f(\mathbf{x})$ | $\frac{\partial f}{\partial \mathbf{x}_i}$. |
| $\mathbf{X}^T$ | Transpose of matrix $\mathbf{X}$. |
| $\text{Diag}(\mathbf{x})$ | Diagonal matrix whose diagonal entries are entries of vector $\mathbf{x}$. |
| $|\mathcal{X}|$ | Cardinality of $\mathcal{X}$. |
| $\text{Span}(\mathbf{X})$ | The subspace spanned by the columns of $\mathbf{X}$. |
| $\langle \cdot, \cdot \rangle$ | Inner product. For vectors, $\langle \mathbf{x}, \mathbf{y} \rangle = \mathbf{x}^T \mathbf{y}$; for matrices, $\langle \mathbf{A}, \mathbf{B} \rangle = \text{tr}(\mathbf{A}^T \mathbf{B})$. |
| $\|\mathbf{x}\|_{\mathbf{M}}$ | $\sqrt{\mathbf{x}^T \mathbf{M} \mathbf{x}}$, where $\mathbf{M} \succeq \mathbf{0}$. |
| $\| \cdot \|_2$ | Spectral norm of a matrix. |
| $\| \cdot \|$ | $\ell_2$ norm of a vector or the Frobenius norm of a matrix, $\|\mathbf{v}\| = \sqrt{\sum_i \mathbf{v}_i^2}$, $\|\mathbf{X}\| = \sqrt{\sum_{i,j} \mathbf{X}_{ij}^2}$. |
| $\| \cdot \|_*$ | Nuclear norm of a matrix, the sum of singular values. |
| $\| \cdot \|_0$ | $\ell_0$ pseudo-norm, number of nonzero entries. |
| $\| \cdot \|_1$ | $\ell_1$ norm, $\|\mathbf{x}\|_1 = \sum_i |\mathbf{x}_i|$, $\|\mathbf{X}\|_1 = \sum_{i,j} |\mathbf{X}_{ij}|$. |
| $\text{conv}(\mathcal{X})$ | Convex hull of set $\mathcal{X}$. |
| $\partial f$ | Subgradient (resp. supergradient) of a convex (resp. concave) function $f$, or the limiting subdifferential of a proper and lower semicontinuous function. |
| $f^*$ | Optimum value of $f(\mathbf{x})$, where $\mathbf{x}$ varies in $\text{dom} f$ and the constraints. |
| $f^*(\mathbf{x})$ | The conjugate function of $f(\mathbf{x})$. |
| $\text{Prox}_{\alpha f}(\cdot)$ | Proximal mapping w.r.t. $f$ and parameter $\alpha$, $\text{Prox}_{\alpha f}(\mathbf{y}) = \text{argmin}_{\mathbf{x}} \left( \alpha f(\mathbf{x}) + \frac{1}{2} \|\mathbf{x} - \mathbf{y}\|^2 \right)$. |
| $D_\phi(\mathbf{y}, \mathbf{x})$ | Bregman distance between $\mathbf{y}$ and $\mathbf{x}$ w.r.t. convex $\phi$, $D_\phi(\mathbf{y}, \mathbf{x}) = \phi(\mathbf{y}) - \phi(\mathbf{x}) - \langle \nabla \phi(\mathbf{x}), \mathbf{y} - \mathbf{x} \rangle$. |

## A.2   Algebra and Probability

**Proposition A.1 (Cauchy–Schwartz Inequality)** *For any* $\mathbf{x}, \mathbf{y} \in \mathbb{R}^n$, *we have*

$$\langle \mathbf{x}, \mathbf{y} \rangle \leq \|\mathbf{x}\| \|\mathbf{y}\|.$$

**Lemma A.1**  *For any* $\mathbf{x}, \mathbf{y}, \mathbf{z}$, *and* $\mathbf{w} \in \mathbb{R}^n$, *we have the following three identities:*

$$\langle \mathbf{x}, \mathbf{y} \rangle = \frac{1}{2} \left( \|\mathbf{x}\|^2 + \|\mathbf{y}\|^2 - \|\mathbf{x} - \mathbf{y}\|^2 \right), \tag{A.1}$$

$$\langle \mathbf{x}, \mathbf{y} \rangle = \frac{1}{2} \left( \|\mathbf{x} + \mathbf{y}\|^2 - \|\mathbf{x}\|^2 - \|\mathbf{y}\|^2 \right), \tag{A.2}$$

$$\langle \mathbf{x} - \mathbf{z}, \mathbf{y} - \mathbf{w} \rangle = \frac{1}{2} \left( \|\mathbf{x} - \mathbf{w}\|^2 - \|\mathbf{z} - \mathbf{w}\|^2 - \|\mathbf{x} - \mathbf{y}\|^2 + \|\mathbf{z} - \mathbf{y}\|^2 \right). \tag{A.3}$$

**Definition A.1 (Singular Value Decomposition (SVD))**  Suppose that $\mathbf{A} \in \mathbb{R}^{m \times n}$ with $\mathrm{rank}\mathbf{A} = r$. Then $\mathbf{A}$ can be factorized as

$$\mathbf{A} = \mathbf{U} \boldsymbol{\Sigma} \mathbf{V}^T,$$

where $\mathbf{U} \in \mathbb{R}^{m \times r}$ satisfies $\mathbf{U}^T \mathbf{U} = \mathbf{I}$, $\mathbf{V} \in \mathbb{R}^{n \times r}$ satisfies $\mathbf{V}^T \mathbf{V} = \mathbf{I}$, and

$$\boldsymbol{\Sigma} = \mathrm{Diag}(\sigma_1, \cdots, \sigma_r) \quad \text{with} \quad \sigma_1 \geq \sigma_2 \geq \cdots \geq \sigma_r > 0.$$

The above factorization is called the economical singular value decomposition (SVD) of $\mathbf{A}$. The columns of $\mathbf{U}$ are called left singular vectors of $\mathbf{A}$, the columns of $\mathbf{V}$ are right singular vectors, and the numbers $\sigma_i$ are the singular values.

**Definition A.2 (Laplacian Matrix of a Graph)**  Denote a graph as $\mathfrak{g} = \{\mathcal{V}, \mathcal{E}\}$, where $\mathcal{V}$ and $\mathcal{E}$ are the node and the edge sets, respectively. $e_{ij} = (i, j) \in \mathcal{E}$ indicates that nodes $i$ and $j$ are connected. Define $\mathcal{V}_i = \{j \in \mathcal{V} | (i, j) \in \mathcal{E}\}$ to be the neighborhood of node $i$, i.e., the index set of the nodes that are connected to node $i$. The Laplacian matrix $\mathbf{L}$ of the graph $\mathfrak{g} = \{\mathcal{V}, \mathcal{E}\}$ is defined as

$$\mathbf{L}_{ij} = \begin{cases} |\mathcal{V}_i|, & \text{if } i = j, \\ -1, & \text{if } i \neq j \text{ and } (i, j) \in \mathcal{E}, \\ 0, & \text{otherwise.} \end{cases}$$

**Proposition A.2 (Properties of Laplacian Matrix)**  *A Laplacian matrix* $\mathbf{L}$ *of a graph with n nodes has the following properties:*

1. $\mathbf{L} \succeq \mathbf{0}$.
2. $\mathrm{rank}(\mathbf{L}) = n - c$, *where c is the number of connected components in the graph, and the eigenvector associated to 0 is* $\mathbf{1}_n$.

**Proposition A.3**  *Given random vector* $\boldsymbol{\xi}$, *we have*

$$\mathbb{E} \|\boldsymbol{\xi} - \mathbb{E} \boldsymbol{\xi}\|^2 \leq \mathbb{E} \|\boldsymbol{\xi}\|^2.$$

**Proposition A.4 (Jensen's Inequality: Continuous Case)** *If $f : C \subseteq \mathbb{R}^n \to \mathbb{R}$ is convex and $\boldsymbol{\xi}$ is a random vector over $C$, then*

$$f(\mathbb{E}\boldsymbol{\xi}) \leq \mathbb{E}f(\boldsymbol{\xi}).$$

## A.3   Convex Analysis

The descriptions for the basic concepts of convex sets and convex functions can be found in [1]. We only consider convex analysis on $n$ dimensional Euclidean spaces.

**Definition A.3 (Convex Set)**   A set $C \subseteq \mathbb{R}^n$ is called convex if for all $\mathbf{x}, \mathbf{y} \in C$ and $\alpha \in [0, 1]$ we have $\alpha \mathbf{x} + (1 - \alpha)\mathbf{y} \in C$.

**Definition A.4 (Convex Function)**   A function $f : C \subseteq \mathbb{R}^n \to \mathbb{R}$ is called convex if $C$ is a convex set and for all $\mathbf{x}, \mathbf{y} \in C$ and $\alpha \in [0, 1]$ we have

$$f(\alpha \mathbf{x} + (1 - \alpha)\mathbf{y}) \leq \alpha f(\mathbf{x}) + (1 - \alpha)f(\mathbf{y}).$$

$C$ is called the domain of $f$.

**Definition A.5 (Concave Function)**   A function $f : C \subseteq \mathbb{R}^n \to \mathbb{R}$ is called concave if $-f$ is convex.

**Definition A.6 (Strictly Convex Function)**   A function $f : C \subseteq \mathbb{R}^n \to \mathbb{R}$ is called strictly convex if $C$ is a convex set and for all $\mathbf{x} \neq \mathbf{y} \in C$ and $\alpha \in (0, 1)$ we have

$$f(\alpha \mathbf{x} + (1 - \alpha)\mathbf{y}) < \alpha f(\mathbf{x}) + (1 - \alpha)f(\mathbf{y}).$$

**Definition A.7 (Strongly Convex Function and Generally Convex Function)**   A function $f : C \subseteq \mathbb{R}^n \to \mathbb{R}$ is called strongly convex if $C$ is a convex set and there exists a constant $\mu > 0$ such that for all $\mathbf{x}, \mathbf{y} \in C$ and $\alpha \in [0, 1]$ we have

$$f(\alpha \mathbf{x} + (1 - \alpha)\mathbf{y}) \leq \alpha f(\mathbf{x}) + (1 - \alpha)f(\mathbf{y}) - \frac{\mu \alpha(1 - \alpha)}{2}\|\mathbf{y} - \mathbf{x}\|^2.$$

$\mu$ is called the strong convexity modulus of $f$. For brevity, a strongly convex function with a strong convexity modulus $\mu$ is called a $\mu$-strongly convex function. If a convex function is not strongly convex, we also call it a generally convex function.

**Proposition A.5 (Jensen's Inequality: Discrete Case)** *If $f : C \subseteq \mathbb{R}^n \to \mathbb{R}$ is convex, $\mathbf{x}_i \in C$, $\alpha_i \geq 0$, $i \in [m]$, and $\sum_{i=1}^{m} \alpha_i = 1$, then*

$$f\left(\sum_{i=1}^{m} \alpha_i \mathbf{x}_i\right) \leq \sum_{i=1}^{m} \alpha_i f(\mathbf{x}_i).$$

**Definition A.8 (Smooth Function)** A function is (informally) called smooth if it is continuously differentiable.

**Definition A.9 (Function with Lipschitz Continuous Gradients)** A differentiable function $f : C \subseteq \mathbb{R}^n \to \mathbb{R}$ is called to have Lipschitz continuous gradients if there exists $L > 0$ such that

$$\|\nabla f(\mathbf{x}) - \nabla f(\mathbf{y})\| \le L\|\mathbf{y} - \mathbf{x}\|, \quad \forall \mathbf{x}, \mathbf{y} \in C.$$

For simplicity, if the constant $L$ is explicitly specified we also call such a function an $L$-smooth function.

**Proposition A.6 ([3])** *If $f : C \subseteq \mathbb{R}^n \to \mathbb{R}$ is $L$-smooth, then*

$$|f(\mathbf{y}) - f(\mathbf{x}) - \langle \nabla f(\mathbf{x}), \mathbf{y} - \mathbf{x} \rangle| \le \frac{L}{2}\|\mathbf{y} - \mathbf{x}\|^2, \quad \forall \mathbf{x}, \mathbf{y} \in C. \tag{A.4}$$

*If $f$ is both $L$-smooth and convex, then*

$$f(\mathbf{y}) \ge f(\mathbf{x}) + \langle \nabla f(\mathbf{x}), \mathbf{y} - \mathbf{x} \rangle + \frac{1}{2L}\|\nabla f(\mathbf{y}) - \nabla f(\mathbf{x})\|^2. \tag{A.5}$$

**Definition A.10 (Subgradient of a Convex Function)** A vector $\mathbf{g}$ is called a subgradient of a convex function $f : C \subseteq \mathbb{R}^n \to \mathbb{R}$ at $\mathbf{x} \in C$ if

$$f(\mathbf{y}) \ge f(\mathbf{x}) + \langle \mathbf{g}, \mathbf{y} - \mathbf{x} \rangle, \forall \mathbf{y} \in C.$$

The set of subgradients at $\mathbf{x}$ is denoted as $\partial f(\mathbf{x})$.

**Proposition A.7** *For convex function $f : C \subseteq \mathbb{R}^n \to \mathbb{R}$, its subgradient exists at every interior point of $C$. It is differentiable at $\mathbf{x}$ iff (aka if and only if) $\partial f(\mathbf{x})$ is a singleton.*

**Proposition A.8** *If $f : \mathbb{R}^n \to \mathbb{R}$ is $\mu$-strongly convex, then*

$$f(\mathbf{y}) \ge f(\mathbf{x}) + \langle \mathbf{g}, \mathbf{y} - \mathbf{x} \rangle + \frac{\mu}{2}\|\mathbf{y} - \mathbf{x}\|^2, \quad \forall \mathbf{g} \in \partial f(\mathbf{x}). \tag{A.6}$$

*In particular, if $f$ is $\mu$-strongly convex and $\mathbf{x}^* = \operatorname{argmin}_{\mathbf{x}} f(\mathbf{x})$, then*

$$f(\mathbf{x}) - f(\mathbf{x}^*) \ge \frac{\mu}{2}\|\mathbf{x} - \mathbf{x}^*\|^2. \tag{A.7}$$

*On the other hand, if $f$ is differentiable and $\mu$-strongly convex, we can have*

$$f(\mathbf{x}^*) \ge f(\mathbf{x}) - \frac{1}{2\mu}\|\nabla f(\mathbf{x})\|^2.$$

*We can further have*

$$\langle \nabla f(\mathbf{x}) - \nabla f(\mathbf{y}), \mathbf{x} - \mathbf{y} \rangle \geq \mu \|\mathbf{x} - \mathbf{y}\|^2. \tag{A.8}$$

*In particular,*

$$\|\nabla f(\mathbf{x}) - \nabla f(\mathbf{y})\| \geq \mu \|\mathbf{x} - \mathbf{y}\|. \tag{A.9}$$

**Definition A.11 (Epigraph)**  The epigraph of $f : C \subseteq \mathbb{R}^n \rightarrow \mathbb{R}$ is defined as

$$\text{epi } f = \{(\mathbf{x}, t) | \mathbf{x} \in C, t \geq f(\mathbf{x})\}.$$

**Definition A.12 (Closed Function)**  If epi $f$ is a closed set, then $f$ is called a closed function.

**Definition A.13 (Monotone Operator and Monotone Function)**  A set-valued mapping $f : C \subseteq \mathbb{R}^n \rightarrow 2^{\mathbb{R}^n}$ (also denoted as $f : C \subseteq \mathbb{R}^n \rightrightarrows \mathbb{R}^n$ for brevity) is called a monotone operator if

$$\langle \mathbf{x} - \mathbf{y}, \mathbf{u} - \mathbf{v} \rangle \geq 0, \quad \forall \mathbf{x}, \mathbf{y} \in C \text{ and } \mathbf{u} \in f(\mathbf{x}), \mathbf{v} \in f(\mathbf{y}).$$

In particular, if $f$ is single-valued and

$$\langle \mathbf{x} - \mathbf{y}, f(\mathbf{x}) - f(\mathbf{y}) \rangle \geq 0, \quad \forall \mathbf{x}, \mathbf{y} \in C,$$

then it is called a monotone function.

**Definition A.14 (Maximal Monotone Operator)**  Define the graph of an operator $\mathcal{T}$ as

$$\text{Graph}(\mathcal{T}) = \{(\mathbf{x}, \mathbf{u}) | \mathbf{x} \in C, \mathbf{u} \in \mathcal{T}(\mathbf{x})\}.$$

For a monotone operator $\mathcal{T}$, if it has the property: for any monotone operator $\mathcal{T}'$, $\text{Graph}(\mathcal{T}) \subseteq \text{Graph}(\mathcal{T}')$ implies $\mathcal{T} = \mathcal{T}'$, then it is called a maximal monotone operator.

**Proposition A.9**  *If $\mathcal{T}$ is a maximal monotone operator, then its resolvent $(\mathcal{I} + \mathcal{T})^{-1}$ is single-valued.*

**Proposition A.10 (Monotonicity of Subgradient)**  *If $f : C \subseteq \mathbb{R}^n \rightarrow \mathbb{R}$ is convex, then $\partial f(\mathbf{x})$ is a monotone operator. If $f$ is further $\mu$-strongly convex, then*

$$\langle \mathbf{x}_1 - \mathbf{x}_2, \mathbf{g}_1 - \mathbf{g}_2 \rangle \geq \mu \|\mathbf{x}_1 - \mathbf{x}_2\|^2, \quad \forall \mathbf{x}_i \in C \text{ and } \mathbf{g}_i \in \partial f(\mathbf{x}_i), i = 1, 2.$$

*If $f$ is closed and convex, then $\partial f(\mathbf{x})$ is a maximal monotone operator.*

**Definition A.15 (Bregman Distance)** Given a differentiable convex function $\phi$, the associated Bregman distance is defined as

$$D_\phi(\mathbf{y}, \mathbf{x}) = \phi(\mathbf{y}) - \phi(\mathbf{x}) - \langle \nabla\phi(\mathbf{x}), \mathbf{y} - \mathbf{x} \rangle .$$

If $\phi$ is convex but not differentiable, then the associated Bregman distance is defined as

$$D_\phi^{\mathbf{v}}(\mathbf{y}, \mathbf{x}) = \phi(\mathbf{y}) - \phi(\mathbf{x}) - \langle \mathbf{v}, \mathbf{y} - \mathbf{x} \rangle ,$$

where $\mathbf{v}$ is a particular subgradient in $\partial\phi(\mathbf{x})$.

The squared Euclidean distance is obtained when $\phi(\mathbf{x}) = \frac{1}{2}\|\mathbf{x}\|^2$, in which case $D_\phi(\mathbf{y}, \mathbf{x}) = \frac{1}{2}\|\mathbf{x} - \mathbf{y}\|^2$.

**Lemma A.2** *The Bregman distance $D_\phi$ has the following properties:*

1. *When $\phi$ is $\mu$-strongly convex, $D_\phi(\mathbf{y}, \mathbf{x}) \geq \frac{\mu}{2}\|\mathbf{y} - \mathbf{x}\|^2$.*
2. *$\langle \nabla\phi(\mathbf{u}) - \nabla\phi(\mathbf{v}), \mathbf{w} - \mathbf{u} \rangle = D_\phi(\mathbf{w}, \mathbf{v}) - D_\phi(\mathbf{w}, \mathbf{u}) - D_\phi(\mathbf{u}, \mathbf{v})$, for any $\mathbf{u}$, $\mathbf{v}$, and $\mathbf{w}$.*

**Definition A.16 (Conjugate Function)** Given $f : C \subseteq \mathbb{R}^n \to \mathbb{R}$, its conjugate function is defined as

$$f^*(\mathbf{u}) = \sup_{\mathbf{z} \in C} (\langle \mathbf{z}, \mathbf{u} \rangle - f(\mathbf{z})) .$$

The domain of $f^*$ is

$$\operatorname{dom} f^* = \{\mathbf{u} \,|\, f^*(\mathbf{u}) < +\infty\}.$$

**Proposition A.11 (Properties of Conjugate Function)** *Given $f : C \subseteq \mathbb{R}^n \to \mathbb{R}$, its conjugate function $f^*$ has the following properties:*

1. *$f^*$ is always a convex function.*
2. *$f^{**}(\mathbf{x}) \leq f(\mathbf{x}), \forall \mathbf{x} \in C$.*
3. *If $f$ is a closed and convex function, then $f^{**}(\mathbf{x}) = f(\mathbf{x}), \forall \mathbf{x} \in C$.*
4. *If $f$ is $L$-smooth, then $f^*$ is $L^{-1}$-strongly convex on $\operatorname{dom} f^*$. Conversely, if $f$ is $\mu$-strongly convex, then $f^*$ is $\mu^{-1}$-smooth on $\operatorname{dom} f^*$.*
5. *If $f$ is closed and convex, then $\mathbf{y} \in \partial f(\mathbf{x})$ if and only if $\mathbf{x} \in \partial f^*(\mathbf{y})$.*

**Proposition A.12 (Fenchel-Young Inequality)** *Let $f^*$ be the conjugate function of $f$, then*

$$f(\mathbf{x}) + f^*(\mathbf{y}) \geq \langle \mathbf{x}, \mathbf{y} \rangle .$$

**Definition A.17 (Lagrangian Function)** Given a constrained problem:

$$\min_{\mathbf{x}\in\mathbb{R}^n} f(\mathbf{x}),$$

$$s.t. \quad \mathbf{A}\mathbf{x} = \mathbf{b},$$

$$\mathbf{g}(\mathbf{x}) \le \mathbf{0}, \tag{A.10}$$

where $\mathbf{A} \in \mathbb{R}^{m \times n}$ and $\mathbf{g}(\mathbf{x}) = (g_1(\mathbf{x}), \cdots, g_p(\mathbf{x}))^T$, the Lagrangian function is

$$L(\mathbf{x}, \mathbf{u}, \mathbf{v}) = f(\mathbf{x}) + \langle \mathbf{u}, \mathbf{A}\mathbf{x} - \mathbf{b} \rangle + \langle \mathbf{v}, \mathbf{g}(\mathbf{x}) \rangle,$$

where $\mathbf{v} \ge \mathbf{0}$.

**Definition A.18 (Lagrange Dual Function)** Given a constrained problem (A.10), the Lagrange dual function is

$$d(\mathbf{u}, \mathbf{v}) = \min_{\mathbf{x}\in C} L(\mathbf{x}, \mathbf{u}, \mathbf{v}),$$

where $C$ is the intersection of the domains of $f$ and $g$. The domain of the dual function is $\mathcal{D} = \{(\mathbf{u}, \mathbf{v}) | \mathbf{v} \ge \mathbf{0}, d(\mathbf{u}, \mathbf{v}) > -\infty\}$.

**Definition A.19 (Dual Problem)** Given a constrained problem (A.10), the dual problem is

$$\max_{\mathbf{u},\mathbf{v}} d(\mathbf{u}, \mathbf{v}), \quad s.t. \quad (\mathbf{u}, \mathbf{v}) \in \mathcal{D},$$

where $\mathcal{D}$ is the domain of $d(\mathbf{u}, \mathbf{v})$. Accordingly, Problem (A.10) is called the primal problem.

**Definition A.20 (Slater's Condition)** For convex primal problem (A.10), if there exists an $\mathbf{x}_0$ such that

$$\mathbf{A}\mathbf{x}_0 = \mathbf{b}, \quad g_i(\mathbf{x}_0) \le 0, \ i \in \mathcal{I}_1, \quad \text{and} \quad g_i(\mathbf{x}_0) < 0, \ i \in \mathcal{I}_2,$$

where $\mathcal{I}_1$ and $\mathcal{I}_2$ are the sets of indices of linear and nonlinear inequality constraints, respectively, then the Slater's condition holds.

**Proposition A.13 (Properties of Dual Problem)**

1. $d(\mathbf{u}, \mathbf{v})$ *is always a concave function, even if the primal problem (A.10) is not convex.*
2. *The primal and the dual optimal values, $f^*$ and $d^*$, always satisfy the weak duality: $f^* \ge d^*$.*
3. *When the Slater's condition holds, the strong duality holds: $f^* = d^*$.*
4. *Let $\mathbf{x}(\mathbf{u}, \mathbf{v}) \in \underset{\mathbf{x}\in C}{\text{Argmin}}\, L(\mathbf{x}, \mathbf{u}, \mathbf{v})$, then $(\mathbf{A}\mathbf{x}(\mathbf{u}, \mathbf{v}) - \mathbf{b}, \mathbf{g}(\mathbf{x}(\mathbf{u}, \mathbf{v}))) \in \partial d(\mathbf{u}, \mathbf{v})$.*

**Definition A.21 (KKT Point and KKT Condition)** $(\mathbf{x}, \mathbf{u}, \mathbf{v})$ is called a Karush–Kuhn–Tucker (KKT) point of Problem (A.10) if

1. Stationarity: $\mathbf{0} \in \partial f(\mathbf{x}) + \mathbf{A}^T \mathbf{u} + \sum_{i=1}^{p} \mathbf{v}_i \partial g_i(\mathbf{x})$.
2. Primal feasibility: $\mathbf{A}\mathbf{x} = \mathbf{b}$, $g_i(\mathbf{x}) \leq 0, i \in [p]$.
3. Complementary slackness: $\mathbf{v}_i g_i(\mathbf{x}) = 0, i \in [p]$.
4. Dual feasibility: $\mathbf{v}_i \geq \mathbf{0}, i \in [p]$.

The above conditions are called the KKT condition of Problem (A.10). They are the optimality condition of Problem (A.10) when Problem (A.10) is convex and satisfies the Slater's condition.

**Proposition A.14** *When $f(\mathbf{x})$ and $g_i(\mathbf{x})$, $i \in [p]$, in Problem (A.10) are all convex,*

1. *Every KKT point is a saddle point of the Lagrangian function.*
2. *$(\mathbf{x}^*, \mathbf{u}^*, \mathbf{v}^*)$ is a pair of the primal and the dual solutions with zero dual gap iff it satisfies the KKT condition.*

**Definition A.22 (Compact Set)** A subset $S$ of $\mathbb{R}^n$ is called compact if it is both bounded and closed.

**Definition A.23 (Convex Hull)** The convex hull of a set $\mathcal{X}$, denoted as $\text{conv}(\mathcal{X})$, is the set of all convex combinations of points in $\mathcal{X}$:

$$\text{conv}(\mathcal{X}) = \left\{ \sum_{i=1}^{k} \alpha_i \mathbf{x}_i \,\middle|\, \mathbf{x}_i \in \mathcal{X}, \alpha_i \geq 0, i \in [k], \sum_{i=1}^{k} \alpha_i = 1 \right\}.$$

**Theorem A.1 (Danskin's Theorem)** *Let $\mathcal{Z}$ be a compact subset of $\mathbb{R}^m$, and let $\phi : \mathbb{R}^n \times \mathcal{Z} \to \mathbb{R}$ be continuous and such that $\phi(\cdot, \mathbf{z}) : \mathbb{R}^n \to \mathbb{R}$ is convex for each $\mathbf{z} \in \mathcal{Z}$. Define $f : \mathbb{R}^n \to \mathbb{R}$ by $f(\mathbf{x}) = \max_{\mathbf{z} \in \mathcal{Z}} \phi(\mathbf{x}, \mathbf{z})$ and*

$$\mathcal{Z}(\mathbf{x}) = \left\{ \bar{\mathbf{z}} \,\middle|\, \phi(\mathbf{x}, \bar{\mathbf{z}}) = \max_{\mathbf{z} \in \mathcal{Z}} \phi(\mathbf{x}, \mathbf{z}) \right\}.$$

*If $\phi(\cdot, \mathbf{z})$ is differentiable for all $\mathbf{z} \in \mathcal{Z}$ and $\nabla_x \phi(\mathbf{x}, \cdot)$ is continuous on $\mathcal{Z}$ for each $\mathbf{x}$, then*

$$\partial f(\mathbf{x}) = \text{conv}\left\{ \nabla_x \phi(\mathbf{x}, \mathbf{z}) | \mathbf{z} \in \mathcal{Z}(\mathbf{x}) \right\}, \quad \forall \mathbf{x} \in \mathbb{R}^n.$$

**Definition A.24 (Saddle Point)** $(\mathbf{x}^*, \boldsymbol{\lambda}^*)$ is called a saddle point of function $f(\mathbf{x}, \boldsymbol{\lambda}) : C \times D \to \mathbb{R}$ if it satisfies the following inequalities:

$$f(\mathbf{x}^*, \boldsymbol{\lambda}) \leq f(\mathbf{x}^*, \boldsymbol{\lambda}^*) \leq f(\mathbf{x}, \boldsymbol{\lambda}^*), \quad \forall \mathbf{x} \in C, \boldsymbol{\lambda} \in D.$$

**Lemma A.3 (Hoffman's Bound [2])** *Consider the non-empty polyhedron*

$$\mathcal{X} = \{\mathbf{x} | \mathbf{Ax} = \mathbf{a}, \mathbf{Bx} \le \mathbf{b}\}.$$

*Then there exists a constant $\theta$, depending only on $[\mathbf{A}^T, \mathbf{B}^T]^T$, such that for any* $\mathbf{x}$
*we have*

$$\text{dist}(\mathbf{x}, \mathcal{X})^2 \le \theta^2 (\|\mathbf{Ax} - \mathbf{a}\|^2 + \|[\mathbf{Bx} - \mathbf{b}]_+\|^2),$$

*where $[\cdot]_+$ means the projection to the non-negative orthant.*

## A.4   Nonconvex Analysis

**Definition A.25 (Proper Function)** A function $g : \mathbb{R}^n \to (-\infty, +\infty]$ is said to
be proper if dom $g \ne \varnothing$, where dom $g = \{\mathbf{x} \in \mathbb{R}^n : g(\mathbf{x}) < +\infty\}$.

We only consider proper functions in this book.

**Definition A.26 (Lower Semicontinuous Function)** A function $g : \mathbb{R}^n \to$
$(-\infty, +\infty]$ is said to be lower semicontinuous at point $\mathbf{x}_0$ if

$$\liminf_{\mathbf{x} \to \mathbf{x}_0} g(\mathbf{x}) \ge g(\mathbf{x}_0).$$

**Definition A.27 (Coercive Function)**   $f(\mathbf{x})$ is called coercive if $\lim_{\|\mathbf{x}\| \to \infty}$
$f(\mathbf{x}) = \infty$.

**Definition A.28 (Subdifferential)** Let $f$ be a proper and lower semicontinuous
function.

1. For a given $\mathbf{x} \in$ dom $f$, the Fréchet subdifferential of $f$ at $\mathbf{x}$, written as $\hat{\partial} f(\mathbf{x})$, is
   the set of all vectors $\mathbf{u} \in \mathbb{R}^n$, which satisfies

$$\liminf_{\mathbf{y} \ne \mathbf{x}, \mathbf{y} \to \mathbf{x}} \frac{f(\mathbf{y}) - f(\mathbf{x}) - \langle \mathbf{u}, \mathbf{y} - \mathbf{x} \rangle}{\|\mathbf{y} - \mathbf{x}\|} \ge 0.$$

2. The limiting subdifferential, or simply the subdifferential, of $f$ at $\mathbf{x} \in \mathbb{R}^n$, written
   as $\partial f(\mathbf{x})$, is defined through the following closure process:

$$\partial f(\mathbf{x}) = \left\{ \mathbf{u} \in \mathbb{R}^n : \exists \mathbf{x}_k \to \mathbf{x}, f(\mathbf{x}_k) \to f(\mathbf{x}), \mathbf{u}_k \in \hat{\partial} f(\mathbf{x}_k) \to \mathbf{u}, k \to \infty \right\}.$$

**Definition A.29 (Critical Point)** A point $\mathbf{x}$ is called a critical point of function $f$
if $\mathbf{0} \in \partial f(\mathbf{x})$.

The following lemma describes the properties of subdifferential.

**Lemma A.4**

1. *In the nonconvex context, Fermat's rule remains unchanged: If $\mathbf{x} \in \mathbb{R}^n$ is a local minimizer of $g$, then $\mathbf{0} \in \partial g(\mathbf{x})$.*
2. *Let $(\mathbf{x}_k, \mathbf{u}_k)$ be a sequence such that $\mathbf{x}_k \rightarrow \mathbf{x}$, $\mathbf{u}_k \rightarrow \mathbf{u}$, $g(\mathbf{x}_k) \rightarrow g(\mathbf{x})$, and $\mathbf{u}_k \in \partial g(\mathbf{x}_k)$, then $\mathbf{u} \in \partial g(\mathbf{x})$.*
3. *If $f$ is a continuously differentiable function, then $\partial (f + g)(\mathbf{x}) = \nabla f(\mathbf{x}) + \partial g(\mathbf{x})$.*

# References

1. S. Boyd, L. Vandenberghe, *Convex Optimization* (Cambridge University Press, Cambridge, 2004)
2. A.J. Hoffman, On approximate solutions of systems of linear inequalities. J. Res. Natl. Bur. Stand. **49**(4), 263–265 (1952)
3. Y. Nesterov, *Introductory Lectures on Convex Optimization: A Basic Course* (Springer Science+Business Media, Berlin, 2004)

# Index

© The Author(s), under exclusive license to Springer Nature Singapore Pte Ltd. 2022
Z. Lin et al., *Alternating Direction Method of Multipliers for Machine Learning*,
https://doi.org/10.1007/978-981-16-9840-8

Printed in the United States
by Baker & Taylor Publisher Services